NANOTALK

Conversations With Scientists and Engineers About Ethics, Meaning, and Belief in the Development of Nanotechnology

NANOTALK

Conversations With Scientists and Engineers About Ethics, Meaning, and Belief in the Development of Nanotechnology

by

Rosalyn W. Berne, PhD

2006

LAWRENCE ERLBAUM ASSOCIATES, PUBLISHERS
Mahwah, New Jersey London

The material in this book is based upon work supported by the National Science Foundation under Grant No. 0134839. Any opinions, findings, and conclusions or recommendations expressed in this material are those of the author and do not necessarily reflect the views of the National Science Foundation.

Lawrence Erlbaum Associates, Inc., Publishers
10 Industrial Avenue
Mahwah, New Jersey 07430
www.erlbaum.com

Cover photograph by Gordon D. Berne: The Scientists, artist, Elaine Pear Cohen. This sculpture sits in front of Marine Biological Laboratories in Woods Hole, Massachusetts.

Cover design by Tomai Maridou

Library of Congress Cataloging-in-Publication Data

Berne, Rosalyn W.
Nanotalk : conversations with scientists and engineers about ethics, meaning, and belief in the development of nanotechnology / by Roslyn W. Berne.
p. cm.

Includes bibliographical references and index.
ISBN 0-8058-4810-X (alk. paper)
1. Nanotechnology—Philosophy. 2. Nanotechnology—Moral and ethical aspects. 3. Scientists—Interviews. 4. Engineers—Interviews. I. Title.
T174.4.B37 2005
174.96205—dc22 2005040158
CIP

Books published by Lawrence Erlbaum Associates are printed on acid-free paper, and their bindings are chosen for strength and durability.

Printed in the United States of America
10 9 8 7 6 5 4 3 2 1

Dedicated to my husband Gordon Berne
our children Kaya, Ari and Zoe;
my parents Roland and Muriel Wiggins;
Beth and Robert Berne

And in memory of John C. Fletcher;
bioethicist, mentor, teacher, and friend

Contents

Foreword

Standard procedure in academic publishing is for a manuscript to be reviewed, which gives the author an important opportunity to say "Oh! I hadn't thought of that" or "Oops," and then make corrections and changes to improve on the original document. The reviewers of this manuscript were wonderfully encouraging and, fortunately for me, detailed in their critiques. One point of criticism, however, warrants mention in a foreword; that is, consideration given in *Nanotalk* to questions of faith, belief, and God as they may be relevant to conversations about nanotechnology. The reviewer wondered why I "danced around the subject of whether or not God exists," and why "I failed to be clear about that." For that reviewer, the matter is straightforward. He wrote "God is what people place in the gap between what we understand and what we experience .... We thought, many years ago, that lightning must be a god throwing down thunderbolts. What else could it be? But now we understand what lightning is, and have no further need for a god of lightning. We still marvel at seeing lightning and shudder at its immense power. But it is not a religious experience anymore. Zeus can rest in peace." That same reviewer said he felt irritation over not being able to be with me and my researchers during our discussion in order to correct us with the understanding that it is a very human desire to invent supernatural beings for things we do not understand. He concluded, "Someone reading Berne's book 100 years from now will find it very amusing, indeed."

My purpose has been to garner from individual nanoscale science and technology researchers a sense of what matters to them, what inspires them, concerns them, motivates them, and instills curiosity in their work toward the research and development of nanotechnology. Whether or not they believe in Zeus (or Yahweh, Allah, Jehovah, or Brahma) and what those beliefs may mean to them in the societal context of developing nanotechnology is only significant here to the extent that it offers a fuller view of who they are as individual researchers in the context of nanotechnology quests. It does not matter how curious, peculiar, or even amusing those beliefs may appear 100 years from now. What matters are the stories we tell one another now, and how the narratives we weave form and ground the basis of nanoscience and nanotechnology endeavors.

A question I have asked of my self while listening to researchers talk about their work is whether and how faith, agnosticism, or atheism might play a role in the pursuit of nanoscaled science and technology. In what way might the human quest to control and manipulate the physical universe be a response to beliefs about the will and existence of God? I sometimes raised these questions in the conversations. As such, the reader will find here and there throughout the text—varied and sometimes abstract—other times explicit, references to that which might be understood as Mastermind and First Mover of our worlds, or God.

As for my own beliefs, I have no other explanation for the incredible complexity and absolute profundity of the universe, especially of life, except for the existence of an Infinite Intelligence and Creative Force, which/who enjoys dancing even more than I! I do not accept as true, as does my reviewer, that God's existence is a human creation for the sake of explaining that which is otherwise frightening or perplexing.

The cover of a 2005 issue of MIT's *Technology Review* magazine reads in bold, "God; But for How Long?" Although the actual subject of the inside story is a computer search engine, the title is clever in that it grabs attention with the haunting notion that God's existence is subject to change. It seems to me that whereas increasing scientific knowledge and technological abilities may change ideas and beliefs about God, these don't necessarily negate the existence of God, at all. Rather, the evolving ability of the human mind to grasp and perceive God's existence as knowable reality will for some human beings depend entirely on how scientific understandings and technological creations mature and evolve. One hundred years from now, those who do laugh at *Nanotalk*'s references to God, may very well be laughing at the limited, primitive, and ignorant nature of our current abilities to perceive and understand God's presence in ourselves and in the Universe.

—Rosalyn W. Berne
Charlottesville, VA

Preface

Many different people are talking about nanotechnology these days. In federal agencies, staff members are talking about funding initiatives for its research. Politicians are talking about new jobs that will be created as a result of its development. Economists speak of its potential for new international markets. Science fiction portrays its potential horrors. Scholars are publishing papers on its social, legal, educational, and policy implications. Industry analysts speculate on nanotechnology as the future for drug delivery, semiconductors, and energy. Transhumanist chat rooms claim it as the answer to human radical life extension. Whether through the media of magazines, academic journals, radio talk shows, novels, the World Wide Web, or legislative session proceedings, nanotechnology is a very popular subject of discussion. But, neither the popular press nor the literature of scholars has yet to adequately deliberate the ethical implications of nanotechnology from the perspectives of scientists and engineers who are themselves the researchers of it. That is the purpose of this book.

Nanotalk is written by an academic scholar, with the hopes of reaching a broad audience. Whether it is read by curious individuals with an interest in technology and the future, or used in undergraduate or graduate school classrooms of philosophy, science & technology studies (STS), engineering ethics, nanotechnology, or science education, the author's intention is to contribute to the public discussion of what nanotechnology may mean to human life. For some readers, the primary interest may be the thoughts and ideas of actual nanotechnology researchers, as revealed through the conversations that are the core of this book. For others, the rhetorical and philosophical interpretation of those conversations are of interest, and become the subject of their reflection and study. Hopefully, all readers will come to appreciate the importance of taking a conscientious approach to developing nanotechnology, with careful deliberation toward humanitarian purposes and uses.

The book begins with the full text of a conversation with a research scientist named Russell. That conversation is placed at the beginning, before the book's introduction, in order to set the tone for the subject of nanoscaled science and technology, in the context of the individual researcher's own thoughts, beliefs, and ideas. It gives the reader an immediate sense of what the conversations are like, what kinds of issues and subjects are talked about, how they flow, and what it is like to "listen" to the scientist speak through the medium of conversation with the author. However, no interpretation is offered. The meaning of the conversation with Russell is left entirely up to the reader.

The introduction to the book is placed after the conversation with Russell. It explains the importance of listening to research scientists and engineers and of including their individual voices in the larger public discourses about nanotechnology. It then details a research project funded by the National Science Foundation, which serves as the basis of this book, to begin to address ethical questions pertaining to nanotechnology research and development. Three main parts follow the introduction. Excerpts of individual conversations are placed and discussed within the individual chapters. Full text conversations stand alone in between the chapters of each part. Those conversations serve to provide a subtext, of sorts. As freestanding texts, the possible meanings and implications of the content of the conversations is determined by the reader. However, the placement of those conversations is strategic, because the content alludes to some of the subjects that are significant to that particular part. For example, the conversation with Caroline is placed inside of Part I, Ethics; because the content of her conversation was pertinent to the subject of moral responsibility for nanotechnology. In fact, it is the conversations themselves which determined much of the content of the book. In other words, it is the author's interpretation and analysis of the conversations that determined the book's organization, and the subjects of each chapter.

Part I, Ethics, considers questions of responsibility for moral leadership of the so-called nanotechnology revolution and suggests that such responsibility is critical. It also offers a particular analytical approach to the formulation of an ethics of nanotechnology, suggesting that no existing normative approach is sufficient to address the unusual elements of a technology whose future is so vast while also being virtually unpredictable. Part II, Meaning, moves toward an analysis of the conversations. It identifies themes and frameworks that appear with some consistency in the conversations and the implicit activity of meaning making that they entail.

The third and final part of the book, belief, shifts away from normative concerns of what nanotechnology might mean, to metaethical considerations of belief. Working with conceptualizations of nature, and the use of imagination, myth, and metaphor in the construction of belief, Part III is concerned with the moral choices that come from recognizing that belief, rather then absolutes about human evolution through science and technology, is an essential feature of nanotechnology development.

ACKNOWLEDGMENTS

At this point I wish to express my appreciation for all those who have helped me with this project. From the very conceptualization of the book to its final editing, the tremendous help I have received has been a testimony to the fact that writing a book is not something that one can do alone. After first thanking the staff of the National Science Foundation who encouraged the initial proposal for this work (especially Joan Siber) there are many others to thank. Absolutely central to this entire project have been the scientists and engineers who kindly gave of their time to meet with me for these conversations. (I promised to keep their identities anonymous so that they could feel comfortable speaking openly with me, otherwise I would thank them by name.) Attorney Philip Lamar explained to me the contractual elements of working with publishing companies. Publishing agent Stan Wakefield presented my proposal to both academic and trade publishers. When it came time to make a decision about which publisher's offer to accept, his guidance was enormously helpful. I have received incredible support, encouragement, and the timely responses from Lawrence Erlbaum Associates editors, Bill Webber and Lori Stone. Everyone I have been in contact with at Erlbaum has been helpful, kind, and attentive.

Midway through this project, I needed to go to a place where I could think and write without distractions, so that I could truly immerse myself in a world of nanoscience research. Leonard Feldman, director of Vanderbilt's Institute for Nanoscale Science and Engineering, made that possible through a visiting professorship at Vanderbilt. While I was there, chemist Sandra Rosenthal's graduate students showed me around, demonstrated the microscopes, and explained to me all about their work. After spending my days on the Vanderbilt campus, I was very fortunate to be able to go "home" to evening writing at the Hermitage Hotel. The very professional hotel staff provided for me a safe, nurturing, and comfortable place to be where people actually knew me by name.

Kay Neeley got me thinking about conceptual roadblocks, which led to a flow of many new ideas for my approach to the book. Joe Pitt made a close read of Part I in its very early stages; by offering both suggested changes and corrections. Bernie Carlson helped me think clearly again when the fog descended. Mike Gorman encouraged me to apply for the original research grant. Deborah Johnson, my department chair, reassures me that being unconventional may actually allow me to see things a bit differently, and from this unusual perspective, is what gives me something of significance to say. She provided the leave of absence I needed for uninterrupted focus on getting this book written. I am ever grateful for Ingrid Townsend's wisdom, mentoring, friendship, and support. Emmanuel Smadja worked as my research assistant. His critical mind and attention to detail have been invaluable. A special acknowledgment goes to Isis Ringrose, whose consultations offered immeasurable insight, and whose encouragement lifted me up from out of more than a few moments of disillusionment, self-doubt, and anxiety. And, finally, I give thanks to my husband Gordon, who not only cares for me, but also assures that family and home are well cared for during all of my travels, writing sprees, and extended periods of total distraction. Thank you all.

Russell

ROSALYN: *Alright, assuming a divine order or intelligence in the universe, is there a connection between that intelligence, and our increasing capacity to manipulate and control the material world, and where we seem to be going with it? If there is one, that's what I want to talk about today.*

RUSSELL: *OK.*

ROSALYN: *I am searching for a reason for this madness, whether it has to do with the convergence of these technologies that are emerging and what that might mean in terms of a radical reconstruction of humanity. That gives me pause to ask, OK is there something cosmologically connected to what we're doing?*

RUSSELL: *That's a very large question. I told you I read* Prey *this summer.*

ROSALYN: *Yes, you did.*

RUSSELL: *And I guess for the first time I would say I understood why it is that some thoughtful people might look at the possibilities of nanoscience and say "thanks but no thanks" and it has to do with this convergence of bio and nano and info in the creation of self-adapting mechanisms.*

ROSALYN: *Right.*

RUSSELL: *But also with the possibility of self-adapting mechanisms that are freed from one of the very important constraints of evolution in the historic past as in the long past. That is, as you discover in reading the book, the problem is there are no natural enemies to this system that has been created and therefore there is no check or balance on what evolves from this and if the creators of the system do not have, or by some means lose their sense of direction about what it is they want to have happen, then you have this situation that ... I think in my very first conversation with you I mentioned this sentence from Hannah Arendt's,* On Human Condition *that has stuck with me. "Then we become thoughtless creatures at the mercy of every gadget that is technologically possible no matter how murderous it is." In that sense, nanoscience is no different from atomic weapons technology, for exam-*

ple, but it is more dangerous. I think it's potentially more dangerous. I worry about it in the sense that I happen myself to believe that there is an order in the universe and that there are certain things that are natural and appropriate and so on and there are other things that are not. I worry about the fact that the scientific community and especially in the live sciences part (and this is true at the intersection of nano and bio as well) is also inhabited by some people who may be among the most thorough going materialists and reductionists in the entire scientific community, and that's a worrisome prospect.

ROSALYN: *I think that's in part where my question is coming from. OK, are you suggesting that there may be limits other than material limits to what we do with nanotechnology?*

RUSSELL: *There are two kinds of answers to that question I can think of. One is the question of whether or not the technologies themselves have the potential to do harm.*

ROSALYN: *Sure, sure.*

RUSSELL: *OK, and that clearly is wrong. But then there's also the question that Freeman Dyson has raised very articulately in recent years and that is, in the face of enormous needs that are far more basic than the issue of whether we can compress a computer to the size of a pinhead, are we justified in pushing ahead and spending lots and lots of money to do this in the hopes of creating economic benefits, perhaps technological benefits, when in fact, some of us who are working on this ought instead to be building houses in Paraguay, or ...*

ROSALYN: *If we would we just get potable water to everyone on the planet.*

RUSSELL: *For example.*

ROSALYN: *Yes we could, so why don't we?*

RUSSELL: *We probably could. And, in fact, it is possible that nanoscience might well contribute to that. As you probably know, environmental issues like that are a major part of the Rice initiative in nanoscience.*

ROSALYN: *That's what I understand.*

RUSSELL: *Perhaps if our focus is on things like that, then ultimately people will say yes, there is something more than just curiosity value or gadget value in what comes out of nanoscience.*

ROSALYN: *What I'm hearing is that fundamentally this is about curiosity and that there is great satisfaction in the hope that it could actually improve the quality of life.*

RUSSELL: *Yes.*

ROSALYN: *OK.*

RUSSELL: *That's fair.*

ROSALYN: *When I ask what are we really up to, and I would love to know, the answers are more varied than that. Do you think about this? In the larger scheme of things, what is it we're up to? Are scientists and engineers doing something for the whole of humanity? You serve a very specific role in terms of the human community. For those of you who are pursuing scientific knowledge and particularly the application of it to nanotechnology, what is that all about? Particularly in terms of that divine order that we have agreed exists?*

RUSSELL: *OK, I think if you ask that question in a university, you're likely to get a different answer than if you ask it in the Naval Research Laboratory or at a pharmaceutical research facility such as Merck, Sharpe, and Dohme.*

ROSALYN: *Sure.*

RUSSELL: *All those places have nanoscience efforts going on.*

ROSALYN: *Merck is a for-profit pharmaceutical, we know what they're doing. They are doing basic research to bring new drugs onto the market which will increase shareholder value.*

RUSSELL: *The executive from Merck next to whom I rode on an airplane recently said that they weren't always attentive to increasing shareholder value in the long run. He felt that in some cases they had neglected basic research over the last 4 or 5 years, increasing shareholder value in the short run but leaving the company in a weaker situation in the long term.*

ROSALYN: *Hum.*

RUSSELL: *Be that as it may, in a university I think the situation is a little different in the following sense. At least in the physics department we are relatively remote from interest in applications, the focus is on trying. In nanoscience I see as one very interesting aspect of the whole question of can we learn to understand very complicated material systems better than we presently do and nanoscale objects, especially as we learn how to fabricate them, give us an opportunity to ask those questions in a way that we never could before and to isolate features of complex behavior that we could not understand before. The long-term view of that and my belief as a researcher, whether with undergraduate or graduate students, is that my contribution to the world revolves less around whatever specific things I am doing at any given time and much more around my capacity, my opportunity to interact with very bright young people and train them in the art of solving complicated problems while learning a certain set of skills, which they apply. That capacity in some sense adds to the store of human potential that is available for solving problems. Some of my students are in the academic world, some of them are in the industrial world, some of them are at national laboratories, so my hope is that they are carrying with them that sense of how to responsibly, creatively, effectively go about applying those skills that you learn in universities to the solution of other classes of problems. But there is nothing in that activity as I see it that relates explicitly to the ethical or moral dimensions of the question that you ask. I mean, the only way those things get developed in our group is through the informal interactions that we have with one another as individuals and not particularly as scientists.*

ROSALYN: *That's understandable. As I have remarked to others, we bring ourselves with us to work. In one conversation I asked, "Was it necessary to check your belief system at the door?" And the response was in effect "yes," because there is not room for those questions in science. They just don't have any relevancy in science, which is about discovery and learning, so.*

RUSSELL: *Yes, but if you, if you water at a public trough as we do, in terms of where our funding comes from.*

ROSALYN: *Yes?*

RUSSELL: *Then it seems to me that implicitly, if not explicitly, you cannot check your belief system at the door because you must have some sense of the value of what you are doing to the people who pay for it.*

ROSALYN: *If you take your water from the trough of the public, but inside you have a belief system where you take your water from the trough of God, then how does this work with nanoscience research?*

RUSSELL: *Well, in our faith tradition, there is a very strong concept of stewardship, of individual stewardship, not only over material sources, but over time, energy, and the sense that all of these things are, well in fact, this idea; this notion is generally referred to in our church as the law of consecration and stewardship. The idea is that fundamentally everything that we have and are or can be is a gift from God and that we as stewards are obliged to both husband it carefully but also to recognize that life itself and all that we do in it, whether it is my work here or time spent with our children or whatever, is in some sense to be lived as a consecration and—the way I put it is that my own personal view of myself is that there is no part of my life, at least to the extent that in my life's activities I am doing things that I know I should rather than things I know I shouldn't—that all of that is part of this idea of a consecrated stewardship; whether it is involved with doing physics or listening to music or being with my wife or whatever. So for me, I don't feel the necessity for checking anything anywhere, it's all kind of a package.*

ROSALYN: *Now whether or not other scientists, nanoscientists, have that belief, would you say nanoscience is a gift from God?*

RUSSELL: *Yes, especially to the extent that it has the potential to relieve suffering, to make the world environmentally or ecologically a better place than what it has been or is now. If nanoscience could, for example, rescue us from some of the pollution created by an industrial revolution which was too little animated by use of stewardship, for example, long-term responsibility, sure, and I think also pure curiosity has a place in that world. I don't think that the idea of stewardship is necessarily bound up entirely with utilitarianism.*

ROSALYN: *Yes.*

RUSSELL: *You know, if Johannes Kepler could look at his planetary ellipses and believe that through this new geometry that he developed that he was getting a glimpse into the mind of God, then why not through nanoscience? The nearest example I can think of is actually from biology (and I am sure that my colleagues in the bio side would say this is an extraordinarily simpleminded view), but one of the things that amazes me about what we've learned about biology at the nanoscale is that what appeared to be extraordinarily complex systems are in fact made from a remarkably small number of very beautiful simple building blocks and they go together in the most amazing ways. Last week I gave a colloquium and in honor of the occasion I fished out one of my very favorite quotations from where Brigham Young says that "man's machinery takes things which are different and tries to make them all alike. God's machinery takes things which appear to be alike and imparts to each a pleasing difference." And so if I think about how people look at nanoscale building blocks of sea animal shells, for example, crustacean shells, and they find that there are these wonderful variegated patterns that arise out of, again, very simple building blocks, and apparently quite simple processes but with little twists and turns, I guess I would say I find room for that in my view of nanoscience just as much as Kepler found room for his ellipses in his astronomy.*

ROSALYN: *So then it all comes together inside of that divine order?*

RUSSELL: *It does for me.*

ROSALYN: *And, what do we do about the stewardship, for the whole of scientific inquiry. Is there anything we can do? Say I want to say that's a good thing and that's a commitment I think should be broadly held, and if we're all swimming around in the soup of meaning making and belief, as it were, I would want that to be a universal principle. (Yes, that's a value judgment.) How do we get nanoscience as an enterprise to be an enterprise of stewardship over a gift and a capacity?*

RUSSELL: *I find it difficult to imagine that you can create a new directorate in the National Science Foundation which does this. However, I do believe that in some sense it is an absolute good that people at the National Science Foundation are asking this question. The formation of a directorate to do this would suggest that either there was a straightforward answer that could be achieved by a bureaucratic administrative technique of some kind, or that there was some terrible problem that cried out for an answer and could not be let alone on political grounds. But I think that to the extent that people are raising these questions, then good things are likely to happen. I mean there are a number of centers that deal with issues like this. There is this center at Berkeley for Theology and Natural Science. Another example is the new initiative at*

Vanderbilt, a Center for the Study of Religion and Culture. I spoke with one of the codirectors about the possibility of trying to build a project around that. That center would be a very good place for a variety of reasons to do something that pulls in bio, nano, and info technology into a group that includes also some very thoughtful, interesting people from philosophy and religion to talk about these issues. They don't teach you this in graduate school. I mean, the feelings that I have expressed to you about stewardship are things that come from my family and my church and from long sessions late at night talking with people about things and from interacting with people who have different sets of values. And I think that the only way to try to make this a component of nanoscience is to make places where people are able to engage one another in dialogue about these issues.

ROSALYN: *That would be a wonderful opportunity.*

RUSSELL: *But see, you're a part of that in the sense that you're going around …*

ROSALYN: *Having these conversations.*

RUSSELL: *Like Socrates asking people these annoying questions.*

ROSALYN: *I know, that's true. Bear with me on this one.*

RUSSELL: *But nobody's passed you a cup of hemlock yet, so this is …*

ROSALYN: *Not yet, heaven forbid. OK, now supposing we're not very good stewards of this gift, is there a chance that, aside from the obvious sort of catastrophic harms that can come to humanity, that the appearance of progress, the appearance of mastery of the material world will take us into a spiritually dangerous place. Such as, oh, nanoscaled circuits lining our neural system and connected to computers or some of these other far out notions of what could be possible for life extension or for transportation of the body. People say you can only do what's within allowable physical law, but we continue to sort of reconstruct our notion of what the universe is and how it functions, so if that's bendable and flexible and our capacity to master it increases, is there any chance that we are embarking on a place that will compromise our fundamental humanity or ourselves, as we were created. Or are we on an evolutionary journey?*

RUSSELL: *Yes and no. I mean, I think that clearly there are dangers. Michael Crichton sees one kind.*

ROSALYN: *The physical danger.*

RUSSELL: *There is the possibility of physical danger. I don't know of anyone who is writing about the spiritual danger. Part of the spiritual danger, of course, can come from the becoming enraptured with this to the point where it excludes other things and say, well this is just too important.*

ROSALYN: *And my concern, my question is whether in fact our rapture will lead us to no longer needing even a sense of there being a God because all the reasons we might have had are no longer meaningful. If we don't die, we don't get sick, you know, we don't feel vulnerable. I personally think there are all kinds of other reasons to have that relationship and belief, but culturally a lot of it did come out of those bodily limitations or perceptions of bodily threat to survival.*

RUSSELL: *I take to some extent a different view of that, and that is that in spite of the urban legend that we tell ourselves that progress is speeding up, if you look over the last century, yes, a lot of things have happened on a lot of different fronts, but it still takes about a generation to move anything from the laboratory to any practical application, and my excuse for not worrying too much about that is to say probably a lot of these dangers are unlikely to happen in my lifetime and for that reason I don't worry about them. But also it's because I believe that even though we may be able to make on a demonstration basis things that appear to show, for example, that we can make an artificial skin for infantry soldiers that would be self-healing, anything that we envision presently in that regard is so hugely expensive that people will still be going to Fort Benning 30 years from now and doing things in very similar ways....*

I think the greater danger is that by coming to focus on these, let me call them man–machine interfaces, things where we are trying to invent artificial substitutes for things, or artificial enhancements for our life and so on, that we can become so preoccupied with those simply because they are interesting and crowd out other things, that we then suddenly discover that we are no longer easily able to resolve issues that can't be solved by technological means. There was a very interesting interview on NPR last night with a Palestinian attorney. He grew up and continues to live in Ramala, which has been the subject of a lot of attacks by Israel and so on. He describes what happened when Israeli soldiers came to his door early one morning to search their house looking for whatever, bomb making equipment, whatever. He said the first thing that struck him about them was how insulated they were in some sense by the enormous amount of gear which they carried, radios and weapons, and flak jackets, and so forth. And then he said that he tried to talk to them as they went around the house and it appeared—he said he read into their behavior— they were recognizing that this man and his wife were not a threat to them and were not engaged in any of the things they were worried about but he said they were constantly in communication with various places and so on and that it was not possible to engage them in a normal human conversation about "What are you finding?" "What are you looking for?" "Can I help you?" that sort of thing. Now, if that's the direction we take, if we become so isolated, if we insulate ourselves from human contact by the development of nanomachinery of whatever kind—that strikes me as extraordinarily dangerous.

ROSALYN: *And the reason why I don't want to let you go with "this is a generation away" is because my perception is it has already begun to happen with technology that's already come, as an extension of ourselves. Here's my example. I walk up and down the halls of my own engineering school and then today here, and I see the same thing in every office—yours is an exception—the door is here, the person in the office is this way, completely focused in on the screen. The back of the head faces the door, 9 times out of 10. So the primary engagement is with this artificial intelligence that now is becoming really attractive, really alluring, it has a wealth of information but it also creates a sense of relationship that's not as difficult as the one we have with the person who comes to the door. It asks less of us. It asks different things of us. I am concerned that because of the appearance of convenience and power, we are not conscious of what our relationship is with the technologies we express and that we will continue to revamp our belief systems to accommodate the things we create. Here's another example. There was a time when the family dinner table was sacred, it was important, it was fundamental. Today we bring our laptop home, we bring our cell phone, we bring our phone, there's the TV and now we have devices that compete for our time and unless we make a real conscious decision that the family dinner table is still sacred to us, the other things become more valuable. I'm suggesting that we weren't conscious when we brought those things into the same domain as family and that the devices became very influential and very powerful, dinner became shorter, the conversations at the table became abbreviated. Already we are changing as a result of the way we express ourselves in technology, so I'm not really going to settle with you on the generation away thing.*

RUSSELL: *I take your point. In fact, if these things come to dominate life in the way that you describe, then in fact this is a manifestation of our turning away, it's a conscious act, I mean, it's a deliberate thing.*

ROSALYN: *It is conscious.*

RUSSELL: *When I set up this office, one reason why I didn't do that is because I wanted for a variety of reasons, a lot of them having to do with teaching and the fact that students come here, to have something that is, when you open the door, looks like I was prepared to have to turn away from my computer screen and to have a conversation with you.*

ROSALYN: *Yes. And you've done that very nicely. All I had to do is walk by and I got your eyes.*

RUSSELL: *Yeah.*

ROSALYN: *This is a conscious effort on your part. It's not common.*

RUSSELL: *OK.*
ROSALYN: *Take a survey.*
RUSSELL: *OK, I will.*
ROSALYN: *You might be surprised.*
RUSSELL: *I've thought about that.*
ROSALYN: *And so, my work is to bring to consciousness our relationship with the things we are creating so that we don't sacrifice elements of our spiritual selves, of our human or communal selves, and of the things we hold as sacred. I fear that if nanotechnology moves quickly that we may.*
RUSSELL: *Is it that they move quickly or that our culture has become so thoroughly materialistic that now as floods of new inventions have come about in the last 20 years, it has just become easier to succumb.*
ROSALYN: *I think maybe that's right. And, you know, there's a reason why I can't walk to work, why the roads are too dangerous to ride my bike. The design is not reflective of a value system of breathing the fresh air and walking your body and taking your time.*
RUSSELL: *That's right, that's right.*
ROSALYN: *So everything we design and bring to fruition as technology has our value system in it somewhere, somehow.*
RUSSELL: *Now let me ask you this. You have just come back from this trip to Germany and we've talked about that. Is it possible that some of this that we're seeing is a peculiarly American trait?*
ROSALYN: *OK, so part of who we are is the explorers, with the new frontier and the domination of new lands and all that. That is our history, defense of liberty and freedom, that's also our history, the sense of independence. We're pretty young, I mean, indigenous peoples aside for a moment, we have been Americans only for a few hundred years. The German culture and civilization is much older than we are. I mean, early German culture is thousands of years old, correct?*
RUSSELL: *That's right.*
ROSALYN: *Yeah. So they have a very different sense of who they are, I think, yes?*
RUSSELL: *There is something else that's different about them and I think this is important, and it is that they have developed in part through frequent, certainly on the cultural time scales, frequent absolutely devastating horrifying conflicts fought on their own soil.*
ROSALYN: *Yes, on their own soil. We've never had that.*
RUSSELL: *At least not since the 1860s.*
ROSALYN: *The Civil War, but that was internal, that was us fighting us, right?*
RUSSELL: *That's right, but the experience, and it was only the southern part of the United States that had the experience of having devastating battles fought on their own soil. I've spent 10% of my life in Germany, roughly, and I have come away with the feeling that some of this path-*

ology that we have results from the fact that we have not experienced this. We have not learned how precious human connections are because in part we have never been deprived of them except as individuals.

ROSALYN: *I think you're right, OK.*

RUSSELL: *And so, now one of the very interesting questions for you to raise in some sense is whether or not this is culture dependent. Are the scientists who are doing superb work in nanoscience in Germany animated by a different set of values? Do they find it easier to escape from this? Are they less tied to cell phones and computers and so on than we are? I don't know.*

ROSALYN: *I do know that they had a European commission meet last summer to look at the societal implications of nanoscience and where it's leading. They had testimony from the ETC group in Canada and Prince Charles and others who are concerned. Germany seems to be and they are taking their time with these questions. I don't know, I'll go next month to the NSF panel and I'll see. I would really be surprised if our engagement has with it their true sense of the value of humanity. I would be surprised. I was afraid to go to Germany; I have to tell you. I was afraid because my husband's family is Jewish from Germany and his family's experience with Germany is not a good one. I had some stereotypes with me. Now they're gone. Part of the reason is because of the sincerity with which people spoke with me about the war, utter and complete sincerity over the horror of it.*

RUSSELL: *This is something that our policymakers have not understood about this, is that there is no family in Germany, you cannot find one that does not have intimate, personal memories from living persons that are tied up with the war that they experienced on their own soil.*

ROSALYN: *I was, well, shattered because I couldn't find a synagogue, and they said, "Well there really aren't so many left." And I asked, "Where are the Jewish people?" and they said, "Well, there are not too many here." And it became stark reality to me, "Oh my God, it really happened!" I've seen people with branded arms in the States, but I really didn't get it that it really did happen. I got to Germany and it became so very real. The city of Darmstadt was all rebuilt.*

RUSSELL: *But there's no synagogue.*

ROSALYN: *There's no synagogue, no, so it was successful, this genocide was successful.*

RUSSELL: *There's no synagogue in Marburg. There is a little sandstone block that shows where the synagogue stood that was destroyed during the Kristallnacht.*

ROSALYN: *Yes, I think, yes as an American doing this research and looking at nanoscience and looking at some of the claims being made and the time frame, the push to move fast, I have to ask, what about the human spirit?*

RUSSELL: *Well, see now I worry about this, you know. This year for the first time in talking with colleagues, especially American colleagues, I am worried that the community that wants to do this is overselling and that there is going to be a backlash and it will come, sure as shooting, and that to me is an interesting and disturbing development in the sense it suggests that people who are working in this area believe in it and believe it is a fruitful area for exploration but who feel this push. Now the people who talked to me about this are all university types, and they are saying, "We are worried that the rush to do this is being driven by something other than what we would consider to be wholesome scientific motives and like it or not, whether materialist or not, there is a set of values that animates the scientific community." We saw it in nanoscience with the Hendrik Schurn scandal at Velepse that there are things that we still won't do or believe we shouldn't do. I hope we're not on the verge of losing that because we are pushing so hard. But I go back and read Michael Polanyi's books,* Science, Faith, and Society, *for example, and read his description of science as a community held together by a certain kind of faith.*

ROSALYN: *Yes, that's right.*

RUSSELL: *And certain community ideals. I hope we're not losing that, and again, we will lose it unless people like you keep asking the questions that you keep asking.*

ROSALYN: *People ask me, "Why do you focus on the scientists and engineers?" This is why. You are where my hope lies. I think I have some issues with what Francis Bacon was up to but I think he was right with holding scientists as a form of the priesthood. I mean, there is something to be said for that role in society.*

RUSSELL: *Have I ever told you about a historian I know who was asked to give the invocation at a summer graduation ceremony?*

ROSALYN: *No.*

RUSSELL: *And whose opening line was, "Oh God, we are gathered here in the robes of a false priesthood."*

ROSALYN: *OK, however, we have assigned you this role, and it comes with certain responsibilities.*

RUSSELL: *It does.*

ROSALYN: *And you all have a very high level of intelligence that for whatever reason is not that common on the planet or in the human condition and so that comes with stewardship responsibilities. OK, so in terms of the motivations and the aggression and the funding and the push—I read the testimonies from this summer to the Senate Subcommittee on Science and Technology and they were all the testimonies that sort of go ahead with funding initiatives ….*

RUSSELL: *Right.*

ROSALYN: *And that's where I get worried because there were a couple of scientists there, which was good but there was a lot of rhetoric that led me to be concerned that really this is about economic opportunity, international competition for new markets, and military power.*

RUSSELL: *Well, but see now here is where I have to pass the buck in the following sense. The Congress of the United States has since Vanavar Bush on the endless frontier has accepted the idea of scientific and technological funding from the federal government as being part of the responsibility to provide for the general welfare in the sense of providing economic future, but I think there has been a fundamental change in the view of that obligation since let us say the 1980s, just to date it in some way, and the belief that market forces should dominate and determine pretty much everything that we do. There is no longer the idea that there is a sense of public purpose that is driven by high-mindedness or feelings of noblesse oblige or whatever you say, and I think that as that idea about the market as the ultimate arbiters of what is worthwhile and what is not worthwhile in our human community, as that idea has come to be dominant, now it has become necessary to make use of that for political cover as scarce federal funds are allocated among competing visions. There is no sense, for example, that our activities in nanoscience are part of some sort of intergenerational compact, something that we owe to our children to try to provide them with a better life. It's solely, as you said, it's the search for new markets in the global marketplace. And that's really scary to me because it just reinforces what are fundamentally selfish materialist motives for political acts which ought to be instead directed at the survival and the expansion of a humane human community.*

ROSALYN: *Well, I'm the choir and you are the preacher here.*

RUSSELL: *Yeah. But look, people in the panel you're going to next week have got to hear this. I mean, Mihail Roco is not just concerned about the next "nanoscience miracle of the month" that goes into his Power Point presentation. As effective and as able as he has been in promoting the National Nanotechnology Initiative, he and others like him who are in those leadership positions are ready and willing to incorporate this if they think it's important.*

ROSALYN: *Absolutely.*

RUSSELL: *I have to say that when I first heard that NSF was planning to devote efforts in programmatic funding and energy to this I thought, man, this is a curious thing. Why not just do more science? I don't feel that way anymore. It strikes me as just extraordinarily important.*

ROSALYN: *I hope it's sincere. I hope the commitment is really there. I hope it's not just to allay public fears or concerns. I said to someone from Foresight*

Institute once, "I have read your guidelines for the development of nanotechnology and I have to tell you I'm suspicious." This person happens to have been one of the authors of the guidelines and he said, "why?" And I said, "You are at the same time claiming that your concern is to take care and have Foresight move forward with prepension while at the same time you're the loudest advocates of rapid nanotechnology development. Seems to me it's a lot of smoke to get people distracted." He started laughing and said, "Well there is that."

RUSSELL: *Now, of course, you certainly understand also that the impetus behind nanoscience comes also from the attempt on the part of NSF and DOE and other agencies to find an umbrella for what otherwise looks to the Congress like a whole array of competing principalities and chiefdoms that were not even close to being on the same page, even though people working in those areas understood that there was a certain kind of interrelationship. So in a way, this is also an important attempt to try to build a community that is capable of telling a story in a way that our very much overstimulated, overpressured political leaders can get their arms around and try.*

ROSALYN: *Sure, and that's good, and also the new collaborations and the new alliances that are forming seem to me also a very good thing.*

RUSSELL: *Sure, I think that's why, I don't even mind the hype, if you will, from Foresight Institute and others who are trying to push the business community together with us if it will, especially if it will generate in the business community some sense that there are actually things that are important out there that might take longer than the next quarter to develop. But if we could change the basic industrial time horizon from 3 months to 18 months, for example, this would probably be a very good thing for everybody.*

ROSALYN: *One of the things I appreciate is that the government is willing to make long-term investments in science.*

RUSSELL: *Yes.*

ROSALYN: *I think that's been true. However, I have a sense that people are feeling very pressured to move fast, so it's like the government's saying take your time, do this research, and at the same time the actual proposals are having to bring results in really quickly, the reports.*

RUSSELL: *Let me tell you about reporting pressure. We are just barely one year into our nanoscience interdisciplinary research team project. I'm on my fourth request for a report.*

ROSALYN: *Oh, come on. In one year?*

RUSSELL: *First one came 3 months after we started.*

ROSALYN: *This is what I've been picking up all over the place.*

RUSSELL: *Really?*

ROSALYN: *Yes.*

RUSSELL: *Yeah. And it's all very innocent but part of it again, there is this vicious circle of policymakers trying to find what they perceive to be political cover for next year's election.*

ROSALYN: *Sure.*

RUSSELL: *We have the permanent election campaign at this point where, no matter what the time horizon is on the election, whatever is being done to justify a particular program has to fit within the context of the permanent campaign.*

ROSALYN: *That's right.*

RUSSELL: *And so there are a lot of things that have nothing whatever to do with science that are driving the time table that we see in the direction that you describe.*

ROSALYN: *Yes. Which for me comes back to the question of how does that time table affect the stewardship, the responsibility, the care with which we need to proceed?*

RUSSELL: *Well, see, if you're a member of Congress and you place high value on staying in office, you might be tempted to forget about noblesse oblige, or intergenerational impacts or any of that stuff. I don't know what the mechanism is to avoid it. As I look around and ask myself who are the people, regardless of what community they happen to be in, for whom I have the greatest respect in this area of driving science, policy, and public goods, it is invariably people who for whatever reason, whether it's the eminence of their scientific position or a Quaker background that will not quit, or the fact that like Jay Rockefeller they are rich enough that they can devote themselves to a cause without worrying about the consequences. It's those people who have long-term outlooks on what they are doing who are in a position to ask the questions that you are asking, or they are people who stand outside science and who therefore look at us and are interested in us as a community but who do not gain or lose by the question of whether or not funding goes up or down this year.*

ROSALYN: *Yes. That's right. OK, thank you.*

RUSSELL: *Well, it's a pleasure, you know, the normal routine of getting out NSF reports does not permit the luxury of this sort of conversation as often as one would like. I'm very excited about the formation of the Center for the Study of Religion and Culture here because there are people in divinity school and other areas who are really primed for this and, so this is really a great opportunity for us …*

There's one thing that is really terrific about the way nanoscience has been elevated and the way it has become thematic and that is it has made it also possible for people to talk about it as a thing, as a cultural object to describe it.

ROSALYN: *That's true.*

RUSSELL: *And in that sense, the nanoscience program has been very valuable in that way. It has made people think in a more global way about what has been traditionally a very fractured, fractious ...*

ROSALYN: *Well, that's a good point. Without this, would I be looking for biochemists or physicists to talk to ...*

RUSSELL: *Or material scientists ...*

ROSALYN: *Material scientists.*

RUSSELL: *You wouldn't be sure where to find them and then so on.*

ROSALYN: *That's a good point.*

RUSSELL: *There wouldn't have been Institutes for Nanoscale Science and Engineering all around the country.*

ROSALYN: *Yeah. And there wouldn't have been an NSF grant for me to do this work.*

RUSSELL: *Right. And in spite of the fact that this is what people would have been doing in any case, probably, you know, as individuals ... we just had our nano forum here 2 weeks ago, and it's fun to be in a room where the speeches were given by an electron microscopist, a biomedical engineer, chemist, and physicist. It was a lot of fun to see these connections form and see students being interested in one another's work and crossing the great divides between departments. This is clearly a good thing.*

ROSALYN: *And you were the physicist on the panel?*

RUSSELL: *Right.*

ROSALYN: *Did you find that you share enough of a language of science that there was understanding?*

RUSSELL: *Well, at least I tried hard to do that and I think everybody who was there was conscious of it. Another one of the good things that has happened because of nanoscience programs, is we've all become more conscious of the fact that we don't speak the same language, but that we are capable of developing some kind of Esperanto that more or less works and brings us together and that I think is very exciting.*

ROSALYN: *This reminds me of one more memory from my trip to Germany. I walked into an antique store in Darmstadt. The vendor was a Turkish man who speaks German and his native language. I speak neither. Despite that, we spent 30 minutes in conversation, he in German, me in English. Somehow, we understood one another.*

Introduction: Narrative and the Voices of Research Scientists and Engineers

> *We live out narratives in our lives, we reconstruct them for our self-understanding, we explain the morality of our actions at least partly in terms of them, and we imaginatively extend them into the future .... It is in sustained narratives that we come closest to observing and participating in the reality of life.*
>
> —Johnson (1994, p. 155)

"The category of narrative has been used to explain human action, to articulate the structures of human consciousness, to depict the identity of agents, to explain strategies of reading, to justify a view of the importance of storytelling, to account for the historical development of traditions, to provide an alternative to foundationalist and / or other scientific epistemologies, and to develop a means for imposing order on what is otherwise chaos" (Hauerwas & Jones, 1989, p. 2). Narrative is one of the most basic tools that human beings have for making sense of perception and experience and to invest those with meaning. Narrative provides access to important but often unarticulated hopes, fears, expectations, and assumptions regarding our relationships to our bodies, to one another, and to the physical world we inhabit. It also brings to light essential, yet otherwise tacit, elements of the human psyche.

As a cultural icon, narrative in the public domain provides a means by which members of society can take part in the development of meaning about technology. There are myriad forces at work inside the development of nanotechnology. One of those forces is the competition to shape the course of human events. Narrative regarding nanotechnology functions as part of the process any technological development entails, to construct an agreed on ethics for its evolution and development. It can also function to illuminate the values, intentions, and belief systems, which are implicit in the nanotechnology initiatives, and varied social responses to them. Language-based stories, (narratives) which are part of the public discourse, reveal the myriad notions of who we believe ourselves to be, what we believe in, and how we wish to live in relationship to one another and to the nanotechnologies being developed. Unfortunately, "in the interest of securing a rational foundation for morality, contemporary ethical theory has ignored or rejected the significance of narrative for ethical reflection," which according to Hauerwas and Burrell (1989, p. 158), has resulted in a distorted account of moral experience.

Whenever a new technology is perceived to have a potentially significant impact on society, narrative emerges in the public discourse to establish the meaning and significance of that technology. For example, developments in in-vitro fertilization, recombinant DNA/genetic engineering, mapping of the human genome, and human cloning have all been subject to intense and critical public discourse. Such public debates over the social impact of technology are made most apparent in the media. This is the primary apparatus for disparate and competing interests to be explored and revealed. As new technologies emerge, so do the disparate and varied voices of talk show hosts, television personalities and news anchors, science fiction authors, science journalists, politicians, commentators, citizen group representatives, community leaders, and scientists who have media access, as they compete to establish the meanings and direction of those technologies. In turn, the public responds with varied, often conflicting expressions of enthusiasm, fear, anticipation, and mistrust, in the effort to control and direct the uses of those seemingly powerful, promising, or threatening technologies. Questions and responses are levied about the effects of particular technologies on the human condition and the ability of those technologies to adequately address alleged material needs of society. Often, ambitions to use technology for improving the human condition are pitted against beliefs that the forces of technology should be limited, and can bring potential harm to the individual and to the society. Narrative also addresses questions of who will have access or be denied access to these technologies. Constructed and expressed within public discourse, these narratives provide society with a platform from which public policy is constructed. They also give private individuals access to the collective making of meaning and the assertion of belief about new technological development. This is good. As technology moves faster and more intensely into our individual and collective lives, excitement and ambivalence run up against outcries of technological doom. Public debate in an open society provides a critical forum for exchange and understanding. My contention is that what is not good is when narrative, which contains significant elements of the meaning from beliefs about technology, is disregarded as irrational or unimportant. As Hauerwas and Burrell (1989) pointed out:

> To live morally, we need a substantive story that will sustain moral activity in a finite and limited world. Classically, the name we give such stories is tragedy. When a culture loses touch with the tragic, as ours clearly has done, we must re-describe our failures in acceptable terms. Yet to do so ipso facto traps us in self-deceiving accounts of what we have done. Thus our stories quickly acquire the characteristics of a policy, especially as they are reinforced by our need to find self-justifying reasons for our new-found necessities. (p. 188)

The tactic described places policy itself as central to the story of our collective lives. The story, such as the one being formed around the imperative of nanotechnology development, becomes indispensable, because it provides us with a conceptual place to be. We are left with ill-informed notions of the intersections of science, technology, and society, or with narratives that are mired in technological determinism.

A common theme of literature and film in contemporary Western society is public mistrust and fear associated with scientists, industrialists, and politicians who spearhead technological change. For example, *Jurassic Park* (Crichton, 1990) tells the story of how scientific ambition, business greed, and political motivations can turn the fruits of good, basic research into technological horror. Mary Shelly's *Frankenstein* (1831, 1992) speaks to the compromises made when scientific curiosity and personal ego supercede personal integrity. The novel *Prey* (Crichton, 2002) points to the horror that can result when business ventures drive scientific research. The moral of these stories is that we will be consumed and destroyed by our own selfish misuse of knowledge and our blatant abuse of nature. One social value of the science fiction genre that puts forth such messages is its ability to engage moral imagination. Thus, literature and film are potentially powerful narrative sources for the public to make and form beliefs about nanotechnology. The problem is when the narrative value is dismissed as irrational, and the content of the story is misconstrued, reacted to as untruth, or used as a source of primary information on which to form cultural belief systems and policy. Ignorance is bred from the reliance on one-dimensional, limited sources of knowledge, bereft of the richness of moral reflection on narrative itself. It is furthered by people's natural tendency to selectively listen to the voices that further their own interests and beliefs, rather than to receive and participate in the creation of new narrative constructions.

Disparate voices have arisen with genuine questions about the goods and harms that may emerge as the result of the ever-increasing ability to manipulate and control matter. Some voices tell stories of nanotechnology as ultimately generating more health/environmental, social, and cultural problems than it eliminates. They eschew rapid development of nanotechnology on the fear that it may release uncontrollable and toxic biological/synthetic new substances, encourage the proliferation of war, exacerbate gaps in access to wealth and resources, increase ecological demise, and even possibly foster social isolation. The ETC Group,[1] Prince Charles (Rhodie, 2003), Michael Crichton, Bill Joy (see next section), and others have been unabatedly bashed by some influential nanotechnology proponents, publicly dismissed as hyped, irrational, and ignorant. On the other hand, public policymakers and scholars appear to be trying in earnest to address the social/ethical questions of "ought this to be done, and if so, how can it be done safely?" with regard to the international quest to manipulate matter at the nanoscale.

Federal funding allocations reflect some of these concerns and provide resources for the study of nanotechnology's health, environmental, as well as societal and ethical implications. Also of concern to some legislators is the potency of public opinion, in its potential to deter or otherwise interfere with nanoscience and technological development. They want to educate the public and inform them in such a way that elicits its support. I question the integrity of pundits who seek to justify their campaigns to keep consumers passive and noncritical about nanotechnology development. This intent is demonstrated by some nanotechnology proponents, who being protective of the nanotechnology initiative seek to discredit the voices of science fiction writers (those fantastical visionaries), as a means to gain public trust and to avert the potentially devastating effect of public

---

[1]ETC Group (Action Group on Erosion, Technology, and Concentration); this is "an international, nongovernmental, civil society dedicated to the conservation and sustainable advancement of cultural and ecological diversity and human rights." See http://www.etcgroup.org/text/txt_about.asp

opposition. Public dismissal of this science fiction by nanotechnology proponents reflects the common failure to recognize the insight into the human condition provided by those narratives. Narrative is a constructive alternative to the standard account (see Hauerwas & Burrell, 1989, p. 176).

Authentic public support for nanotechnology development can only be earned through ethics, as attained from narrative. There is an important role for the public to play in the negotiation and determination of nanotechnology's appropriation. To this end, the U.S. government has funded citizen panel programs based on the European models, which provide a forum for the involvement and inclusion of the public in dialogues about nanotechnology concerns. Here is one likely means to authentically educate and include our citizenry toward a commonwealth in nanotechnology development. This is the way to garner authentic public support, if it is warranted. But the narratives of individual research scientists and engineers should also be included, not solely as the voices of professional experts,[2] but as interested citizens with a story to tell. Successful public trust and understanding warrants the inclusion of individual laboratory researchers as persons, who might be willing to contribute their own stories to the wider public discourse, along with their understandings, ideas, beliefs, and perspectives, as they pertain to the nanotechnology initiative.

RESEARCHERS AS EXPERTS

Public statements made by researchers about newly emerging technologies have the potential to influence both public conceptualizations and political decisions about those technologies. For example, Bill Joy's (2000, former chief scientist for Sun Microsystems) "Why the Future Doesn't Need Us" stimulated a public debate into an emotive, provocative, and far-reaching discourse on the convergence of newly emerging technologies: nanotechnology, genetic engineering, information technology, and robotics.[3] When Eric Drexler said before the 1992 Congressional Subcommittee on Science that "this technology will clearly have broad applications. If you can work with the basic building blocks of matter, you can make virtually anything, producing a much wider range of products than can be

[2]The role of experts in nanotechnology research and development is discussed by Sarewitz and Woodhouse in their essay, "Small is Powerful," (2003). In D. S. A. Lightman & C. Desser (Eds.), *Small is powerful. Living with the genie* (pp. 63–84). Washington, DC: Island Press.

[3]For further information on the converging new technologies, see Roco, M., Ed. (2004). The co-evolution of human potential and converging technologies. Annals of the New York Academy of Sciences. New York: The New York Academy of Sciences.

made by processes that lack this direct control of the fundamental pieces ..." (Drexler, 1992). His words apparently had an influence in policy decisions, which led ultimately to funding of the original National Nanotechnology Initiative. It is not apparent how those who testified during that session were selected to do so. One thing they seem to have in common is "expertise," as demonstrated through professional, public notoriety. But Drexler's science is not laboratory based—it is speculative. Some researchers in the science community say they are perplexed by Drexler's early influence on public policy. He has come to be a controversial figure among research communities. By some he is considered to be an instrument of public and political interest in nanotechnology. Others consider him to be a nanotechnology visionary and spokesperson on the potential for nanotechnology to effect profound material change, and still others reject his ideas about "molecular manufacturing" (programmable molecular machines called assemblers that can build any molecular structure from the bottom up—atom by atom—and molecular scale replicators that can copy themselves and self assembling)—(see Drexler, 1986) as scientifically implausible. Early on in the government's consideration of funding nanotechnology research, Drexler was invited to give testimony in federal hearings. Drexler was not included in the 2003 hearings on nanotechnology appropriations (a time during which there was a significant increase in funds being proposed and in a climate of highly sensitized public perceptions about nanotechnology) and he felt personally offended by the blatant exclusion.[4] Other futurist-scientist-visionaries, such as Ray Kurzweil, did testify. During the spring 2003 congressional hearings, MIT's Kurzweil claimed, "Our rapidly growing ability to manipulate matter will transform virtually every sector of society, including health and medicine, manufacturing, electronics and computers, energy, travel and defense" (House Committee on Science, 2003). Disparaging those who object to human manipulation of the "natural world," he proclaimed, "The increasing intimacy of our human lives with our technology is not a new story, and I would remind the committee that had it not been for the technological advances of the past two centuries, most of us here today would not be here today" (House Committee on Science, 2003).

The words of IBM's physical scientist, Thomas Theis, were equally encouraging about the importance and significance of nanotechnology. His

[4]This sentiment was expressed directly in both formal and informal remarks to author and others attending a meeting in March, 2004 at the University of South Carolina on the use of imagery in nanotechnology.

testimony to Congress included the statement: "Nanotechnology is key to the future of information technology." Theis explained:

> Nanotechnology allows us to characterize and structure new materials with precision at the level of atoms, leading to materials as superior to existing materials as steel was to iron, and iron was to bronze in earlier eras. Nanostructured materials hold the promise of being stronger and lighter than conventional materials. This would have innumerable beneficial impacts from more fuel efficient and safer airplanes and cars, to luggage that can withstand baggage handling at airports! But strength is just one property. Designing materials with atomic precision allows unprecedented control of their electronic, magnetic, optical, and thermal properties—in fact, any property that we want to enhance. (House Committee on Science, 2003)

Dr. Vickie L. Colvin, director for Biological and Environmental Nanotechnology at Rice University, opened her testimony at the hearings with reference to the novel, Prey. She warned the Committee on Science that public fear could bring nanotechnology to its knees. She suggested that nanotechnology needs strong public support in order to proceed, and further, that there are still many unanswered questions about the effects of nanomaterials on human health and the environment.

Drexler, Kurzweil, and Joy are well published and popularized in the public discourse. Their voices are persistent and influential. Theirs and the voices of other laboratory scientists such as Colvin, Theis, and others have clearly been a critical source of expertise for the determination of public policy. They reinforce the political aspiration and hopes for grand and powerful material goods to result from nanotechnology. On May 1, 2003, the House Science Committee approved legislation that would authorize a national nanotechnology research initiative. Six months later, in November 2003, the U.S. Senate passed by unanimous consent a version of the 21st-Century Nanotechnology Research and Development Act (S. 189; see Appendix D). That bill authorized $3.7 billion over 4 years for the program. It requires "the creation of research centers, education and training efforts, research into the societal and ethical consequences of nanotechnology, and efforts to transfer technology into the marketplace" (S. 189, 2003). The act provides a substantial amount of support for the participating researchers and the major centers that they either lead or of which they are a part. Although some have private or industrial funds, the main investor in nanotechnology research in the U.S. is the government, through such agencies as the National Science Foundation (NSF), the National Institutes of Health (NIH), the Department of Energy (DOE), or by military agencies such as the Defense Advanced Research Projects Agency (also known as DARPA, the major research branch of the Department of Defense, DOD).

LIMITATIONS ON THE VOICES OF RESEARCHERS

Communication scholar Chandra Mukerji (1989) voiced concern about what she considers to be the prescriptions and limitations on individual research scientists to engage their voices in the public domain. Her research looked at the relationship between marine biologists and oceanographers and the federal government that largely supports them. Her analysis of 74 interviews and various written documents led her to conclude that whereas individual scientists do have intellectual autonomy, it comes at the price of relatively little freedom of external expression about the topics of their research, especially its larger social / policy implications. This is because they have made a tacit agreement to serve as what Mukerji called, "an elite reserve labor force for the interests of the state." These are an elite group of people who are supported with public funds, so that "their well-honed skills will be available when they are needed (by, for instance, the military in case of war, by industry in case there are major changes in the direction of the economy, or by the medical community if there is an outbreak of some new and threatening illness" (Mukerji, 1989, p. 6). Comparing research scientists to Army or Navy reservists, Mukerji viewed them as having in common being paid by the government in order to sustain the skills that might one day be needed. According to Mukerji (1989), "What seems sociologically significant is that they are paid money to keep their research skills sharp on the agreement (for scientists, implicit) that they are 'on call' to be mobilized when their services are needed. Members of government can ask them to apply their scientific expertise to practical problems, and the norm of reciprocity requires that they agree to do so" (p. 7). If Mukerji's model is correct, then it is apparent that the forces have been mobilized again, not under the banner of war per se, but under the imperatives of the National Nanotechnology Initiative. It could be that the nanotechnology initiative is primarily concerned with economic growth in the face of a weak economy. It also appears to be occupied with national security under the shadows of terrorist threats, international prowess in military, global alertness, and health crises in the costs of treating persistent diseases such as cancer and heart disease.

Mukerji traced the development of the relation between science and its federal funding agencies back to World War II, when the government had to scramble to immobilize a workforce of scientists and engineers to address the immediate and critical military and economic problems of the nation. The researchers enjoyed the status and financial support they received in exchange, but quickly became dependent on the military and on business for their ongoing support. After the war, the government had a continuing interest in having access to a highly trained source of scientists and engineers. Today, that elite workforce has the freedom to set its own research agenda and strategies, but must compete for federal funds to pay for expensive equipment, laboratory space, graduate students, and so on. The sociological problem Mukerji (1989) identified is that their mobilization by the state is hidden within the reward structure of science: "They cannot see their value to the state, so they tend to discount the significance of science to national policies.... They do not see themselves as having a political role, per se" (p. 12). One of the physicists in my study adamantly disagreed with Mukerji's assessment. He explained that at least in physics, the profession has a robust public voice when necessary, with strong moral convictions, and one that has ample historical precedent for politically opposing particular technological applications of scientific knowledge.

Although engineer and scholar Samuel Florman indicated that scientists are thinking more introspectively about their work than in the past, Mukerji suggested that the very system that supports them in their research is unreceptive to their individual opinions outside of what is elicited from them as professional experts for the interests of the state. Mukerji (1989) explained, "What is said by the scientists that seem to retard a desired policy can be kept confidential because it is so effective, just as it can be used publicly when its rhetorical power seems to fit the interests of the bureaucrats who contracted to hear it ...." (p. 201). The *voice of science*, as Mukerji called it, is harnessed for achieving lasting knowledge, detached analysis, and thoughtful reflection for the political purposes of the government. Mukerji concluded that the power scientists have is in the hands of the modern state, but little power is accorded to them as individuals; they have tacitly agreed to give it over to their supporters, a state requiring personal detachment from their work. This, claimed Mukerji, makes the scientist vulnerable. She concluded that, "the scientific establishment routinely gives away one of its greatest assets; its voice" (p. 203).

My own encounters with individual scientists and engineers in the nanotechnology initiative reveal most to be deeply introspective in private, openly expressive of their personal ideas outside of the "expert" voice they are sometimes asked to carry. It may be true that their individual participation in the public discourse is not institutionally supported or encouraged. It may also be true that, as Mukerji pointed out, the government solicits their expert testimony, and asserts some controls over the flow of the information the scientist provides as an expert. There are, however, examples to the contrary, such as the response of key scientists in the 1970s in calling for a moratorium on "recombinant DNA research," and following that call with the presentation of carefully construed, precautionary guidelines. A few scientists are speaking and writing about nanotechnology to non-expert audiences in the public domain. There are nonetheless many researchers who are very highly regarded by their respective science and engineering communities, and are guarded and insulated from participation in the public domain. Although potentially important, those individual voices are disengaged from the public discourse. What is needed is for more behind-the-scenes experts to voice their concerns or their thoughts about the corporate-political motivations, societal dimensions, and ethical implications of nanotechnology. Why is the majority so silent? Perhaps because the enormous time pressures of securing funding, managing laboratories, and teaching students keeps them internally focused. Or, maybe no one wants to bite the proverbial hand that provides one's sustenance? Of course, there may be less dubious reasons for the relative silence of the individual scientist, as a person. Introversion, personal preference, professional isolation, or cultural boundaries may also keep them at a distance.

SCIENTIST AS PERSON

As a group, research scientists and engineers have a long history as agents of social awareness.[5] They have spoken out into the public and against government policies when reasonable moral limits are crossed. On occasion, individuals have done likewise, such as when Noble Prize winner Dr. Henry W. Kendall spoke out on missile defense (Society, 1997). As such, their beliefs about the ethics and possible meaning of nanotechnology, and about the ends and purposes of the initiative, could potentially play a significant role in policy, regarding the direction and outcomes of the

[5] For example, the Union of Concerned Scientists, public stand on global warming, available at: (http://www.ucsusa.org/).

nanotechnology quest. But so far, as individuals they have generally been publicly uninvolved in the ethical guidance of nanotechnology development. The explanation, however, may lie beyond Mukerji's research findings. The "voice of science" she referred to is often a collective voice, which arises from within a highly formalized system of critical, peer review. Most individual researchers seem to prefer to abstain from expressing their own individual voices, relinquishing that role to the designated public spokespersons of the science community, but not necessarily under the thumb of the government that sponsors them. Researchers participating in this study readily acknowledge that they sometimes tailor and adjust their own research proposals to the aims and specifications of particular federal agencies, industries, and corporations. Responding to the question of their relative absence from the public discourse, some have offered the following explanation: They feel they haven't got the expertise or understanding to speak out individually on matters of public policy and ethics in the nanotechnology initiative. Their primary skills and principal interests in nanotechnology lie with the business of their own research groups, inside of their own personal laboratories.

When I inquired of these same researchers about the possibility of their participation in the public discourse over the societal and ethical dimensions of nanotechnology, some initially asked why I am so interested in the individual researcher when the science community takes good care of representing itself through professional organizations, lobby groups, and its publications. One answer I have given is that there is not yet any such professional organization specifically for nanotechnology. There are organizations for researchers with an active interest in nanotechnology, such as the American Society for Mechanical Engineering (ASME), the American Association for the Advancement of Science (AAAS), and the National Academies of Engineering and of Science. There are also think tanks, such as the Foresight Institute, which annually recognize outstanding research in nanotechnology. But, at the time of this writing, no professional organization has made its primary mission the ethical leadership of humanitarian, nanotechnology development.

There is too much at stake in the development of nanotechnology in terms of ethics, humanity, and the well-being of the planet, for individual researchers to cloister themselves inside of professional associations and affiliations. And further, society needs scientists to contribute to the public discourse as persons (not only as experts in the interests of the State), thereby helping to frame the meaning and significance of their work toward a democratized process of knowledge creation and technological development. The expertise of researchers, and their core beliefs, could be a rich source of understanding toward the social agenda of conscientiously directing the unfolding of nanotechnology. Without clear, definitive moral leadership offering the articulation of absolute humanitarian aims, the development of nanotechnology especially needs the engagement of individual scientists, who are unencumbered by the boundaries of the collective, professional voice. Again, there are a few very powerful and audible voices of individual researchers out here. But those are primarily the voices of proponents, and not necessarily voices of discerning, critical moral leadership. And, furthermore, it is the voices of lesser known, behind-the-scenes, individual researchers, speaking not for or from "the community of science" or on behalf of their sponsors, but speaking for themselves, which are especially needed at this time to contribute their perspectives to the understanding and wisdom of the wider community.

NARRATIVES IN THE PUBLIC DISCOURSE

Narratives allow for entering into the life of a story, identifying with the characters, and participating in their moral deliberations as the plot unfolds. Technology is expressed through, and embedded in, personal and collective narratives that function as part of larger and often competing discourses. Humans use discourse to make meaning of their lives, as well as their relationships to the physical world and to one another. This is one means we have of constructing and expressing essential reflections about moral inquiries. The distinctively human endeavor to make and share meaning, to express and reflect belief, to formulate and establish ethical understanding, and to respond to varied competing sources of motivation and inspiration are embedded in discourses about nanotechnology.

Under the auspices of nanotechnology, science, (the observation and contemplation of the material world) partners with engineering (the transformative, design focused reconstruction of the material world). That process is amorphous, however. Where this merging is taking us, the teleology (design and purpose in the material world) of nanotechnology, is yet to be determined. I believe, however, that its quest is value laden and intentional. The question at issue here is whether the voices of individual researchers might offer support to the public's observations of, and participation in, the conscientious development of nanotechnology. Unfortunately, research scientists and engineers are often portrayed in the public as amoral and taciturn about seeking to understand and solve problems through technology. Generally speaking, the language and literature of science are inaccessible to outsiders. When researchers communicate among themselves formally and informally at professional meetings, through journals or to government officials, they are only partially and indirectly engaged in the larger public discourse.[6] They are stereotyped as people who have difficulty connecting to humanity because of their obsessive interest in the mundane elements of the material world of observation, processes, gadgets, and devices. There are few public mechanisms to correct this false portrayal or to reveal their own individual struggles to make meaning of their work and contribute directly to the benefit of humankind. And yet that is precisely what many of them aspire to do. One hope I have is to make a contribution to what Daloz et al. (1996) called "a different public mirror" (pp. 7, 8). Her listening-based research describes the lives of "those who are committed to the common good, who seek to align themselves to the life of the whole, and work on its behalf." Certainly there are individual nanoscale science and engineering researchers who could be identified with Parks' group, but who are not reflected in the public mirror. The intention of my project is to gather and reflect voices of individual researchers toward an open, honest, barrier-free dialogue with an otherwise inaccessible world of other, and in so doing, to see that they are included in the reflection of the public mirror of the nanotechnology initiative.

One interesting aspect of the nanotechnology initiative is the challenges it presents to ethics, meaning, and belief, which arise with the changes that come from the development of any new and novel technologies. No one really knows where the nanotechnology initiative is leading, what it will mean, and how it may affect human life. Nevertheless, we are moving at full speed into that unknown. One of my book reviewers suggested that the rapid movement toward who knows where is endemic to engineering; there are many things engineers do just for fun, without knowing the consequences, believing their efforts will be beneficial, but not really considering the negative consequences. Further, this enthusiasm comes with no malicious intent but rather from simple ignorance. This is most often the case. But, in the case of developing nanotechnology,

---

[6]Christopher R. Toumey has a very helpful discussion about why this is so in his book, *Conjuring Science.* The section entitled, "The Cultural Failure of Popularization," explains that the scientific ethos has not been understood by the public, which led to a troublesome 'conjuring' of science in the public domain.

such undirected movement also represents a technological leap of faith that necessarily calls for recognizing the imperative of real time, imaginative, and conscientious engagement with the future being created. If somehow the scientists' voices can be brought into the public domain, as they describe and tell stories about their ideas, experiences, perceptions, beliefs about their work, nature, change, and the future, we might better respond to the moral inquiry that asks: why, and to what end, are we pursuing nanotechnology development? This is an inquiry the society needs to address with laboratory researchers, not without them. The 21st-Century Nanotechnology Research and Development Act (see Appendix D)[7] includes the requirement of studies about the societal and ethical consequences of nanotechnology. Individual research scientists and engineers could be called on to help society seek to establish the social significance and ethical domains of the possible nanotechnology futures.

RESEARCH PROJECT: MEANING AND BELIEF INSIDE THE DEVELOPMENT OF NANOTECHNOLOGY

Nanoscale science and engineering, generally and commonly referred to together as "nanotechnology," represents a broad range of research, distinguished no so much by subject matter, but rather by the novel properties presented at that scale, and the technical applications that manipulations at that scale make possible. It is a rapidly developing, relatively recent scientific and technological endeavor. Nanoscale science pertains to interrelated processes that occur at the scale of one billionth of a meter. One definition of nanotechnology reads, "Working at the atomic, molecular, and supermolecular levels, in order to understand, create, and use new materials, devices, and systems with fundamentally new properties and functions because of these small structures."[8]

Nanoscale science and engineering researchers work inside of a larger social context of technological development. They participate inside of the wider network of culturally constructed notions of the good, of government and business motivations for technological ingenuity, of large organizations and the public domain, of religious and popular belief systems

[7]Which put into law programs and activities supported by the NNI (National Nanotechnology Initiative), continuing to support activities aimed at assessing the societal implications of nanotechnology, including ethical, legal, public and environmental health, and workforce related issues.

[8]www.nano.gov.omb_nifty50.html

about human life, of political pressures and personal desires, and of all the other myriad elements that comprise this web of social-technical evolution. What is it that they believe they are doing with nanoscale science? What do they imagine is possible as a result of their work? What personal hopes, aspirations, beliefs, and fears do they have? What do they believe motivates their work? What, if any, kinds of studies might they reject from their own labs? As with any new technological development, nanotechnology is an expression not just of the intellect, but of other dimensions of human psyche as well. Like art, music, dance, and literature, scientific exploration and technological development is an external expression of internal longings, curiosities, struggles, desires, fears, and possibilities. The curiosity to know, to understand, and to control our material universe, which fuels nanoscale scientific inquiry and technological development, is a source of passion and motivation for most of these researchers. These voices are valuable, even perhaps essential, to any well- informed public discourse on nanotechnology development.

This book presents narrative as a essential source of ethical analysis and reflection about nanotechnology. The focus here is on conversations with research scientists and engineers who are at the heart of nanotechnology development. The hope is that through discourse, we might better be able to garner understandings about the inevitable changes nanotechnology will bring to humanity, toward the development of an ethical analysis of nanotechnology. Indeed, making scientists and engineers the focal point is a curious undertaking. Even the individual researchers who are participating see themselves as having little to say in the public domain about the societal or ethical implications of their work. Furthermore, they feel their skills and expertise lie in the research laboratory, and not so much in expressing their views on the meaning and social significance of their work. So why engage researchers in conversations over questions of ethics and values about nanotechnology, and their own personal visions and beliefs? Because as experts, they have formed certain kinds of perceptions about the material universe to which most of the general public has little access. Their stories are potentially a powerful force, influential in shaping the wider ideas and beliefs about nanotechnology's direction and purposes. Through interdisciplinary collaboration efforts and shared expertise, nanoscale science and technology researchers are in large measure the designers of our emerging world. It is these individuals who are in the laboratories, framing the basic research questions, making the observations and discoveries that open new possibilities for society through knowledge and technology. As such, they have the unique capacity to offer moral leadership to the pursuit of the imagined, technological future.

Generally, sharp distinctions are made between scientists and engineers, but those distinctions are becoming increasingly vague in the nanotechnology initiative. Both scientists and engineers work as principal investigators within laboratory settings. Both are deeply curious about novel phenomenon and "interesting" problems. Both research scientists and engineers receive funding from federal sources. Furthermore, because of the increasingly applied nature of basic science in the nanotechnology domain, both are motivated by the allure of material applications as outcomes of their work: "The core of nanotechnology is its inter-disciplinarity."[9] Under the umbrella of "nanotechnology" flourishes an intriguing diversity of formerly distinctive and often fractious fields of science and engineering research, with researchers increasingly collaborating across disciplines, toward the acquisition of new knowledge, the observation of newly perceptible phenomena, and the creation of new devices and processes at that fantastically small scale. Biologists; chemists; physicists; biochemists; theoretical and applied mathematicians; material scientists; computer scientists; mechanical, civil, chemical, biochemical, and biomedical engineers; and researchers from other specialized and distinctive fields exchange a plethora of findings and engage fascinating problems to take on under the rubric of nanotechnology.

Nanotechnology centers tend to be made up of individual investigators from varied academic departments and schools in one or multiple institutions, who are in some way working to inform and cross-fertilize one another's work. For this reason, both research scientists and engineers are included in my research, and are referred to in this writing more generally as researchers. The focus of my own research is on them—the individual scientists and engineers whose research is sponsored both privately and federally under the national nanotechnology initiative. It reaches for researchers as individuals: who they understand themselves to be, what they aspire to achieve, and how they are conceiving the use of their research in nanotechnology in the design and determination of future human and technological evolution. What personal dreams, imaginings, hopes, and motivations are inspiring them? What role, if any, do they ascribe to their work in the evolution of humanity within the evolu-

[9]As stated by the chair of the ASME Nanotechnology Institute Advisory Board, in his welcoming to the Third Integrated Nanosystems Conference, Pasadena, CA, 2004.

tion of the world? How are they able to maintain a clear sense of purpose amidst the fierce external pressures that accompany this quest? What internal beliefs may be enhancing, or perhaps convoluting their moral commitments? Most importantly, what pseudovalues might need to be unmasked in order for them to assure that their part in this "technological revolution" will be pursued ethically, and with a genuine regard for human and earthly life?

This project entails yearly, informal, private, and confidential conversations with individual male and female scientists and engineers who are working at the nanoscale in their research, in major collaborative or independent efforts. From these conversations come a rich and fascinating body of textual material, which provides the basis for a dialectic process of engaging and deliberating the meaning and significance of nanotechnology development. What do individual researchers believe about nanotechnology, their own research, and their role in the future that nanotechnology will bring? How might those beliefs and understandings, their unique perspectives and perceptions, be embedded inside of the research itself? What is the nature of the beliefs are at work in their perceptions? These types of questions are considered in the hopes of bringing into the public domain, the personal commitments and beliefs held by some of the very people on whom the nanotechnology initiative depends. The intention has been to elicit their ideas, concerns, beliefs, fears, and motivations, as those pertain to their work as researchers in nanoscale science and technology. The aim here is to help "disparately interested parties overcome their language differences in order to join in a common cause."[10]

My studies follow these scientists over a 5-year period, as they move deeper into their own abilities and understandings, and as they make more discoveries, broaden their collaborations, and facilitate the development of new technologies. This book comes midway through that 5-year study. Excerpts from multiple conversations with the 35 researchers are woven throughout the following chapters. In addition, eight full text conversations standing alone, for the reader to unravel, without comment or explanation from the author. The National Science Foundation (NSF) sponsored this research as a career award, formally entitled *Meaning Making and Belief Inside the Development of Nanotechnology*. The participants were principal investigators who are conducting nanoscaled research in their own laboratories at universities across the United States. They and

[10]See Fuller (1993) for discussion on the role of rhetoric in Fuller's approach to the social epistemology of science.

their institutions' names are held in anonymity, with identifiable indicators removed from the text. Otherwise, the essence of the interviews is presented in an unadulterated form.

The scientists and engineers have been quite generous with their time and ideas. And, as participants in this study, they have given richly to the quest for understanding the meaning and significance of the development of nanotechnology. They have spoken from their roles as primary characters in the unfolding story of new scientific knowledge and novel technical abilities, and as individual persons with their own unique ways of seeing the world in which they live and work. The words and ideas expressed in these dialogues suggest that there are deep and complex relations between nanotechnology explorations and its meanings to humanity—meanings that are, to some extent established and conveyed through discourse itself.

Originally, I asked over 50 individual researchers to participate. Thirty-five said "yes," and met with me once. Twenty-three have met with me twice, and I anticipate that by the time of this writing, 18 of those will have completed or be scheduled for a third conversation, and one will have had a fourth. It could be argued that the group of 23 continuing participants is a self-select group. It is likely that those who continue to make themselves available for these discussions probably have a genuine interest in reflecting on the meaning and ethics of their work in nanotechnology. They may have been predisposed to participate.

From my own perspective, the relationships have become more trusting and open with each subsequent interview, which means the researchers and I are likely adapting to one another in ways that may be altering both the types of questions asked and the answers given. Nevertheless, there is valuable data in the interviews in that the language used and stories told belong uniquely to the researchers. Although there may be some self-conscious maneuvering on both my part and that of the individual scientist or engineer to provide what we believe is expected of one another, there is ample evidence to suggest that honesty pervades these discussions. There is another feature worth noting. That is, changes seem to be taking place. As a researcher, I was initially somewhat skeptical about the research scientist or engineer's commitment to consider ethics as it pertains to their work. (I now perceive a consistent and genuine concern on the part of most of those with which I have spoken.) Some researchers in the study seem to be changing as well. In the beginning, they were participating out of politeness and answering my questions guardedly. Now, most are coming across as personally engaged and actively interested in these discussions. Perhaps in the end, personal growth for all of us will be an unanticipated consequence of this basic research project.

The aims of this research project, with this book as an interim report, are threefold. First, I wish to contribute to the quest for the ethical, humanitarian evolution of nanotechnology by stimulating further thought and conversation among and between students, scholars, policymakers, private investors, corporate managers and executives, and the interested individuals in the general public. It is a common but mistaken belief that science and technology cannot be directed, that is, that they take on a life and course that are independent of human volition and will. Directing the course of nanotechnology development toward humanitarian aims may be an unwieldy, even implausible, undertaking. Attempting to do so is, nevertheless, a moral obligation. To that end, the voices of researchers provide a critical vantage point from which to engage the conscientious process of setting nanotechnology development on an ethical course. If left to the random influences of corporate market incentives, institutional ingenuity, personal curiosity, and national struggles for global dominance and economic power, then the nanotechnology quest is likely to be indeterminate (vulnerable to an uncontrollable, boundless course of evolution). If public policymakers, industry leaders, politicians, venture capitalists, the lay public, and laboratory-based researchers will engage in an open, honest dialogue toward the negotiation and determination of nanotechnology's course of direction, then there is hope for humanitarian ends. This kind of dialectic has the powerful capacity to "focus an otherwise indeterminate reality." It can offer a critical means through which the socio/cultural process of meaning making about nanotechnology's influence on the future of our civilization might occur.

Second, it is my desire to stimulate another level of public discourse among and between scientists who function from inside of somewhat cloistered communities, and the broader public communities (students and scholars included) who otherwise might have relatively little access to (or interest in) the personal thoughts and ideas of individuals who are integral to the nanotechnology initiative. Third, I hope that by addressing what may be some of the elements of meaning and belief that are enmeshed in the nanotechnology initiative, the project will demonstrate a multidimensional, nanotechnology ethics analysis. It would be especially wonderful if this project might also lend some moral support to those research scientists and engineers whose intent it is to contribute conscientiously to the ethical development of our nanotechnology future.

Here and there, some of the excerpts offer a bit of technical description of the researcher's work, but mostly they reflect conversations about their personal values, hopes, beliefs, and worldviews. These elements of individual reflection seem inextricably linked to the uniquely human quest for control over destiny and the material universe in which we perceive ourselves to live. One task in thinking about the stories researchers tell about their work and, more generally about the enterprise of nanoscience, is to identify any conceptual roadblocks, which may stand in the way of an ethical, conscientious evolution of nanotechnology. A second task is to decipher elements of perception, value, and belief that may be useful in the public and policy-level formulation of an ethics for nanotechnology.

As with all discourse about technology, the researcher's discussions here include powerful, cultural symbols that play a key role in the searches for meaning. If reality itself is constituted through and mediated by symbols, then those symbols themselves create structures of meaning. To understand belief, it is necessary to decipher how meaning is being ascribed. To establish ethical foundations for the development of nanotechnology, it is necessary to identify both the explicit and tacit working assumptions embedded in the nanotechnology quest. Researchers are at its core. Therefore, their assumptions, beliefs, and personal perspectives are integral to its direction and development. Individual researchers are integral to the process of meaning making (a process that is constantly underway in the human mind, and communally, in the public domain) as important voices in the conscientious development of nanotechnology.

I hope and believe that some trust has been built between myself and the researchers who are at the heart of these conversations. I also trust that there has been honesty and openness on both our parts. However, given my formal training in the humanities and my limited exposure to the sciences or engineering, there is a boundary that keeps me from ever truly penetrating their world. On the other hand, our sharing of the values of knowledge and scholarship may create a sense of mutual respect and appreciation. Perhaps our mutual interest in what is "good" has some influence as well. There at least seems to be permeability between our otherwise very distinct worlds.

Dialogue is the primary method of inquiry in this project, wherein listening is central. Intentional listening puts me in a position, with respect to the researchers, which offers them a chance to indicate a certain amount about themselves through me to others. I ask very specific questions in the interviews (see Appendix C), such as:

- When did you first get interested in science? (Or, engineering?)
- How did you first get involved in nanotechnology?
- What are you ultimately hoping to accomplish?
- What would it mean if you were successful?
- What makes your research most interesting to you?
- What do you imagine could go wrong with your research?
- What is the best outcome you can imagine from your research? The worst?

The discussion questions are open-ended, with follow up responses to solicit deeper reflection: Why is that important to you? What do you mean by that? Why did you use that word? Why do you care? But it is their stories that are the central matter. The goal is to move beyond the simple transcription of interviews to deliberately shaping the material into stories that reveal certain life themes. The method is inspired by the "participant observer" work of Robert Coles, who acknowledged its risks as including intrusion of the observer's subjectivity and biases. My own subjectivity and biases are clearly present in the discussions. However, when I am aware of them or they are perceived by the researchers, I try to acknowledge them openly. I try also to remain tentative in making judgments, and to engage in deeply respectful listening. My effort has been to honor and respect all researchers who have given their time to the project through conversation with me, and to validate and make audible their wisdom and understanding about the world they inhabit and the beliefs they hold as their own, as researchers of nanoscale phenomenon.

A common closure researchers make in these conversations is to say something like, "This is great. I rarely get the opportunity to talk about these issues in this way. Is that simply because they don't have time?" Yes, in part. Is it also because society at large is disinterested in the voices of scientists and engineers who are philosophically reflective or critical about the nature of their work? No, absolutely not. Is there something about the scientific method of research that inhibits such reflection? I don't think so, but the scientific method itself is commonly, incorrectly touted as an amoral endeavor, and therefore wholly objective. Does the science-state relation preclude this kind of introspection and open dialogue? Perhaps, but I cannot say with certainty. What I can proclaim with conviction is that the diverse voices of individual researchers need to be heard more often, and engaged more openly in the public domain. To that end, I present these voices, along with my own interpretation of what their words might imply or suggest about ethics, meaning, and belief in the development of nanotechnology.

Part I

ETHICS

Well, you can plant seeds every step of the way. You have got to be careful what grows from them. We used to have this argument, and I don't know if this is a good example, but talking about global warming I always said, "Well you know you have got to really consider it, think about it and make sure we understand it scientifically." One perspective is, "Why should you do anything until you understand, really scientifically understand the impact of changing things?" And my argument was that we have already done that. We introduced fossil fuel consumption as a source of energy before we understood it. Now our whole society is based on it. No one thought in the beginning when there were very few cars driving around and a few homes being heated that every family would have two cars, almost every home would be using fossil fuel. Industrial plants would depend on it. When they first started, no one ever really thought about it that way. It was hard to imagine how such a little change could make such a big change in our lives.

—Shalini

CHAPTER ONE

The Nanotechnology "Revolution"

ROSALYN:	*Do you think we have any control over where we are going to end up, where we are going in the process of discovery?*
IRVING:	*As a species?*
ROSALYN:	*No, as a species endeavoring in nanotechnology.*
IRVING:	*I think that in the end we will do what is possible. We will do what is possible and we'll then have to deal with the consequences. If something is discovered to be possible, someone will do it. The fear of someone doing it and gaining a competitive advantage is enough to drive people who are able to at least find out what is possible.*
ROSALYN:	*So is this inevitable, that we will do what is possible?*
IRVING:	*Yes. It's our nature as a species to be curious. It's the nature of life.*

Scientists and engineers who work at the nanoscale are not the only ones who are intrigued with the possibilities that arise from exploring that dimension of the material world: Science fiction writers are also interested in the future possibilities; the military is looking at nanotechnology's potential to effect the protection of soldiers and to increase efficiency of warfare; venture capitalists are positioning for competition in its newly created markets, while banking on its promise of phenomenal returns on investment; and policymakers debate the allocation of funding it, preparing for global domination with it, and try as well to anticipate and manage potentially debilitating public perceptions. Economists, psychologists, historians, philosophers, and sociologists of science and technology, rhetoricians, and religious ethicists grope with how to abstract the material, social, cultural, and ethical significance of nanotechnology. Political leaders all over the world have made the development of nanotechnology a national priority. As spoken by the director of the U.S. Office of Science and Technology Policy (Marburger, 2003),

> Not until recently have we actually had the instruments to make atomic level measurements, and the computing power to exploit that knowledge. Now we have it, or are getting it, and the implications are enormous. Everything being made of atoms, the capability to measure, manipulate, simulate, and visualize at the atomic scale potentially touches every material aspect of our interaction with the world around us. That is why we speak of a revolution-like the industrial revolution-rather than just another step in technological progress.

Documents from the U.S. National Nanotechnology Initiative (National Science and Technology Council, NSTC, 2002) describe nanotechnology as "the ability to work at the molecular level, atom by atom, to create large structures with fundamentally new properties and functions." Nanotechnology is also spoken of as the search to understand and mimic nature's own mastery of the atoms. Nanotechnology's very purpose is revolutionary: to put into human hands the ability to "manipulate and control matter with precision, and to specification." Some nanotechnology proponents say that this technological revolution could generate amazing and very powerful capabilities, such as dramatic improvements in accurate diagnostics in health care; more precise delivery of drugs and other forms of medical treatment; development of highly sensitive, sophisticated military capabilities; use of increasingly small and highly sensitive surveillance devices and other tools of security; opening major new markets through the development of new materials; rapid processing and distribution of information; storage of vast amounts of information in exceedingly small devices; breakthroughs in strong artificial intelligence; abundance in energy sources and bolstering the semiconductor industry by compensating for Moore's law.[1]

A number of the laboratory-based researchers participating in this study (particularly chemists) maintained that they have been working at nanometer scales for over two decades. They explained that the term nanotechnology functions to cast a unifying net around increasingly interdisciplinary fields of scientific inquiry. And, to their benefit, it opens new but highly competitive opportunities for support of their own ongoing research interests, and greater potential for applications of their results. One of these researchers, a physicist, speaks of the origins of nanoscale science in this way:

---

[1]Moore's Law: The exponential rate at which computer circuit size has been decreasing is predicted to slow down, thus drastically limiting earning potentials that depend on new product development. A primary assumption is that this trend will only be abated if and when an entirely new computing architecture is developed.

> When physical chemists and chemical physicists learned to make cluster beams, they rapidly discovered that matter in very tiny clusters with up to 100 atoms did not behave like bulk matter. "Buckeyballs" were one outcome of this research, but more generally cluster physics and cluster chemistry triggered a search to find out where the "cross-over" is between atoms or molecules and bulk materials. At the same time, developments in semiconductor physics, and in particular the ability to make extremely thin films of semiconductors, created a science of "reduced dimensionality." In a real sense, these are the two intellectual forces that created nanoscience.

Nascent nanotechnology development has already produced nanoscaled devices, such as nanoscale storage and nanotube transistors, molecular transistors and switches, atomic force microscopes, focused ion and electron beam microscopes, novel materials, nanowires and nanostructure-enabled devices, nonvolatile RAM, nano-optics, nanoparticle solubilization, and nanoencapsulation for drug delivery. Products already on the market include sunscreens, fabrics, sports equipment, house paint, and medical devices. So far, national governments have been the major investors in nanoresearch, with the United States, Europe, Japan, and the United Kingdom with the highest financial commitments. Private investors, major companies, and start-up technology companies in those regions of the world have also invested large amounts of funding for research leading to potentially revolutionary breakthroughs and spin-offs of nanotechnology in intellectual property, instrumentation, novel materials, modeling, platform techniques, security, surveillance, and nanobiotechnology.

FROM VISION TO INITIATIVE

Nanotechnology is coming increasingly into public awareness. But creative imaginings of its potential applications began to surface years ago. Early inspiration and vision for the pursuit of nanoscience and nanotechnology is widely credited to physicist Richard P. Feynman, who in December 1959 delivered a presentation to the annual meeting of the American Physical Society at the California Institute of Technology entitled, "There's Plenty of Room at the Bottom." He concluded that speech with an offer of two $1,000 dollar prizes, one to the "first guy who can take the information on the page of a book and put it on an area 1 / 25,000 smaller in linear scale in such a manner that it can be read by an electron telescope." Two decades later, Drexler (1986) popularized the use of the word *nano-*

technology in *Engines of Creation*. Back when the U.S. government was still focused on funding other promising, "revolutionary" technologies such as cold fusion, this nanopioneer was writing about molecular assemblers that could make possible low cost solar power, cures for cancer and the common cold, cleanup of the environment, inexpensive pocket supercomputers, accessible space flight, and the limitless acquisition and exchange of information through hypertext. At that time, Drexler's ideas were considered novel.[2] More recently, his ideas have fallen to harsh criticism by some policymakers, and research scientists and engineers, who reject those visions as fantastical. Nevertheless, there are those who embrace his visions with hope.

One of the revolutionary aspects of nanotechnology is the tools it uses to see, study, and manipulate material phenomenon at the nanoscale. Decreasingly, "our intellectual command over the fine structure of matter (whether living or non-living) has at every stage been limited by the tools at our disposal—both our laboratory instruments and our intellectual ones" (Toulmin, 1962, p. 338). Jeffrey, a nanoscience researcher, would probably still agree with Toulmin's statement, over forty years later:

> *JEFFREY:* *In some respects, things can get simpler at the nanoscale. An example is quantum confinement where as you shrink a nanoparticle down, say a semiconductor nanoparticle, the band gap between the balance bands on the conduction band becomes higher in energy .... They in some respect are much simpler phenomena to study than the bulk material where you have so many states and these broad, continuous states. You can explain them using much simpler physics, but the nanoparticle is more of a molecule. OK, so you are approaching the molecular behavior, which in some level one can think of as being simpler, at least simpler quantum mechanically. I think that the phenomena are actually going to be quite simple, and we have some ideas as to what these phenomena are, which actually govern the behavior that we are seeing, but we don't have really good tools to study it with. That's the main stumbling block, trying to develop tools that allow us to really have an understanding of what's going on at that length scale.*

At the time when Toulmin wrote in the early 1960s, nanoscience was in its infancy. The tools required for observation at that length scale had not yet been developed. In 1982, Gerd Binnig and Heinrich Rohrer invented the scanning tunneling microscope (STM), making Feynman's earlier hypo-

[2]David Barube of the University of South Carolina is writing a rhetorical analysis of nanotechnology, which includes the study of Eric Drexler and his ideas.

thetical challenge (and perhaps some of Drexler's visions) more technically feasible. IBM patented the Bennig–Rohrer invention, and then demonstrated the microscope's incredible power to the world's scientific community and beyond, by writing the initials "IBM" with 35 individual xenon atoms. In essence, development of the STM marked the beginning of nanoscale science and technology research. Throughout the 1990s, various U.S. government agencies began working together toward the formulation of what was to become the first National Nanotechnology Initiative (see Appendix B). In January 2000, 30 years after Feynman's speech, also at the California Institute of Technology, U.S. President Bill Clinton (President Clinton's Remarks, 2000) announced "a major new national nanotechnology initiative worth $500 million." Later that year, the U.S. National Nanotechnology Initiative was formalized. Meanwhile, other countries were launching their own government-sponsored efforts. Nanotechnology initiatives are underway all over the globe, including in Taiwan, Japan, Singapore, Germany, South Africa, the United Kingdom, South Korea, Brazil, Hong Kong, Australia, and Switzerland.

Worldwide, billions of dollars are being poured into nanotechnology research and development.[3] Government appropriations point to the significant political and economic motivations to fuel the pursuit of nanoscale scientific knowledge, and to accelerate and advance technical understanding and control of the material world. Some political rhetoric about nanotechnology references a "technological race" with tremendous winning stakes in economic gain. Incredible economic promise is certainly one of the claims that have been made. According to the NanoBusiness Alliance (a U.S. trade group), nanotechnology's annual global revenues are estimated to be U.S. $45.5 billion, including microelectronic devices. The National Science Foundation (NSF) has projected that the global market for nanotechnology products could reach $1 trillion by 2015.[4]

In the United States, the pursuit of nanotechnology is being engaged with vigor. On December 3, 2003, President Bush signed into law the 21st-

[3]Examples of public funding for research and development in nanoscience and nanotechnology: Europe, 1 billion Euros; Japan, funding rose from $400M in 2001 to $800M in 2003 and is expected to rise by a further 20% in 2004. The USA's 21st Century Nanotechnology Research and Development Act (passed in 2003) allocated nearly 3.7 billion to nanotechnology from 2005 to 2008. This compares with $750M in 2003. With the launch of its nanotechnology strategy in 2003, the UK government pledged $45M per year from 2003 to 2009. (European Commission, 2004, taken verbatim from the Royal Society's Final Report on Nanoscience and Nanotechnologies, Dowling, 2004.)

[4]As reported by Phillip J. Bond, Undersecretary of Commerce for Technology, U. S. Department of Commerce, in a speech delivered at the National Nanotechnology Initiative 2003, Washington, DC. Available at http://www.ta.doc.gov/Speeches/PJB_030404.htm

Century Nanotechnology Research and Development Act. This legislation authorizes $3.7 billion for nanotechnology research and development, and puts into law programs and activities supported by the National Nanotechnology Initiative. Claims are that nanotechnology is leading to "dramatic changes in the ways materials, devices, and systems are understood and created," and lists among the envisioned breakthroughs "orders-of-magnitude increases in computer efficiency, human organ restoration using engineered tissue, 'designer' materials created from direct assembly of atoms and molecules, and the emergence of entirely new phenomena in chemistry and physics" (NSCT, 1999). Policymakers, investors, and researchers hope for the possibilities of miniaturized drug delivery systems and diagnostic techniques, positive environmental impacts through drastic reductions in energy use and the rebuilding of the stratosphere, extending and repairing deficits in the human senses, and security systems smaller than one piece of dust. The incredible promise is that through nanotechnology, previously unimaginable and inaccessible material possibilities are now coming within our reach. In the words of one U.S. senator; "We are poised to take the next major leap into the future where the possibilities are endless" (Mikulski, 2001). If, in fact, humanity is embarking on yet another major shift in our technological capacities, what might that "revolutionary" shift mean socially, economically, and otherwise for individual and collective human life?

RIDING THE WAVE OF RESEARCH FUNDING

Nanotechnology is also sometimes referred to as the "next big wave," Why not? For people with the financial means, there are promising, exciting investment possibilities with nanotechnology. For consumers with purchasing capital, there are fun and thrilling new products on their way to market. For those researchers whose interests happen to fit the description of nanoscale science, there is a wonderful (albeit competitive) opportunity to jump in on the momentum and gain funding support. But only if their current research area is popular or pertinent to the nanotechnology initiative stated goals and challenges do they have a chance at catching the wave. Otherwise, they have to hope to catch the next wave. Certainly, researchers involved in nanotechnology initiatives are much more deeply committed to their work than an ocean wave metaphor might suggest. What is a problem is that having to catch the wave in order to be able to participate may preempt the careful and deliberative moral reflection that is needed to inform a conscientious pursuit of nanotechnology research and development. The wave may just be too powerful and swift for that. Ryan seems to recognize this:

> ROSALYN: *Are we moving in some kind of trajectory or some kind of particular direction in the nanoscience inquiry or nanotechnology pursuit?*
> RYAN: *Oh, you are asking a philosophical question.*
> ROSALYN: *Well, I can't help myself.*
> RYAN: *It's an inquiry for those who have had time to reflect on these things.*
> ROSALYN: *Well, this is that time.*
> RYAN: *I'm just enjoying what I do, too much to think about what should be my ultimate goal. I think most of us should be happy and that means different things to different people. So professionally, to do what you like to do is happiness, I think.*
> ROSALYN: *And this is what you are doing?*
> RYAN: *I think so, most of the time.*

VENTURING INTO UNCERTAINTY

Each new technological era is ushered in on claims of great promises and hopes, along with competing expressions of concern and alarm, doubt and disbelief. And, nearly always, new technological eras are propelled by futuristic dreams, fantasies, and ardently stated human values. The discovery of nuclear fission came with proclamations of a carefree life, made possible by this tremendously powerful new source of energy. The space program came with promises of discovery, which would exploit the resources of heretofore unexplored worlds, and perhaps open the doors for future inhabitation. The arrival of electricity brought with it dreams of the end of human labor, and the beginning of a life with the complete fulfillment of material needs. In the case of nuclear fission, American engagement in war at the time and the desire for military dominance fueled an enormous expenditure of social and economic capital toward its aggressive and rapid development.

The explosion of nuclear bombs and its shocking moral implications led to the redirection of nuclear science toward nuclear power, a more socially acceptable technology. The early vision and excitement of the nuclear age that followed was fueled by proclamations of great promise for improvements to the quality of living. Eventually, its promise was subdued by the stark reality of tragic nuclear accidents, by the irrationality of fear, and by political and environmental challenges related to disposal of nuclear radioactive waste. The nuclear power technological promise rose with great hubris, but fell (in the U.S.) to economic woes and accidental horrors. With the original space program, fear and propaganda about communism, competitive notions about "getting there first," and the eternal promise and allure of new frontiers instigated a tremendous outlay of capital. But those ambitious dreams of the original space program were dampened by financial as well as technological constraints, and by the somber reality of multiple tragedies. Recently, new possibilities and new leadership have rekindled the hopes and dreams of space exploration. Political leadership in the U.S. has once again pointed the nation toward the skies with ambition and aspiration, with new initiatives proposed for further exploration of Venus, Mars, and beyond.

What about nanotechnology? What will come of the great proclamations being made about the material and economic gains to be had in its development? What will be the actual results of current commitments and its ramifications to society at large? What will it mean to human life, and to the Earth, if the development of nanotechnology comes to fruition? A sense of urgency abounds.

In some of the testimonies given in the congressional hearings on nanotechnology and proceedings of the Senate Subcommittee on Science and Technology, speakers talked about the need for expediency and of the urgency of a race for international competition in nanotechnology development. Nationally, the articulated "stakes" are high for economic gain, military superiority and surveillance advances in human health, and for worldwide, competitive dominance in scientific knowledge. Individual researchers at universities, national labs, in corporations and start-up firms, are competing globally and professionally to acquire resources to support their work, accelerate basic research, and quickly bring to bear the results of that research in the form of new knowledge, materials, processes, and technologies.

The rapid emergence of nanoscience and nanotechnology is an incredible, technological undertaking with profoundly potent, yet undetermined possibilities. While fueled by scientific ingenuity, it is also motivated by political pressures, competition for new international markets, political competition, venture capital ambitions, and culturally construed conceptualizations of progress. One issue warranting consideration is whether nanoscience is moving forward too quickly, before adequate health, safety, societal implications, and ethical concerns can be adequately assessed. Some concerned individuals believe the research and development of nanotechnology ought to be decelerated before it is too late for society to respond effectively and proactively, and to avert any consequential and irreversible harm.

When Joy (2000) reflected on that potential writing, "These possibilities are all thus either undesirable or unachievable or both. The only realistic alternative I see is relinquishment: to limit development of the technologies that are too dangerous by limiting our pursuit of certain kinds of knowledge," he unleashed vigorous and emotional debate from various sectors in the public discourse, and especially from nanotechnology proponents such as the Foresight Institute. The narratives were vehemently opposed to one another. Foresight has put forward self-regulation guidelines for the development of nanotechnology. That institute's leaders reject outright any notion of relinquishment. Rather, the story they tell is that guidelines should suffice in addressing concerns about the safe development of nanotechnology if adopted by research scientists and institutions doing nanotechnology. Other advocates defend the continued pursuit of nanotechnology on moral grounds, claiming that if U.S. scientists don't do it, scientists from other countries will, that it is a relatively benign enterprise representing a good and natural evolution in scientific enquiry, and further, that any hold on development of nanotechnology means keeping humanity from its rightful self-improvement.

The Canadian-based ETC Group (Action Group on Erosion, Technology and Concentration), a nanotechnology watchdog organization, is among those who are concerned that nanotechnology development is moving too quickly. The ETC demarcates four phases of nanotechnology development. The first phase (which is already well underway) involves bulk production of nanoscale particles for use in sprays, powders, coatings, fabrics, and so forth. In these applications, nanoparticles contribute to lighter, cleaner, stronger, more durable surfaces and systems. In the second phase, the goal is to manipulate and assemble nanoscale particles into supramolecular constructions for practical uses. The third phase would be mass production, possibly self-replicating nanoscale robots, to manufacture any material, on any scale. Finally, according to ETC's assessment, nanomaterials will be used to affect biochemical and cellular processes, such as for engineering joints, performing cellular functions, or combining biological with nonbiological materials for self-assembly or repair (Group, 2003).

Many varied and often contradictory ideas about the future of nanotechnology have been articulated. When Jim Urh, the founder of the nanotechnology venture called Zyvex, shared his view with me informally in a personal conversation, he spoke of the nanotechnology future in terms of "African people being able to use their plentiful air and soil as raw materials for the assembly of the material needs of life." He spoke of every one on the planet being able to take a pill a day for the assurance of good physical health and longevity. He envisioned a world without want. Stephenson's *The Diamond Age* also portrays a future with molecular assemblers, which can take air and soil and convert them into material goods. But Stephenson's world is not without want. In fact, it has massive starvation in some regions and brutal wars in others, while spiritual meaningless pervades all regions of the world. Why, in Stephenson's fictional account of the future, do assemblers not bring the promise of plenty to all, which is now alluded to in some of the narratives about future social scenarios possible through nanotechnology? Some researchers and other proponents of nanotechnology debunk visions such as Urh's. But Urh is not alone. In fact, Urh was an observer and guest in the Oval Office with President Bush during the signing of the 21st-Century Nanotechnology Research and Development Act. There are plenty of other intelligent, high profile people putting their resources of time, energy, and money into such visionary efforts as Urh's.

One of the difficulties of accurately determining the meaning, significance, and direction of the possible nanotechnology futures is the confusion and disagreements over what claims are dubious and which are reasonable. It also doesn't help that there is a relative dearth of research on its possible health and environmental impacts. Moreover, there is not enough understanding of the behavior of nanoscale phenomena to know for sure which imagined applications are plausible and which are fantastical. There is a great deal of speculation and debate over future outcomes and applications of nanotechnology. For example, in a now well-publicized and contentious dialogue, Noble Laureate Richard Smalley challenged Eric Drexler's notion that scientists will one day be able to create self-replicating, self-assembling devices (Baum, 2003). Whitesides (2001) pointed out that we have no sense of how to design a self-sustaining, self-replicating system of machines. He believed that the most promising possibilities of nanoscience lie in the imitation of biological systems (p. 81).

Given today's knowledge, Drexler's vision for the future is highly suspect, but who can really know for sure if one day nanotechnology will actually make possible some of his imaginings:

> In short, replicating assemblers will copy themselves by the ton, then make other products such as computers, rocket engines, chairs and so forth. They will make disassemblers able to break down rock to supply raw material. They will make solar collectors to supply energy. Though tiny, they will build big. Teams of nanomachines in nature build whales, and seeds replicate machinery and organize atoms into vast structures of cellulose, building redwood trees. There is nothing too startling about growing a rocket engine in a specially prepared vat. Indeed, foresters given suitable assembler "seeds" could grow spaceships from soil, air, and sunlight .... Assemblers will be able to make virtually anything from common materials without labor, replacing smoking factories with systems as clean as forests. They will transform technology and the economy at their roots, opening a new world of possibilities. They will indeed be engines of abundance. (Drexler, 1986, p. 63)

Claims of this kind about fantastical nanotechnology futures have spawned excitement and enthusiasm from some groups, but ambivalence, concern, even horror from others. Michael Crichton's *Prey* is an example of the latter. It portrays "nanobot" swarms aggressively and intelligently organized to eat living human flesh. According to the moral of the story, there is the potential for horror if research is rushed toward the development of nanotechnology processes and devices. Beyond science fiction writers, there are journalists, industry leaders, and members of Congress, scholars, environmental activists, a European royal figure, and scientists who urge careful consideration of the possible risks and harms embedded in precise, atomic manipulation of matter by humans. They warn as well about the possible inability to reverse any resulting harmful effects. Some have called for a moratorium on nanotechnology research and development until critical questions of safety and long-term effects can be definitively answered. Others say that a moratorium would be irresponsible and ineffectual. I am not a moratorium advocate at this point. There is still too much to learn and too much to understand before putting the brakes on what may actually prove to be a wonderfully beneficial development for humankind. That being said, I do assert the imperative to carefully reflect on the basic assumptions held about the inevitability of nanotechnology's development. I believe that there are still many choices to be made; that much of what will come in the future of nanotechnology is taking root in the values, ethics and commitments that are embedded in nanotechnology development, now.

Where nanotechnology may be leading, whether it will bring us into danger, and whether it can and will make good on the promises that have been made is unknown. Varied and divergent answers to those questions are coming from a number of different points of view. For some, there are no answers at all. Seven preeminent research scientists and engineers who were panelists at the 2002 National Science Foundation Symposium for Nanotechnology were asked, "What is the single most exciting potential of nanotechnology?" The resounding response from each of them was essentially, "I have no idea." This is an understandable response (albeit ironic, coming from the very experts on whom we are depending for the initiative's success). The fields of inquiry are so immensely broad, with so many multiple disciplines involved, that it is difficult even for researchers to project very far into the future where the nanotechnology quest is leading. A few who are confident in their visions can be heard among the voices of public discourse, actively speculating and debating what might be outcomes and applications of nanotechnology. But most research scientists and engineers seem more inclined to avoid making futuristic predictions in public, or do so with deep reservations. Lawrence completely dismisses any value in attempts to anticipate the outcomes of nanotechnology research and development, preferring to see his role as a scientist as proving wrong core assumptions in the world:

LAWRENCE: *Having conversations about predicting the future is guaranteed to fail for one of two reasons. Both you can predict the future, and anything that you can predict is probably relatively speaking, uninteresting. Or, you can't predict the future in which case the conversation hasn't gotten you very far. Well, even in things in which you know the consequences, because of the law of unintended consequences you don't get it right, so we don't know the consequences of anything. What are the consequences of this perfectly innocent conversation? Let's go at it another way and not take what we are doing and try to predict where it will lead. Rather, let's look at the core assumptions in the world and then ask, what is the probability that science will prove those assumptions wrong? One of our assumptions is that we are mortal.*

ROSALYN: *Yes. That is an assumption we have made.*

LAWRENCE: *What happens if it's wrong? What happens if science allows us to live to be over 200 years old? I think the chances of science allowing us to live over 200 years are very likely. Human life is valuable, that's another assumption. That one is much frailer around the edges because we already know that human life is a commodity that you spend in various ways for various purposes. I think you can see the processes in*

> *science that are going to challenge that very substantially. It's interesting to make that kind of list of assumptions and sort through how science may challenge those assumptions. I find it much more interesting, and more productive to go that way than to ask, "Where is nanotechnology going?"*

A few of the researchers participating in this project have explained that many wonderful new discoveries and technologies have been wholly unanticipated, emerging accidentally without predetermination or intention. One individual related her own experience:

> *It's serendipity because it was not what we intended to prepare. We wanted to prepare just simple metal particles, without the graphite. But some graphite carbon evaporated from around the particles. We wanted to do a chemical analysis to make sure that there was no contamination of our materials. The procedure was no different than welding. You evaporate the metal and collect the vapor. Then dissolve it in acid, which is supposed to dissolve anything .... Months went by and they didn't dissolve. We couldn't figure it out, and I went and said, "Why don't you throw it in the microscope, see what it is?" The student threw it in the microscope on some Friday. I remember, he was on his way to go on vacation. Then I saw what he had and I said, are you sure you want to go on vacation with this? Because I had nanotubes and so forth, I could see that shells were on these particles and so it was like elements. You have a seed inside with a hard shell outside. So it protected the inner particle from the acid, yet it was magnetic. I found that to be very exciting. What was intriguing is why nothing happened to these metal particles when treated with acid. It was because graphite was acting to protect the particles. So, I immediately realized that we had some new material with magnetic properties.*

Commonly, the researchers consider the element of the unknown to be one of the beauties of science.

A number of nodes of ethical concern are at issue in questions about the direction and future outcomes of nanotechnology. One concern is potential environmental accidents and abuses, or threats to human health and safety, which arise from the intentions to introduce nanoscale devices and nanoparticles into the atmosphere, waterways, the food chain, and the human body. (Rice University's Vickie Colvin focuses her scientific research on nanotechnology safety in these areas. And Günter Oberdörster of the University of Rochester received a $5.5 million grant to research whether nanotechnology poses a health risk. One study already completed shows that inhaled nanosized particles accumulate in the nasal cavities, lungs, and brains of rats.) The potential for placing nanoscale computing and other electronic devices into almost any material raises concern over diminishing privacy rights of citizens. The potential for nanotechnology to produce powerful and precise new weapons raises multiple concerns, such as the aftermath of war, the rapidity with which it is fought, and the limits we could be removing regarding when, how, where, and why war is engaged. Miniaturization and hybridization of commonly used electronic devices calls into question the assumption that faster and cheaper is equal to better, and raises issues about how market imperatives could supercede other social goods and deeply held values. Then there are the metaethical challenges of meaning making and belief, regarding the purposes of nanotechnology development and the overlapping purposes of human existence.

As its proponents explain, nanotechnology points to the desirability of humans to manipulate and control matter, with precision and to specification, at the atomic level. The implications of this potential are monumental in terms of technical applications in human health; environmental cleanup; food safety, preservation, and production; alternative sources of energy; novel exploitation of natural resources; military strategy; national security and surveillance; computer processing, miniaturization, and speed; and, at bottom line, grand new market opportunities projected to be $1 trillion by 2015. Theoretically, nanotechnology seeks to take things apart and to build things from the atom up, to rearrange matter and to control its very structure, such as David describes:

DAVID: *But, the success for us with respect to the materials we work with is really going to come when I can sit down with a student in my office and we can start talking about a specific material that we want to design. We want a material that will do X, Y, and Z, and we can sit down with a pad and paper and write down almost a recipe of how to design that material.*

ROSALYN: *Wow.*

DAVID: *That's really what we strive for. What we also work a lot with is creating multifunctional materials, or multiresponse of materials. How do you take one of these particles, say response to temperature, and then add stimuli response to that. Let's say we have a material that responds to both temperature and a protein binding event. How does one stimulus perturb the behavior of the other stimulus?*

ROSALYN: *Um hum.*

DAVID: *So building up multifunctional material is a huge part of our work and what we are really striving for is how do we design that material*

> *or how do we understand enough about the material to sit down and write down a recipe for making a certain nanoscale machine. Let me give you an example. One of the students is working on targeted gene delivery. He is trying to make materials that will find a cancer cell, get into that cancer cell and release genomic material where it can do its job, that is, at the nucleus of the cell. So what does that particle have to do? That particle has to have a surface that can bind to the surface of the cancer cell and only the cancer cell. It has to fool the cell into taking it up into an endosome. It then has to fool that endosome into rupturing and letting the particle back out and then it has to find the nucleus of the cell and when it reaches the nucleus it has to release its genomic material of the nuclear surface or into the nucleus so that it can incorporate itself into that cell's chamber. OK, so that is a complex series of steps and we know how to do some of those steps, but what we need to do is be able to make a particle that has all of that functionality built into it where it can do that task in a repeatable and predictable fashion. So, that's what I talk about, that's what I mean by writing down a recipe for making multifunctional materials.*

At present, the quest to manipulate and control matter, and its consequential effects on humanity, largely is a venture into uncertainty. Perhaps a useful metaphor for thinking about the pursuit of nanotechnology would be a competitive, exploratory hike through a previously impenetrable, vast, and wondrous forest made up of groves of unclassified trees, and unfamiliar species of life living in habitats never before seen. The venture into this dense forest is tremendously exciting because of its novelty, and the great new knowledge that it offers to the explorer. And yet, individual hikers enter from many different points of the forest and blaze their own trails to places they do not know, because there are no signposts to guide the navigation. The exciting journey has a great deal to offer, because its pioneers have varied perspectives, are exploring different aspects, and have many different lessons to learn. Explorers stand to enjoy the delight of discovery, and to forge for themselves the meaning of the experience. Some perceive themselves to be in a race where the winners get the glory of significantly contributing to the creation of new knowledge, and thus to significantly influence its meaning and significance. But will each pursue this venture with enough foresight to know how to recognize danger, when to change directions, whether to proceed, and when, if ever, to turn back?

WHERE IS THE MORAL LEADERSHIP OF THIS REVOLUTION?

Other recent "technological revolutions" had a huge impact on human life. But they have largely been without clear moral direction, taking on a life of their own, with no apparent management, control, or ethical responsibility for their effects. The Internet is one example. Although created with good intentions for open-sourced information, its actual effects have been unwieldy. It has become an amorphous, seemingly intelligent force, which is significantly larger and more powerful than originally anticipated. Despite recent legislation and corporate efforts to control such elements as spam, theft identification, Internet blackmail, gambling, child pornography, and so on, it continues to evolve independent of ethical leadership, and to grow unbridled. Persons who do not even own a computer are identified by search engines and listed in multiple Web sites. It is not possible to maintain privacy any longer, as the Web can provide access to myriad bits of information about nearly any citizen of the industrialized world. Although it has provided fantastic, free, nearly instant open access to vast amounts of information on a seemingly infinite array of subject matter, private citizens have paid a very high personal price for its use. It has made possible the connection of people from all around the globe, who would otherwise be isolated from one another, and it has also solicited consumerism and imposed values of the materialistic West on societies that have little means of participation in the free market. There is no one to blame or to hold responsible for what this technology has become, how it has been abused, and what it is yet to be. Might the same be said of nanotechnology one day? Currently, there is no clearly conscientious moral leadership at the helm of nanotechnology development. So where to turn? Perhaps the myriad, diverse, concerned individuals, who participate in the larger discourses of the socio-cultural world, could learn a great deal from these research scientists / engineers speaking as individuals, on the subject of responsibility for nanotechnology.

RESPONSIBILITY FOR AN UNPREDICTABLE TECHNOLOGICAL REVOLUTION

Development of nanotechnology is moving very quickly, and without any clear public guidance or leadership as to the moral tenor of its purposes, directions, and outcomes. Nanotechnology initiatives of various forms are in place and continuing to emerge all over the world, with active and competitive involvement especially robust in countries that are highly developed technologically, such as Japan, the United States, and Germany. Projections for the future of nanotechnology are varied and sometimes contested. No one can really say what it will mean to incorporate nanotechnology products, devices, and processes into systems of national security and surveillance, warfare, medicine, food, electronics, and other consumer goods. Where nanotechnology is leading and what impact it might have on humanity is anyone's guess. Nanoscience researchers can imagine and explore the possibilities, but they are inclined to suggest that knowing is simply not possible:

ROSALYN: *What about thinking in terms of the future and the effects nanotechnology may have on our societies?*

SHALINI: *I think that right now people can't imagine living to the age of 150. They can't imagine having their organs replaced. They can't imagine having no pain. They can't imagine being able to communicate with anybody on the planet, instantly. They can't imagine these things that perhaps in 30 years will be taken for granted. I tell my daughter that there were no personal computers when I grew up. We didn't know what a computer was. She can't imagine being without one.*

ROSALYN: *So, alright. Then, is there any reason to even think about the future if its so far off, if we can't even imagine it?*

SHALINI: *Well, you can plant seeds every step of the way. You have got to be careful what grows from them. One perspective is "why should you do anything until you understand, really scientifically understand the impact of changing things?" And my argument was that we have already done that. We introduced fossil fuel consumption as a source of energy before we understood it. Now our whole society is based on it. No one thought in the beginning when there were very few cars driving around and a few homes being heated that every family would have two cars, almost every home would be using fossil fuel, that industrial plants would depend on it. When they first started, no one ever really thought about it that way. It was hard to imagine how such a little change could make such a big change in our lives.*

ROSALYN: *Well that's the question. Why can't we think about it?*

SHALINI: *OK. This whole thing about "pathological science" is too strong a word, because people probably were thinking about it to some degree, but just didn't think quite as far ahead or as broad reaching as they possibly should have. It is an interesting example of how at the start of a technology, your society becomes based on that technology and all of a sudden you find out the technology may possibly not be what you imagined it would ....*

To my mind, the most pressing moral imperative at this time is for those involved with nanotechnology research, development, policy, and funding to commit to conscientiously addressing questions of what nanotechnology might and ought to be, in terms of what it may mean to be human in a nanotechnology-driven world. Of course, this suggestion is vexed on a number of counts, not the least of which is what it means to be human is a matter of constant and evolving social negotiation, as well as a matter of private, personal investigation. Furthermore, conscientiousness is subjective, not easily obtained, and not of apparent value in the competitive race for nanotechnology development. It is a moral imperative, nonetheless.

The question of "Who is responsible for the ethical development of nanotechnology?" is exceedingly difficult to answer. Who shall be held responsible for the course of its developments, for its various and multiple impacts on human and other life, or for its possible disruption to the stability of human societies? Where might there be the commander and chief of this enormous machine of nanotechnology development? If it is similar to previous technological revolutions, then there is no such person, or any apparatus by which any individual persons in particular may be held responsible or accountable for its evolution. Oppenheimer is often blamed for the development of the atomic bomb, when in fact he was a relatively insignificant part of grand undertakings, the roots of which were deep, and fanned out in many different directions. It is a similar story with nanotechnology; it is a competitive, multinational undertaking, with many varied players representing multiple personal interests and business and political agendas.

If nanotechnology emerges as even a remote threat to human and environmental health, then responsible parties may be identified through litigation. Meanwhile, nanotechnology will likely be appropriated in the form of consumer goods and government tools, and gradually come into use in socially fulfilling ways. Acceptance will parallel consumer satisfaction, and nanotechnology will take its place in the larger social consciousness as inevitable and essential to human life, just like with many new technologies. If the various nanotechnology initiatives meet their stated goals, then nanotechnology products will be appropriated and consumed voraciously and their presence will be ubiquitous. A paradox may then haunt humanity. If the fantastic claims about the potential of nanotechnology are in anyway realistic, then what is done with the outcomes of the deep financial investments, and tremendous outlay of human capital with the resulting new knowledge and abilities, will likely affect the substance of human material, social, cultural, economic, moral, and perhaps even spiritual lives. In other words, if humans succeed in the projects of nanoscience research and nanotechnology development now underway around the globe, then human life may be facing radical, perhaps even wonderful, possibly unalterable, but surely unpredictable changes. The socio-cultural meanings of nanotechnology's impacts may hardly be noticed, but human life will be fundamentally changed once again. Embracing the advantages and averting the dangers of serious unanticipated consequences, the moral imperative is for nanotechnology development to proceed democratically, and with perpension, by actively engaging and validating the voices of its proponents as well as those with concern or reason for hesitation. We, as prosperous thinking societies, have the capacity and moral responsibility to pursue both a conscious and a conscientious relationship with nanotechnology.

There is responsibility on the shoulders of those who have allocated resources for nanotechnology, and those who are at its core—the researchers themselves. Equal responsibility belongs to the individual consumer for intelligent and thoughtful consumption of nanotechnology products. And trust has been given to those who are in positions of political and industrial leadership who are themselves following a beacon of morality. What needs to happen in order for nanotechnology to be developed ethically, in pursuit of authentically humanitarian ends? It will require that the individual researcher, policymaker, venture capitalist, and other proponents honestly confront questions such as: Do we know enough of the basic science to proceed toward a safe and responsible development of nanoscale devices and applications? Will it lead to decreases in human suffering and will it address yet unmet human needs? Might it be capable of supporting or contributing to the well-being of vast human communities all over the globe? Could it unintentionally contribute to the degradation of our societies, the environment, and the sense of cultural meaning and human connections? Will it be honoring and respectful of the human body, and of Earth? For all concerned, there is a tremendous, creative challenge on the table to discover and embrace the possibilities of directing nanotechnology toward genuinely humanitarian ends.

CAROLINE

CAROLINE: *November first.*
ROSALYN: *That's the big date?*

CAROLINE: *Yes, renewal.*

ROSALYN: *That explains the tension I heard in your voice when I called from the airport. Its in the voices of many of the researchers I listen to.*

CAROLINE: *Distraction.*

ROSALYN: *Is that what it is?*

CAROLINE: *Yes, this week has been crazy. I've had barely any room to breathe all week, its crazy.*

ROSALYN: *What do you account for that? Everyone I'm talking to is saying this.*

CAROLINE: *I've been traveling probably twice a month on average.*

ROSALYN *To give talks?*

CAROLINE: *To give talks and to do NSF study sections; things like that. Consulting, service, now, traveling on the average of twice a month, and this is hiring season, so we have all of our committee meetings, going through those files, a lot of things. And right now I've got a bunch of grants that are expiring, so I'm going like mad trying to write renewals. It's the same thing, balancing research, service, and teaching, and whoever you're beholden to in any of those instances think that you've got 100% of their attention.*

ROSALYN: *And you're also publishing your research in journals?*

CAROLINE: *Um hum.*

ROSALYN: *So on average per year, how many papers are you trying to crank out?*

CAROLINE: *We do about a paper a month.*

ROSALYN: *A paper a month?*

CAROLINE: *Yes.*

ROSALYN: *And you've got people in your group writing, or you're doing most of the writing?*

CAROLINE: *My students usually write the first drafts, and then we go through a revision process.*

ROSALYN: *That involves you?*

CAROLINE: *Yes.*

ROSALYN: *So your grants expire at the end of this academic year, or next?*

CAROLINE: *I have some that have already expired that I'm trying to replace that were nonrenewable. I am trying to replace them with other funding lines through other agencies.*

ROSALYN: *I see.*

CAROLINE: *And then I have some expiring like a June time frame next year, so now is the time to write renewals for those. NSF funding cycles, as you know, has 3-year funding cycles; 18 months into the project you're writing a renewal.*

ROSALYN: *How much time do you get to spend in the lab yourself?*

CAROLINE: *Oh, I don't do any experiments anymore. I go through the lab, and talk to students probably twice a day. I take a little loop through.*

ROSALYN: *See how they're doing. Do you yourself establish the experiments they're going to do from week to week, or is it really sort of up to them to decide how much to keep going, when to change focus?*

CAROLINE: *That changes from student to student. Obviously, new students get a lot of direction from me, but usually the way I guide students is I give them some goal and then if they come to me with a failure or a success on their first few attempts at the small step they're supposed to go through, then you start providing them with options, "Here's one direction you should go, this is the way I would do it, here's a few other things you might try," and that lets them see how the process works, instead of you dictating to them exactly what they need to do. They do the thing that you think is most probably successful and if its not successful then they try to reevaluate, "what are these other approaches going to get me?" and then they've got to make a decision from there.*

ROSALYN: *So you're simultaneously teaching through the lab and pursuing your own research interests through the lab?*

CAROLINE: *Um hum.*

ROSALYN: *Like balancing the two, is that sort of what you're doing?*

CAROLINE: *Right, right. I mean, you're teaching the graduate students the scientific process, how to go about writing, etc., and how to do research, and at the same time you're trying to direct the research in a fashion that pleases the funding agencies, of course.*

ROSALYN: *I may have asked this before, but if you could get an agency to give you money for anything you wanted to do and it didn't have to fall into any particular category of current interest among agencies, would your work look any different than it does now?*

CAROLINE: *My group is pretty diverse, so I think the way it would change is the emphasis that currently exists in the group might be reprioritized, so you could probably divide our group into four separate subheadings and the one that's number four right now might rise to one or two.*

ROSALYN: *Oh, OK. But nothing would go out the door, nothing new would come in?*

CAROLINE: *I don't think so.*

ROSALYN: *Really?*

CAROLINE: *Yes.*

ROSALYN: *So, you're that close to being funded exactly for what you personally want to study?*

CAROLINE: *Yes, I'm pretty happy in that regard. I've been fairly successful and hopefully will continue to be in getting money for things I want to.*

ROSALYN: *So how much of what you want to do, do you think is influenced by what they designate money for?*

CAROLINE: *That's certainly an influence, right. So you're always making compromises. I'll give you an example. One of the things that we do in my group is we're designing nanoparticles that we hope can directly target tumors for directed chemotherapy. The grand idea that I want to pursue, the 20-year plan, if you will, is to create these multifunctional, multiresponsive particles that, as I describe to my students, interact with biological systems to have a dialogue, rather than just kind of grunting at a cell and the cell responding in some way. Really interacting in a designed and programmed fashion and it starts sounding almost like this nanorobotry, or the nanorobotics that people talk about, but, with soft materials.*

ROSALYN: *Polymers?*

CAROLINE: *Using polymers, and, or bioinspired-type materials.*

ROSALYN: *Bioinspired, I like that expression, bioinspired materials. OK, go ahead, I'll ask you about that later.*

CAROLINE: *So the long-term plan is, is I think in my mind not predicated upon one particular application. In my mind, its more of an esoteric pursuit and wouldn't it be wonderful if one understood the design rules for building such things such that you could pick and choose what function you wanted to place into the material and design it from first principles. And I don't care what the application is, I just want to be able to do that. I want to be able to have that kind of control over the synthetic system; the carrot that you dangle early on for the students with the case I've chosen is this drug delivery or targeted chemotherapy. That's commensurate with the funding agency's goals and its something that can motivate students because they don't necessarily appreciate my long-term ideas, they want something that's more tangible, that they can wrap their fingers around.*

ROSALYN: *So if you pull back to talk about just wanting to get this to work, to getting the control over the synthetics, that's just the scientist talking.*

CAROLINE: *Right.*

ROSALYN: *The curiosity, the intrigue, and the mastery of that very specific function?*

CAROLINE: *Correct.*

ROSALYN: *And then the application is gravy, that's what's delightful, but its not necessarily the driving force?*

CAROLINE: *Exactly.*

ROSALYN: *OK, so let me see how to put this. You've got scientists such as yourself who have these sort of ideas about what could happen on a very focused dimension, whether its bioinspired materials or with the polymers, whatever. Its like, I want the X, Y, Z operation, I want to be able to predict it and control it and contain it, which is the scientific mind. Then you've got people like me wandering around the halls saying, well what about society and ethics? Then you can say, well this could be targeted toward tumors and so it does something, it speaks to society's needs, but it almost sounds like one could interfere with the other.*

CAROLINE: *Um hum.*

ROSALYN: *Do you think, if you had to spend much time asking, "what are the implications of this beyond the laboratory, what does this mean in the larger scheme of things?," would it in any way detract from the very focused intention you have to have to get control over that domain?*

CAROLINE: *If I was directed to really explore the societal implications of everything we do, then certainly that's going to intrude upon what we do, but by the same token, there's a long history of academicians really working on fundamental issues, very esoteric issues, that have just fantastic, positive and negative societal implications that are far beyond anything they could have ever imagined. I think we can spend an awful lot of time doing that, thinking about the societal implications and we should, especially in light of how good we're getting at making really functional materials and really functional chemical systems. But just like we have to collaborate scientifically with people in other disciplines, I collaborate with engineers, biologists, etc., in order to make what we think are positive impacts on the science. Maybe its worthwhile having these kinds of dialogues more regularly and thinking of those as collaborative efforts, so that people who are really attuned intellectually to how society works and what the real societal implications are, as opposed to me just kind of trying to educate myself on the fly, because we live in a society where everything is centered around what we need, right? I mean, that's my view of society; its Caroline-centered. No, I shouldn't have said that, but—*

ROSALYN: *But you're being honest, and I appreciate that.*

CAROLINE: *You know that's the—*

ROSALYN: *Its human nature.*

CAROLINE: *Human nature, right. I'll give you an example. Nanoscience is famous for this. Someone will have a system that they perceive to be useful for another field that's outside of their own, they'll read half a dozen papers on that field and then start making just wild claims as to how their system can solve that field's problems without having a true appreciation for what's involved. I think the same thing is true for the societal implications. I can go and I can do something like study the toxicity of the particles we make and I can extrapolate that to societal implications, they're toxic or not toxic, good or bad, that's black or white.*

ROSALYN: *Don't pour it down the drain.*

CAROLINE: *Right, that's not really as far reaching as what you're getting at. It wouldn't be genuine for me and it wouldn't be effective for me to think so broadly in terms of the societal implications that I'm not educated classically to think about.*

ROSALYN: *But you did just mention what could be considered a professional ethics obligation.*

CAROLINE: *Sure, sure, of course.*

ROSALYN: *To know what you're dealing with.*

CAROLINE: *And that's firmly ingrained in all well-trained scientists, but that's taking it down. I think that that's one component of what the real societal implications are.*

ROSALYN: *Right, right.*

CAROLINE: *I gave a lecture this morning on a particular spectroscopic tool that can be used for looking at what are called single nucleotide polymorphisms.*

ROSALYN: *Singular—*

CAROLINE: *Single nucleotide polymorphisms.*

ROSALYN: *OK.*

CAROLINE: *They're just genetic defects. So, the students were baffled, and these are graduate students. I was disappointed because they really didn't get it. They didn't understand why this apparently esoteric spectroscopic tool I was describing had such clear societal implications, and I went on a long tirade, actually, more than half of the lecture I spent talking about how this is important for gene chip technology and genetic screening and how—*

ROSALYN: *Gene chip?*

CAROLINE: *Gene chip technology and genetic screening where people who are doing gene chip development right now are proposing that a few years from now we'll be able to do full genetic mapping of people and tell them whether they're predisposed to certain types of cancers, etc., and that's wonderful and scary at the same time, right. So, you know, I really beat the students up a little bit about exactly what you're talking about; not really thinking more broadly about why its important to get this right. Because if the science isn't done correctly then that means that the technology isn't as functional as we would hope, and the societal implications for having huge numbers of false positives or false negatives in that kind of screen are tremendous.*

ROSALYN: *Which we know from other poor tests that came to market too fast, right?*

CAROLINE: *Exactly, yes.*

ROSALYN: *And you talked about this in your class?*

CAROLINE: *Yes, I did.*

ROSALYN: *And did they have a reaction?*

CAROLINE: *I think they had a reasonably positive reaction.*

ROSALYN: *What do you mean?*

CAROLINE: *I think most of the students saw it as something that was important. I think they thought I probably went a little bit over the top. I was being a little bit apocalyptic in the way I was describing it, just to drive home the point, like, what is the worst case scenario here?*

ROSALYN: *What was it?*

CAROLINE: *Well, this probably isn't the worst case scenario, but the way I proposed it was, what if these tests don't work the way we think they do, and there's huge numbers of false positives for a particular genetic defect that can be treated with known therapies, either drug therapies or gene therapies, etc. And there's a large segment of the population then that is taking these therapies, that induces a heavy financial burden on the health care system, and for a huge segment of the population it changes the quality of life.*

ROSALYN: *Like chemo, for example.*

CAROLINE: *Right. So again that's a little bit over the top, I don't think that the FDA is so stupid and incompetent as to let something like that propagate.*

ROSALYN: *Well, there's the other question, like what if it tests for diseases that are not treatable? Well, then who wants that information? What do you do with it?*

CAROLINE: *Exactly.*

ROSALYN: *Well, you can sell the information to the insurance companies.*

CAROLINE: *Exactly. I didn't talk about that, but that's another aspect of it that's very important.*

ROSALYN: *The other question I have is, will these technologies be applicable to genetic testing of embryos, of fetuses?*

CAROLINE: *Right.*

ROSALYN: *And then, boy do you have a mess of questions about what you do with that information.*

CAROLINE: *Correct.*

ROSALYN: *I had a pregnancy once where I was offered something called the AFP, alpha-fetoprotein test.*

CAROLINE: *Yes, I'm familiar with that.*

ROSALYN: *For neural tube defects.*

CAROLINE: *Um hum.*

ROSALYN: *And it was only because I was already into my 30s. I said, "No thank you," because at that time 50% of the results were false positives. I carried the baby full term to 38 weeks and then learned she would be born with no brain, anencephalic. Then I went through this whole process of talking to the ethics committee of the hospital to have her organs donated, since its always a fatal condition. It was just a week before her delivery, and one of the physicians on the ethics committee got really angry and asked me, "How the hell could you have turned down the AFP test?"*

CAROLINE: *Wow.*

ROSALYN: *He was angry because I didn't use the technology. I had to explain why I believed that technology would not have helped me. We can't put a brain where there is no brain. The test results would have offered very little at the time when the test was done. So, how do you factor those questions into the science? I mean, how is that even possible? Do you continue with the science or do you wait, do something else?*

CAROLINE: *I don't know the answer to that and that's I think up to the individual researcher, I don't know that ethics of one path versus the other really make the decision very clear.*

ROSALYN: *So there's a tension. There's the interest in the science and then there's the science becoming technology, and then you're smack up against society questions. Short of that, maybe its just about knowledge and exploration other than pouring the toxic stuff down the drain.*

CAROLINE: *Its becoming much more important that we think about or at least pretend to think about (for some of us) the societal implications. But I'm not sure how much of that actually changes what we do in the lab. I mean, we think about it and we pay lip service to it because the funding agencies are demanding of academicians that the time scale on which we make an impact on society is shorter.*

ROSALYN: *Yes, I know. The reporting turnaround is amazing.*

CAROLINE: *Its no longer the case that I can propose to study this fundamental process and then say, "This is important fundamental knowledge because 20 years down the road it might lead to advances in these fields." Now I need to say that in the second grant period I'm going to make an impact on these fields. Very often that's the way it is. It changes from funding agency to funding agency but that's definitely often the case.*

ROSALYN: *So this sounds like working for a corporation where they require quarterly reports.*

CAROLINE: *Yes.*

ROSALYN: *And they change everything because they need to get the profits back up.*

CAROLINE: *Uh huh.*

ROSALYN: *And maybe the government's relationship with the funding agency sort of asks the scientist to be somebody different from what the scientist once was.*

CAROLINE: *Of course.*

ROSALYN: *Are the funding agencies generally federal?*

CAROLINE: *Most of my funding comes from federal agencies.*

ROSALYN: *We're not even talking about venture capitalists who want to see a return on their investment quickly.*

CAROLINE: *No.*

ROSALYN: *We're talking about government, public money.*

CAROLINE: *Right.*

ROSALYN: *So what's the rush?*

CAROLINE: *Well, their budgets are dictated by whatever Congress says their budget is. And senators and congressmen have no reason to understand the fundamental science and why its beautiful, right? They want to see their NSF nuggets that show particles inside of cancer cells, and targeted drug delivery and chemical warfare detection agents and sensors being set up on Mars probes so we can see if there was ever life on Mars. They want to see those kinds of things, which in their world justifies giving a 10% increase, or whatever it may be to a particular funding agency.*

ROSALYN: *So, to me there are some ethics problems there. It looks to be that nanoscience and nanoengineering are becoming very closely intertwined, so that devices and technologies are starting to drive the basic research.*

CAROLINE: *That's true for a lot of people, yes.*

ROSALYN: *OK. This is enormously helpful, by the way.*

CAROLINE: *Well, I'm enjoying this, I always enjoy this.*

ROSALYN: *Oh good, me too. So now I want to be very careful about your time. If we go until 1:30, what does that do to you?*

CAROLINE: *I think I'm OK.*

ROSALYN: *Having had two to three conversations now with 23 or 24 people and one conversation with 35, one of the themes that's starting to bubble up for me is that I don't see how what is coming out of nanoscience and nanotechnology is going to have any effect whatsoever on large pockets of the world. I'm thinking about Haiti, Nicaragua, Bolivia, Botswana, I could go on and on and on; the countries where there is, not only is there no initiative because there's no money for it, but there are no labs. I talked to someone in Washington who is involved in nanotechnology policy and I said, "This is bothering me," and he said, "Well, the NSF invited people from all over the world to send representatives to a meeting on nanotechnology. Those countries could have sent a representative." And I said, "You can't expect people to come who don't have any real resources." What can we possibly do? What will the world look like if nano in all its various realms is successful? Its going to mean that Japan, Europe, the U.S., England, Taiwan and maybe Brazil are going to have access to incredible new technologies and resources. What about the rest of the globe? Do you ever think about that? Can you think about it?*

CAROLINE: *Now, that's an excellent point, and before I answer that, may I ask you a question?*

ROSALYN: *Oh, sure.*

CAROLINE: *This feeling that you're getting about this bleakness in terms of nano having a broad impact, or the lack of impact you perceive it will have, is that something that you're hearing from people or something you're construing from what they're telling you?*

ROSALYN: *Right.*

CAROLINE: *So what you're saying is most of the people in the field in your opinion, don't necessarily share the same opinion?*

ROSALYN: *One person I've talked with is really strong about this.*

CAROLINE: *OK.*

ROSALYN: *The rest are happy to discuss it but it doesn't come up unless I bring it up.*

CAROLINE: *OK.*

ROSALYN: *So I'm construing it from, well, I'm looking at where all the funding is, I'm looking where all the national initiatives are, I'm looking at some of what the grants are about, I'm looking at the rhetoric, the national rhetoric about why we're doing this, and I don't see most of the world participating. It's being talked about like another technological race, sort of like the race to the moon was, and I want to know, well, who's in the race and for those who aren't what does that mean? And when you win what does that mean? And does this mean that, yet again, we're going to have more of a technological divide, because if, in fact, even not needing labor for building things which will soon be built by nanomanufacturing, then we won't even be able to say, well that X-Y-Z part of the world takes care of our cheap labor needs.*

CAROLINE: *OK, so?*

ROSALYN: *So I'm, OK, sorry, I get very—*

CAROLINE: *No, this is something that's very important to me. Let's see how I can phrase this most delicately, and then I'll get less delicate as I get into it.*

ROSALYN: *OK.*

CAROLINE: *The divide that you're talking about between us and them, and we know who are in those populations, exists with or without nanotechnology.*

ROSALYN: *Absolutely.*

CAROLINE: *OK? That's obvious.*

ROSALYN: *Absolutely.*

CAROLINE: *And, in my opinion, if the divide gets bigger because of nanotechnology, well, I think that divide is going to get bigger, even if there wasn't nanotechnology. If there wasn't such a thing as nanotechnology there would be something else that the government and scientists … frankly, nanotechnology didn't start with the government, nanotechnology started with scientists. We're the ones who are at fault, and the government was convinced that nano and now nanobio and nanomed, are good things to support. I think there are huge numbers of research programs that are being supported through those mechanisms that are very important and very worthwhile, but the divide, the long-term socioeconomic changes that occur, are as artificial as daily swings in the stock market.*

ROSALYN: *Yes.*

CAROLINE: *They're made up and created by what the government wants to do. There's no substance there. Nanotechnology, you could even go so far as to say nanoscience and nanotechnology, and to steal part of a quote from a colleague, "is a sham repackaging of whatever discipline is taking part in it" and that sham repackaging almost speaks to a desperation. The scientists and the engineers are desperate to maintain funding levels, maintain government support. We weren't making adequate progress toward social implications such that the government thought we should be supported at the million dollar level as opposed to the hundred thousand dollar level. We wanted the million dollar level, therefore it was required for us to repackage ourselves and resell ourselves in a way that the government understood. So in my opinion, the ends sort of justified the means, in that it's allowed us to get more stuff done and a lot of it I think is very good (and a lot of it is crap). But even without nanotechnology, that divide, that microdivide, would still occur. There would still be good science and very bad science, that doesn't change. So coming back to your original question, I think that if nano is even real, which I don't think it is, the stuff that people are doing today they have been doing for years and it's just been repackaged. It's going to increase the divide between us and them in a very artificial fashion, and I don't think that our core industries are going to be affected by this in any way that's really tangible. I was asked to give a talk at an annual international board meeting for a company that's stationed here in the Midwest. I was speaking to a bunch of nonscientists who were the VPs and presidents of the different international divisions. There were people from UK, Australia, and Asia and the United States and they all wanted to hear about nanotechnology. This is a company that's not heavily invested in nano right now; they have almost no R & D effort in anything you would call nano. I was asked to tell them about what nano is and what we can expect, where the grand challenges are. I told them its basic science; that it doesn't have to be called nano, just a repackaging. And they asked me very bluntly, well, "What's going to happen to a company that isn't involved in nano?" "What's our motivation to get into nano as a multinational corporation?," and I said, "You know, if in 10 years none of your products are using nano stuff, I'm sure that outside of a few niche markets, that's not going to be a problem. It won't fundamentally hurt your bottom line if nothing changes. However, we already see it in our culture. What's going on is that companies are already promoting their products as nano, whether that's a genuine statement or not and whether it's meaningful scientifically or technologically or not, it doesn't matter. If you're not going to be playing the game as a corporation, the public perception will be that your product is inferior to your competitors because it doesn't have this buzz word associated with it." How artificial is that, you know? That's Madison Avenue advertising agency artificial.*

ROSALYN: *For trillion dollar new markets.*

CAROLINE: *What's that?*

ROSALYN: *For trillion dollar new markets.*

CAROLINE: *We are getting to the point where the company doesn't even have to worry about whether any of this stuff, which we've already established is artificial, is going to help their product to be better. What they have to worry about is if whether they have visibility in such a field, whether they risk on being outmaneuvered by their competitors simply through advertising. Their competitors can say, "Our product has nano, your competitor does not, therefore, our stuff is better," and there could absolutely be no scientific basis for saying. So, that was the picture I painted for the company and I think they weren't happy to hear that, but I see that as being a reality, that this is all artificial at the scientific level and the only way to cover that up is to layer more little layers of artificial on top of it.*

ROSALYN: *So what would happen if this were talked about honestly and openly among funding agencies, politicians, congressmen, proponents of nanoscience—if we could just cut it all out and back up and regroup around what we value in basic science? A pipe dream, isn't it?*

CAROLINE: *Oh totally, yeah, I don't think that's ever going to happen.*

ROSALYN: *Hum.*

CAROLINE: *I can talk to you about this. I can talk to a few of my select colleagues about this. But there are certain people who would rebel wholeheartedly against what I'm saying. They might see a little bit of the logic in it. They might come down on my side in terms of it being repackaging, however the end justifies the means. I think there's a huge segment of the scientific population who in their hearts know that this is disingenuine.*

ROSALYN: *Yes. Actually I'm hearing it from others who feel like they can say that to me.*

CAROLINE: *So. I'm not going to lie to you. Still, I and a bunch of other people are through the first round of selection for one of these federally sponsored nanomedicine centers. I mean, I don't know what else to do.*

ROSALYN: *I understand. I fully understand that institutionally, you don't have other options.*

CAROLINE: *Right.*

ROSALYN: *That I can see.*

CAROLINE: *Right.*

ROSALYN: *So the divide is there with or without nano. Do you think that nano will accelerate it because of the repackaging and the labeling?*

CAROLINE: *To a certain degree I think it will, but not necessarily because of repackaging and labeling in and of itself. Simply because it's been an effective mechanism for enhancing government support for certain segments of the science and technology community.*

ROSALYN: *Right.*

CAROLINE: *And it's been an extremely effective mechanism for generating venture capital.*

ROSALYN: *Right.*

CAROLINE: *So I think it all comes down to the financial situation, that is, the money is being redirected from biotech to nanotech, the next wave. Biotech was a sham too.*

ROSALYN: *Yes. It appears to have been.*

CAROLINE: *A friend of mine was talking about how in academic circles, institutes and universities, resources are being redirected to those of us who are being multidisciplinary and playing nice with engineers and biologists, etc., and kind of spreading our research out so we can quote unquote, make a greater societal impact. I think as researchers we actually want to do that. Very often I think we just want to do good science, and that's the mechanism by which we can fund science, but because this is being supported so wholeheartedly, being directed by deans and provosts, this friend of mine characterized it as the academic equivalent of corporate reconstruction or reorganization. The universities are undergoing a corporate reorganization.*

ROSALYN: *Because they're following the money.*

CAROLINE: *Right. And they don't know what else to do. That's why corporations reorganize, as a last ditch effort when they don't know what else to do.*

ROSALYN: *So then what does it mean to be a scientist now?*

CAROLINE: *Oh, gosh.*

ROSALYN: *That's my last question.*

CAROLINE: *You know, I don't know. I spend probably 80% of my mental effort thinking about and writing grants.*

ROSALYN: *80%!*

CAROLINE: *At least over the last 6 months that's where my mind's been. I'd say at least 80% of my mental effort has gone into grant writing, 10% into writing papers, and 10% into teaching my classes.*

ROSALYN: *Oh, gee. And do you miss the laboratory?*

CAROLINE: *Well, I knew when I got into this I was never going to see the inside of a lab again.*

ROSALYN: *Do you miss it?*

CAROLINE: *Of course. So, what it means to be an academician is, for me, I am constantly trying to balance being a scholar with being a salesman, and with being an educator, and that balance gets completely out of whack. I find myself every 4 to 6 months, kind of taking a step back and trying to re-center myself and figure out where I really want to be. But then you find that, you know, there is no free will. What you do in this job is you have freedom with respect to the ideas you pursue and the science we do is something that I enjoy and I really want to do, and I couldn't have that if I worked for a company. But by the same token in order to do that science I've got to work incredibly hard, too.*

ROSALYN: *It sounds like you work for the government.*

CAROLINE: *Oh sure, to a certain degree I think that's true.*

ROSALYN: *So if you work for the government, what's your real job?*

CAROLINE: *I still classify my real job as educating graduate and undergraduate students. I still firmly believe that's what I do. The way I do it is by going out and trying to get the funds to support a research program that I think provides the environment and the infrastructure needed to educate good students.*

ROSALYN: *I accept that because I think that's what you're doing here and what you care about, but to the extent that you're paid for by government grants, and that the students are here because of the government grants, you work for the government. So, what is it you do for the government?*

CAROLINE: *To be really cynical, I think they hope that I'm doing work that they can hold out and show to the taxpayers; "We allocated this money to this funding agency and this is what's come out of it. Isn't this wonderful?"*

ROSALYN: *But that's empty, there has to be more than that.*

CAROLINE: *Of course. But don't you think that's what it is?*

ROSALYN: *Then that's just about reelection?*

CAROLINE: *Of course. I mean, what else is there? I think along the way we do good work and I really believe that a lot of the things we're doing in this group and a lot of the things that are going on around the world, will have direct positive impacts on medicine and communication technology, etc. I really fully believe that. That just makes the government look better, because it's government-funded research. But I don't think that's really the immediate goal of the people who give us the money.*

ROSALYN: *OK. So that explains why in the end it's not going to have any impact on Haiti or Nicaragua because nobody in the constituency is going to care if there is less malaria in Nicaragua and cleaner water in Haiti.*

CAROLINE: *People do that kind of research in the United States and they're working for the government, but now what does the government get to say? They get to say "look, we're funding research that doesn't even directly impact you but it impacts third world countries and aren't we wonderful?"*

ROSALYN: *And that actually happens to some extent, is what you're telling me.*

CAROLINE: *Sure it happens, that happens.*

ROSALYN: *So we're willing to allocate a percent of science research to that, something like that?*

CAROLINE: *I don't know what the percentage is, but it's not a huge amount of money.*

ROSALYN: *OK.*

CAROLINE: *A lot of that gets funneled through Centers for Disease Control.*

ROSALYN: *I see, right, right.*

CAROLINE: *Their directive is not just domestic.*

ROSALYN: *So in the end what I want is to see a level playing field so that people in those places can do research themselves and develop technologies themselves that speak to their own needs as opposed to getting stuff trickled down from us so that we feel better about taking advantage, or whatever we do. I've asked people in the NSF, how come we couldn't pair up scientists in Nicaragua with scientists here to look at problems together?*

CAROLINE: *Well, there are small programs that fund those kinds of things.*

ROSALYN: *There are?*

CAROLINE: *Very small programs.*

ROSALYN: *OK.*

CAROLINE: *There are international research fellowships.*

ROSALYN: *Yes?*

CAROLINE: *Yes, through the NSF, but those are not big programs, usually.*

ROSALYN: *I think I've got the picture. I'm glad that what you're doing will probably make a contribution to people's well-being. I am, I'm really glad. At some point, I question how much more quality of life do we need, how many more gadgets do we need?*

CAROLINE: *Sure, sure.*

ROSALYN: *But, cancer is a horrible thing.*

CAROLINE: *Yes, it absolutely is.*

ROSALYN: *And if we could treat that, which would be great.*

CAROLINE: *Yes, there are still some very important pursuits other than those that improve the quality of life like, that will my wrinkles go away, or make it so that I will be able to avoid grey hair.*

ROSALYN: *I hope so.*

CAROLINE: *I think there are grander challenges than that.*

ROSALYN: *I hope so. Although, I hear in California now, pet owners are getting plastic surgery done on their animals.*

CAROLINE: *Oh, my lord.*

ROSALYN: *So, OK. We did it, that's an hour.*

CHAPTER TWO

Three Dimensions of Nanoethics

ROSALYN: *What seems to be driving nanotechnology, and why? And, do we have any control over it its direction? Those are my two primary questions.*

JARED: *Well, in my opinion, there are a few factors at work. First, there's commercial interest. Everybody wants to get the new technology and to build the newest company from their research. Materials and biotechnology and all of these areas represent hundreds of billions of dollars to industry, so there is a lot of commercial impetus behind it. Then, in the university setting, there are needs to be the one to advance the area. Ego drives people who want to do great things. Also, I think we have some basic motives that aren't as sinister as those two. There is the interest in understanding things. A lot of people just want to learn something; they want to see something new and have the answer to the question "why." So we have got all that embedded in this.*

One way that humanity attempts to capture and understand the human condition is through science, and its quest for knowledge of the laws that govern the physical universe. Another is through religion, and its attempt to garner and obey any laws that might govern the domains of the soul. Ethics, a third attempt to capture the human condition, seeks to identify and understand principles or laws that might govern human moral choices and behaviors or, in the absence of discernable or existing moral laws, to capture and reflect on the way in which human communities construct and agree on values about how it is they want to live together. Technological development is not so much concerned with understanding the human condition, but rather it is geared toward solving perceived problems and making improvements to the material conditions in which humans live and function.

Potentially, nanotechnology development can have a tremendous influence in re-shaping and rearranging the evolving material world. How might nanotechnology affect the condition of human life? What is it that most consumers of this complex new technology should be able to understand? What is it that most policymakers ought to assure? What is it that most business leaders must make a commitment to if these new technological developments are to be incorporated into the society in healthful and ethical ways? What might those technologies mean for human health, to the environment, to the structure of society and the common good? Who may be harmed, and who stands to benefit from its use? How might the economy be affected? What social goods will come as a result? Will its appropriation in any way challenge well-established moral or religious beliefs? These are some of the fundamental questions of any ethics of technology, seeking to both ascertain and influence its effect on the human condition.

Some scholars who have made preliminary assessments about the need for ethics in nanotechnology have concurred that ethics need to be considered, now, while it is developing (Dowling, 2004; Feeler, 2001; Weil, 2001). The Report of the Royal Society (Dowling, 2004) suggests that most of the social and ethical issues arising from applications of nanotechnologies will not be new or unique to nanotechnologies. My assertion is that ethics particular to nanotechnology is needed in order to guide nanotechnology development towards humanitarian aims; a challenge which may be beyond the scope of classical ethical imperatives to anticipate and then minimize harm, or in general terms, to maximize "the good." It calls for conscientious pursuit of nanotechnology, which is only possible if there is authentic attention given to the nature, tenor and function of human values and beliefs embedded in the current initiative.

As such, nanotechnology ethics complements other efforts to explore and direct the moral dimensions of scientific and technological transformations in human action. It is similar to nuclear ethics, which deals with the challenges of very large-scale power generation, and to biomedical ethics, which focuses on professional decision making and behavior, as well as bioscientific/technological aspects of medicine. Nanoethics are also similar to computer ethics, with its emphasis on the technological redefinition, processing and exchange of information. Its difference is in extending a vision of what is morally desirable while also tending to the motivations of greed, power and control which may obscure or repress otherwise humanitarian interests.

ETHICALLY CHALLENGING CHARACTERISTICS OF NANOTECHNOLOGY DEVELOPMENT

Formulating ethics for a technology that has yet to develop is a daunting quest, to say the least. There are at least three factors that make it so. First, new instruments have made possible the precise movement of atoms by human intention, so the potential magnitude and potency of technology applications is increased with nanotechnology. Second, in its present, early form, nanotechnology products appear to be no different than familiar products made from macro- or microscaled substances, but its elemental characteristics are indiscernible to the unaided human eye. When its products are eventually, fully appropriated into the marketplace, nanotechnology may be ubiquitous yet invisibly buried in computer processors, energy systems, medical technologies, fabric, skin and other body organs, and in food. Third, nanotechnology has the potential to profoundly touch virtually every facet of human life, but how so is still futuristic and cannot be predicted with any accuracy. Therefore, ethical nodes of concern about nanotechnology products and devices are largely hypothetical.

It would appear that at present, there is no clear direction for nanotechnology. It is developing willy-nilly, subject to political and market forces, as well as to the vagaries of individual curiosity and will. I find it very unsettling that the stated but ambiguous intention of using nanotechnology to control matter with precision is being pursued without explicit ends in mind. If nanotechnology development is to be conscientiously guided and directed in humanitarian and Earth-respecting ways, then ethics of nanotechnology need to be rooted in the commitment to do so.

APPROACHING AN ETHICS OF NANOTECHNOLOGY

Some say that nanotechnology development needs no more than conventional approaches to ethics, such as:

i. Identifying and responding to any stakeholders who may have an interest or be affected.
ii. Identifying issues of justice that may arise.
iii. Establishing guidelines, governing moral rules and principles.
iv. Assessing rights (human and/or animal), resulting harms, and risks that may be associated with it.

v. Determining any associated values and beliefs that may be threatened or preserved.
vi. Ascertaining potential societal impacts or social costs and benefits.
vii. Outlining duties and responsibilities that might be associated with nanoscale science research and development.

I suggest that whereas each of these approaches might be useful for analysis of the ethical development of nanotechnology, the conscientious development of nanotechnology requires a multidimensional approach to nanotechnology ethics. Consider the following distinctive features of nanotechnology:[1]

1. Nanotechnology has no clear moral leadership directing its course.[2]
2. Because of its dimensional features, nanoscale components will be unseen, inaccessible, and provide no sensual, human-to-technology interface (e.g., like the computer that sits atop one's desk or under a car hood, and has contact with one's fingers; or a head set which brings awareness directly to the ears).
3. Nanotechnology products are being targeted for use in many and multiple applications in a vast array of environments from food to clothing to cosmetic products to medical devices to medicine to surveillance systems to the very air we breathe. Therefore, it has only ambiguous focal points for ethical consideration.
4. In the past, technological revolutions have come swiftly and dangerously without moral deliberation. For example, the computer revolution we are still experiencing has unwittingly hit hard on society's values and on human psyches, bodies, beliefs, and social interactions without any warning or preparation. Consumed by the allure of new gadgets, new markets, and the rapid transmission and processing of information, the profound ethical implications of computer technology were anticipated by only a few. As a result, benefits aside, humanity is living with a powerful new industry that has few controls, despite efforts to regulate it, and is daily impacting the lives of billions of individuals. Had individual scholars, politicians, consumers,

[1]Some of these features exist with other technological "revolutions," such as the computer revolution, after which computer technology has become almost ubiquitous, buried in unanticipated places such as the automobile, and used for unanticipated purposes such as distance learning.

[2]Unlike the project of the National Human Genome Research Institute, for example, which under the leadership of Director Francis S. Collins since 1993, set out very clear directions and ethical precepts.

industrial leaders, and scientists been conscientious early in the evolution of the computer revolution, humans and their various societies may today have very different relationships with the consumption of computer technology (i.e., disposal of toxic computer trash), and therefore more adept at maneuvering our way through the quagmire of complexities of computer use in daily life.

5. Personal belief, cultural myth, and the amorphous yet powerful role of the human imagination are elemental in the conceptualizations and purposes of nanotechnology, as expressed in narrative.

Mirrored Pyramid of Nanotechnology

Narrative raises important issues that are central to ethical reflection. But because of what might be described metaphorically as the mirrored pyramid of nanotechnology,[3] many of those issues are opaque.

The bottom pyramid represents humanity's struggle to survive, accompanied by the quest for knowledge of and mastery over the material world. The base is humanity's origins, the pinnacle of the pyramid represents humanity's evolutionary rise to technological development through knowledge of the material world. Mythically, this pyramid is represented in the biblical book of Genesis' story about the expulsion of Adam and Eve from the carefree life to one of struggle. The story also promises humanity dominion over the Earth. But as history reveals, that dominion is in and of itself a struggle, waged through the quest of science and the evolution of human-built technology. On top of the bottom pyramid is the reflection of a second pyramid, which holds inside the story of freedom from nature's constraints, made possible through the mimicry and mastery of nature itself. Narratives about nanotechnology tell of humans being within reach of controlling and manipulating matter with precision.

Symbolically, the top pyramid represents the potential for humanity to control and recreate the physical world in which we live. As a mirror reflects the image of a human form, the second pyramid reflects images of human desire, longing, imagination, myth, and perception. There, at that point in human/technological evolution, nanotechnology makes possible the absolute dominion over and control of matter; the consequences of which are wholly unpredictable. Because of the absolute

[3]As conceptualized by Gordon Berne, unpublished.

power that is represented in the reflected pyramid, the development of nanotechnology may mean profoundly radical changes to human life and to the qualities and constitution of matter itself. Therefore, the intention to attain precise control and mastery over matter comes with unprecedented moral responsibility. It is for this reason (and because of the distinctive features listed previously), that both tacit and explicit ethics issues must be addressed; which is only possible through understanding the multiple dimensions of meaning in narratives about nanotechnology. In theory, a multidimensional analysis of nanotechnology narratives will make apparent the relative complexity of the issues at stake, and reveal the multiple dimensions to be penetrated and understood in order for reflection about nanotechnology ethics to be effective in guiding its humanitarian development.

A THREE DIMENSIONAL FRAMEWORK

In order to address the rather amorphous, unwieldy nature of nanotechnology development, I suggest a three dimensional framework[4,5] for thinking about nanotechnology ethics. The first dimension of analysis makes assessments at the level of practical ethics. The second dimension of analysis moves beyond practicality to normative questions about how we ought to proceed, and what counts as ethical in the development of nanoscaled systems. By penetrating the tacit level of meaning in nanotechnology, the third dimension addresses meta-ethical concerns; an essential level of inquiry for conscientious development of nanotechnology.

FIRST DIMENSION NANOETHICS

Moral assertions about nanotechnology that are apparent, explicit, commonly held, and widely accepted might be categorized as First Dimension Nanoethics. There is common agreement within and outside of research communities that research scientists and engineers have a moral responsibility to be conscientious in their research. For example, basic nanoscale

[4]Note that although useful as an intellectual exercise for purposes of analysis and discussion, the Three Dimensions are actually not separable, practically speaking. Each overlaps the other. Alone, none of these levels of analysis is sufficient. This approach augments conventional approaches that might be taken to elaborate an ethics for the guidance of nanoscience and its development into nanotechnology. As such, it may be useful for the ethical development of any new technology, not just nanotechnology.

[5]There may be more than three dimensions to be considered, such as the forth and fifth dimension of ethics as suggested by Dyson (1997).

research demands that investigators be accurate in their reporting of procedures and results. When human subjects are included in their research protocols, it is expected and assumed that care is given to those subjects in terms of informing them of all known hazards and of respecting their personal rights. Falsifying data and knowingly doing harm to subjects are widely held within the profession to be unethical, unacceptable behavior. When there is a breach of either, investigators are held culpable for their actions. Professional ethics dictates that researchers adhere to rigorous self-regulation.

Normative ethical intentions to avoid doing harm to persons or to the environment also fall under First Dimension Nanoethics. Scientists and engineers who are working with various nanosubstances and doing different kinds of experiments with those substances are expected to take care in storage and disposal, and in exposing the atmosphere, water, and human beings to anything that might be hazardous. Potential dangers (e.g., freely migrating carbon nanotubes penetrating plant, animal, and human cells, or uncontrollable "self-replicators") morally obligate nanoscale science and engineering researchers to learn how to respond effectively and proactively to avert any consequential and irreversible social or environmental harms.

The health and safety of laboratory assistants, as well as of the general public and environment, is to be guarded. One problem here is that there are no clear indications of which substances may be harmful. Preliminary research results on the health and safety hazards of nanotechnology research substances are yet inconclusive. Nanoscale science and engineering fundamentally entail risk taking with novel, unpredictable, relatively untested new materials and devices in the realm of public and environmental safety. But most investigators recognize that one of the occupational hazards of science and engineering research is exposure to unknowns. One scientist explained in our second conversation that she proceeds on the hope and trust that no harm will come to her or her graduate students as a result of exposure to the carbon nanotubes she uses in her laboratory.

Another area for consideration inside of First Dimension Nanoethics is public policy regarding the potential for private individuals to gain access to the raw materials of nanotechnology, such as the carbon nanotubes already mentioned, or perhaps eventually, to self replicating nanodevices. Who is to oversee and monitor the use individuals might make of those materials, such as for the building of experimental devices or weapons of mass destruction? To protect society from other kinds of possible harm, external controls may have to be put in place to regulate and govern industrial uses of various nanotechnology components. Public policy also needs to address concerns over management of nano-related toxicity, release and control of nanoscale, self-replicating artifacts, and subtleties of nanoscale surveillance mechanisms, inequities in access to power, educational resources and support, and other nano-related implications for society.

Generally, guiding principles, codes, and laws will be sufficient to address issues that fall under the category of First Dimension Nanoethics. The Foresight Institute guidelines offer one such example.[6] Foresight is interested in what they call molecular manufacturing, which involves nanoscale components in manufacturing processes, including self-assembly of those components. Their goal is to provide safe opportunities for the development and commercialization of the type of nanotechnology they call "molecular manufacturing." They believe that if adopted by research scientists and the industries involved, their guidelines should suffice in addressing some of the ethical concerns over the safe and responsible development of nanotechnology. Another example would be what Joy (2000) called on scientists and engineers to adopt: a strong code of ethical conduct resembling the Hippocratic Oath, along with the courage to enforce this code on others. Codes of conduct, principles, and guidelines are important, because they can serve as a foundation for an agreed on system of expectations about professional behavior in the realm of nanotechnology development. But, by their very nature, they are not wholly capable of addressing abstract, philosophical questions such as ideological tensions between the imperatives of pursuing capital gains and the imperatives of pursuing justice. Codes, rules, and principles may dictate and control human action, but only reflect current human conditions; they do not challenge or reconsider its basic assumptions (especially when the codes seem to conflict with personal reasons, desires, or beliefs). And, they are not sufficient for aiding understanding of such notions as whether or not the precise control of our material existence is a good to be pursued, over and above other possible scientific or technological aims. When professional codes, principles, and guidelines are in place, the moral agent has simply to learn and understand them, acknowledge their validity, and abide by them, despite the personal moral quandaries and professional perplexities that following them sometimes may create.

[6] See http://www.foresight.org/guidelines/

Consider the following example. Two nanoscientists speak about how wonderful it would be if their research could contribute to eliminating mental depression from the human condition. Various kinds of professional codes might offer these scientists some direction as to how to proceed ethically, with the study and actual development of a direct chemical intervention of the neurological causes of depression. But codes are not designed to offer insight into the deeper, profound social, cultural, and perhaps spiritual implications of such treatment. Maybe the treatment will alleviate painfully difficult and life-threatening symptoms, and treat biological elements of depression, but it may also inhibit the kind of fundamental changes needed in a person's mental constitution for addressing the root, psychological causes of the depression. Or, it could flatten human expressions, such as in the arts, which sometimes comes from states of melancholy, struggle, and despair. How might these two researchers grapple with this level of inquiry regarding their research? Should they (or any laboratory scientist) concern themselves with such philosophical questions? Attempts to understand why nanotechnology is being pursued as it is, what meaning it has for whom, and what deep-seated beliefs and ambitions are stimulating its development move us into Second and Third Dimension Nanoethics.

SECOND DIMENSION NANOETHICS

In the category of Second Dimension Nanoethics are those moral claims that are negotiable and subject to change under the vagaries of will, power, and perception. They incite competing interests and are disputable. Whereas First Dimension Nanoethics can entail discourse (e.g., in the process of interpreting and explaining laws, principles, rules, or codes), Second Dimension Nanoethics extends that discourse to a dialogic, dialectic process of discovery and construction, where stakeholders are engaged in a living, dynamic competitive process of the exploration and negotiation of values. In technological, pluralistic, multicultural societies that predominate in some regions of the world, it can be exceedingly difficult to find agreement over ethical precepts about technological development and use. Second Dimension Nanoethics is where moral consciousness draws interested members of the general public, policymakers, researchers, investors, scholars, and others, to engage in a communicative process, such as that suggested by Jurgen Habermas. In the introduction to Habermas' *Moral Consciousness and Communicative Action* (1990), McCarthy commented:

> Matters of individual or group self-understanding and self-realization, rooted as they are in particular life histories and traditions, do not admit of general theory; and deliberation on the good life, moving as it does within the horizons of particular life-worlds, and forms of life, does not yield universal prescriptions .... Habermas' discourse model, by requiring that perspective-taking be general and reciprocal, builds the moment of empathy into the procedure of coming to a reasoned agreement: each must put himself or herself into the place of everyone else in discussing whether a proposed norm is fair to all. And this must be done publicly; arguments played out in the individual consciousness or in the theoretician's mind, are no substitute for real discourse. (pp. xii–xiii)

Who is to control the emerging nanotechnologies, and for what purposes? What happens to personal privacy when information can be gathered from invisible sources? Using nanotechnology to design sophisticated, ubiquitous surveillance may exacerbate existing concerns over conflicts of values between national security commitments and the civil liberties of citizens. What is to come of the notion of privacy in a world that is driven by nanotechnology? Will undetectable observation become a government right? What about freedom? What will it entail, and for whom? Second Dimension Nanoethics issues such as these are most democratically and honestly addressed through such discourse.

Education is another node of ethics concern falling under the domain of Second Dimension Nanoethics. Questions about access to education and technical training include: Who will pay for and provide the specialized retraining needed for teachers, or for the equipment, facilities, and supplies needed for the schools? How will society assure democratic inclusion and full public access to the products and services that come from nanotechnology developments? Can a racial/socioeconomic nanotechnology education divide be averted? Who will have rights of access and what economic and quality of life opportunities will nanotechnology training afford, and for whom?

Nanotechnology is often spoken about in terms of a national race. Much like the space race of the 1960s, political leaders especially point to the great economic opportunities that can be gained if the United States "wins" the nanotechnology race. The very notion of a race raises important ethics questions:

1. What the stakes are in "winning"?
2. Who gets to participate?

3. Ought the scientific method or technological development ever be rushed?
4. Race? To what end?

The questions of which institutions or nations are likely to "come in first" and how competing world powers will implement and control the applications of nanotechnology are questions for Second Dimension Nanoethics, and the discourse it entails.

The potential for nanotechnology to produce powerful, precise new weapons of destruction calls into question the purposes of advanced and refined forms of military combat and intervention. Although on the one hand, sophisticated materials for military use may assure fewer casualties and greater intelligence for our soldiers, it may also mean swifter, more efficient, and precise modes of destruction and death for others. It may also lead to the removal of all the traditional demarcations of the fair battlefield. A provocative case study for ethics reflection in Second Dimension Nanoethics is the MIT Institute for Soldier Nanotechnologies. Here is a nanotechnology project that seeks to create materials to protect soldiers and to improve their survivability. The technical problem, as it is defined, is that current army soldiers carry way too much weight while having insufficient protection from ballistic, chemical, and biological threats; physical injury; and climatic, environmental, and terrain difficulties. The institute researchers hope to create strong, lightweight structural materials for soldier systems and system components. They hope, as well, to enhance ballistic and blast protection while maintaining soldier mobility. They want to create novel detection systems and create materials that will do many things, including to remotely treat local wounds, address injury triage, and emergency treatment systems enroute and in the battle place.

In First Dimension Nanoethics, the primary nodes of ethical concern might be over minimizing harm and enhancing good for the soldiers themselves. Second Dimension Nanoethics opens up the ideological struggle over what conditions make war just, if any. Such questions as who should be regarded with respect and who counts as a moral agent must be considered in terms of decisions made about destruction, aggression, and the taking of life. Who gets to make those decisions? Also to be negotiated are the resources societies will elect to place in nanowarfare, and at what cost and for whose benefit? There has been many millions of nanoscience research dollars allocated to military applications, but none yet for nanotechnology research toward the elimination of war or for technological solutions to the causes of war. Clearly, such nodes of concern are always present in ethical considerations of war, irrespective of nanoscale material development. However, there are, in fact, issues that arise only because of nanotechnology capacities. Those have to do with the increased efficiency and sophistication of war. Each new technological development brings into human hands greater capacities for deeper and more profound destruction.

There is another unresolved, conceptually variable area for consideration in Second Dimension Nanoethics, subject to competitive social negotiations: The use of nanotechnology in bodily implants for mind and physical body enhancement, drug delivery, or remote control of limbs and other movements and functions. Questions are unresolved in the public domain as to what constitutes bodily integrity and under what conditions that integrity may be morally compromised. Notions of the sacred in terms of certain bodily functions such as brain activity are continually being renegotiated as new technologies offer to humans the capacity to restructure the body at will. For example, the eyes, which for centuries were defined as windows into the soul, are now sliced without any moral trepidation with laser incisions for improved vision, and covered with colored composites for changes in eye color. Concepts of beauty, longevity, and strength are malleable as technology offers alterations and enhancements to formerly permanent body features (e.g., nose shape, breast size, buttocks shape, belly fat deposits, skin color, hair texture, and ear shape). What meaning the body holds in terms of right and wrong in its treatment and care will shift dramatically as nanotechnologies bring yet unimaginable possibilities to what the body may become.

Humans tend to affirm as true and good that which we believe to be possible, in the negotiation of improvements to who we are and how we wish to live. At our best, new technological developments are guided by enlightened self-interest. At our worst, many of us blindly accept, pursue, and consume any technology that promises an improved life, or more control over that life, irrespective of the resources it commands or the values it may compromise. What is called into question in Second Dimension Nanoethics is not the desire for or even the worthiness of change, but rather the purposes, directions, and intentions of it, and whether it actually will bring the improvements it promises.

Gerson (1976) spoke of the "constraining aspects of participation in any situation" (p. 979). People make their contributions to that "situation," and are also bound by its limitations. Yet, the situation is enriching and offers resources and opportunities not otherwise available. What resources? Gerson suggested beginning with four: money, time, sentiment, and skill. Each, when understood as Second Dimension considerations, offers enhancements to "qualities of life" that make nanotechnology so alluring.

Money

Whose money is being used to bring financial benefit to whom in the development of nanotechnology? In the research phase, principal investigators and their institutions are the primary recipients of financial resources. Once development moves to market appropriation of new goods and services, and research findings are materialized in marketable products, then the funding agencies of the federal government, industry, venture capitalists, and principal investigators and their universities whose patents are registered will all begin to see returns on their investments. Eventually, nanotechnology start-up firms will be publicly traded and individuals who invest in them may reap financial returns as well. Individual citizens will purchase the goods and services of nanotechnology, and hopefully benefit from their use. Whether those benefits will mean improvements to their quality of living depends greatly on the answers individuals give to the questions of meaning and purpose in their lives.

Time

Faster and smaller semiconductors will mean on-demand information access and processing. This, in turn, may mean less personal time, not more, as we have already learned from the computer revolution. How might time and money need to be allocated to industrial retooling and retraining, and to public education and consumer information exchanges?

Sentiment

How do we feel about privacy? There may be substantial loss of privacy, but some increase in feelings of national security when invisible surveillance mechanisms become ubiquitous. How do we feel about intimacy? As human communication shifts increasingly from touch, smell, and other sensual perceptions to electronic media, the meaning and expression of intimacy will also shift under new domains of knowing and experiencing other. How do we feel about power? Our insatiable desire for it may bring us face to face with our true and increased frailty, as a result of dangerous liaisons of power accumulation.

Skill

Most of our aging population is without the skills or resources to participate fully in the benefits of computers. How will any of us, especially the aged, acquire the skills needed to manage intellectually, physically, and emotionally in the strange, new nanotechnology environment? Who will have access to those opportunities for training, and at what cost?

Second Dimension Nanoethics provides a mechanism for determining how communities, nations, and individual citizens of the planet might participate in the ethical development and use of nanotechnology, through the dialectic-dialogic method of meaning making and negotiation. It requires the engagement of research scientists and engineers, policymakers, philosophers, social scientists, investors, business leaders, and any other stakeholders who are willing to work toward making explicit the values, intentions, and belief systems, that are at stake in the nanotechnology initiatives. This process includes identifying and acknowledging narrative structures, which are framing the nanotechnology initiatives.

The nanotechnology "revolution" is launched and, one way or another it will take its course. Potentially, it will provide for many new and wonderful opportunities for human health and well-being. Just like with any technology revolution, the coming nanotechnology era will reflect human nature and characteristics, including our ignorance, selfishness, insecurities, hostilities, greed, and hatreds, as well as our tremendous capacity for creativity, wisdom, compassion, generosity, and agape. There are still many choices to be made in negotiating that future. So now the question is posed again: How might that future be directed, ethically, toward humanitarian, Earth-respectful ends?

An ethical nanotechnology initiative will need to be supported in the public domain, threaded into the social fabric of the persons and communities that have a stake in its appropriation. To this end, the processes of negotiation about what nanotechnology might mean to humanity are now underway, with values and beliefs being reordered and redefined under the weight of competing interests and emerging new demands.

THIRD DIMENSION NANOETHICS

Third Dimension Nanoethics seeks metaethical understandings of the purposes of human living, beliefs about existence, and about the way meaning is created as it pertains to conceptualizations of selfhood and purpose. Its access is through myth, art, and other symbolic languages. The defining features of Third Dimension Nanoethics are particularly apparent in imagery, fantasy, and science fictional accounts of possible nanotechnology futures.[7] Through those highly metaphoric, otherwise symbolic and imaginative rhetorical forms, the public encounters the strangeness of what is imagined to yet be possible in the nanotechnology world, and explores what those imaginings could mean if materialized in human individual and social life.

Johnson claimed that classical, rule-based moral reasoning, which consists primarily in discerning the appropriate universal moral principles that reveal what behavior is right and good, presupposes a way of reasoning that is incongruent with actual human thought processes. That is why, he explained, there is so often such a deep tension between the view of one's moral task, on the one hand, and the way people actually experience their moral dilemmas, on the other. For Johnson (1994), "the quality of our moral understanding and deliberation depends crucially on the cultivation of our moral imagination" (p. 1). He cited the now infamous 1978 Pinto case as a sad example of how our conventional ethical reasoning failed us by replacing metaphor-based reasoning with the illusion of an infallible source of moral reasoning—that of rule-based cost-benefit calculations. Johnson argued that it is metaphor, that "lies at the heart of our imaginative, moral rationality, without which we are doomed to habitual acts." And because metaphor is one of the principal mechanisms of imaginative cognition, he wanted us to expect our common moral understandings to be deeply metaphorical, too. According to Johnson (1994), "Since our experience is never static, and since evolution and technological change introduce new entities into our lives, we are faced with novel situations that simply were not envisioned in the historical periods that gave rise to our current understanding of certain moral concepts. Metaphor is our chief device for extensions from prototypes to novel cases" (p. 195). Nanotechnology counts as a novel case. Its ethical development requires

---

[7]As was the subject of an International and Interdisciplinary Conference at the University of South Carolina, Columbia on March 3–7, 2004 entitled "Imaging and Imagining Nanoscience and Engineering."

recognition of the metaphoric basis of moral inquiry. Language both conveys and constructs meaning. As such, it has the powerful capacity to take the otherwise indeterminate reality that nanotechnology represents and focus it toward determined visions and goals. The domain of Third Dimension Nanoethics considers language use such as metaphor in the process of meaning making. It looks at myth in conceptualizations of life, matter, and self. It considers the role of imagination in the search for a sense of place and purpose in living. Tacit awareness, internal and external motivations driven by biological promptings as well as by purely psychological states of mind, are also important in reflections at the Third Dimension of nanoethics. *Design*, *progress*, and *revolution* are three examples of symbolic language used in various narratives about nanotechnology. In each case, their meaning is significant to ethics.

Design

The National Institutes of Health (NIH) is interested in nanotechnology because it operates at the same scale as biological processes, and offers the opportunity to enable understanding of the actual design of biological systems and processes, toward the quantitative modeling of biology. Third Dimension Nanoethics inquiries might consider the stated aims of the NIH, questioning what the meaning of the word and concept *design* may entail. What is assumed and believed when biology is described in terms of systems and design? How does the idea of design intersect with ideas about the order of the physical universe? If that universe is there by design, then is there a designer? If so, who or what is it? And, who is the human in terms of the capacity to take a role in that design? Another query might be into the meaning of "synthetic biology," a rapidly emerging concept in nanotechnology which inspires ideas such as using synthetics as antibiotics.

Progress

The rhetorical strategy of appealing to nanotechnology's potential for material and personal *progress* is one means proponents and political leaders have of cultivating a sense of meaning and conviction about the purposes of nanotechnology. The meaning of the notion of progress, as Ellul (1990) pointed out, is a phenomenon we can neither contest nor grasp. Here is an illustration to support Ellul's point: How can progress be contested or grasped if nanotechnology quests are successful in shrinking the size of transistors say 100-fold, giving humans the incredible capacity to place computing devices in and around the body in order to gain access to information and to control certain physiological functions? There is no contest to the moral good of tending to the health, care, and well-being of the body. But there is no grasping the moral meaning of progress when it challenges the relatively stable emotional and psychological comfort most humans generally have over being alive inside of the body. In order for persons to adjust to the information access and physiological control made possible through such devices, radical subconscious and conscious reconstructions of cognitive processes about selves within the body will have to occur. There are many hundreds of thousands of things going on inside of and around our bodies about which humans have no current conscious knowledge. Gaining access to even a few of those processes, such as particular biochemical changes, temperature changes, exposure to viruses and bacteria, breathing rate and heart rate changes, and so forth, will require that the mind make meaning of that information. If we do not successfully make meaning of that information, then we risk our mental health. But to do so will require reconfiguration of who we believe ourselves to be, and what it means to be alive in the body.

Matters of spiritual orientation will also have to be reconsidered. As technologies increasingly lead us to trust that our bodies and its processes are at our command, the sense of trust and faith in an omniscient being beyond ourselves, which many human beings possess, will need also to be reconciled. New meanings will have to be made of sickness, health, and reliance on a divine other. In other words, as revolutionary material changes take place in the larger society, change will also come to individuals. Technological transformations are interrelated to internal, cognitive processes in searches for meaning, and of the human need to establish a sense of self and purpose in life. Compelling notions of technological progress leave little negotiating room.

Revolution

As a trope of American determination and vision, the word *revolution* is often used to build value and construct purpose around the nanotechnology initiative. For example, John Marburger, director of the Office of Science and Technology Policy (2004), proclaimed:

> Not until recently have we actually had the instruments to make atomic level measurements, and the computing power to exploit that knowledge.

> Now we have it, or are getting it, and the implications are enormous. Everything being made of atoms, the capability to measure, manipulate, simulate, and visualize at the atomic scale potentially touches every material aspect of our interaction with the world around us. That is why we speak of a revolution—like the industrial revolution—rather than just another step in technological progress.

Use of the word *revolution* invokes deep ideologies, and is rooted in an old sense of identification for many people. It also goes to the core of American democracy, and is often invoked in political talk about the nanotechnology initiative. Large-scale changes that become evident over a short period of time are variously defined in terms of "revolutions." But all revolutions, including technological revolutions, require the allocation of tremendous resources at the burden of the society. At the level of Second Dimension Nanoethics, they depend on wide-scale social participation, which is difficult to enlist. Therefore, they must be negotiated in the public domain before they can get underway. In Third Dimension Nanoethics, the role of symbols—linguistic and otherwise—is central. For example, the basic concept of research as movement along a path from ignorance to knowledge is replete with metaphors of control which, if understood symbolically, might free recalcitrant reasoning to broaden the meaning of research in very intriguing ways. What would happen if different metaphorical constructions are used? Might nanoscale science become a different enterprise, or reveal different types of knowledge? Even the concept of knowledge itself has metaphorical roots. Learning becomes a social imperative toward mastery of one's material world when metaphors of increase, power, and capability are associated with it.

Third Dimension Nanoethics also entails engagement with the symbolism of art, which engages imaginative expression, toward envisioning various possible futures. When the film *Gattaca* was released, it sent out a shock wave of horror about the possible true intentions and directions of genetic engineering in our culture. Its viewers were challenged to reconsider the hopes and dreams of mapping and engineering the genetic code. Notions of physical perfection so passively accepted in popular culture, and protected by the classical domain of rule-following ethics, become sources of philosophical concern when portrayed in the drama of bodily life under meticulous genetic control. Those who watched the film, and saw what might be, had to wonder about other possible outcomes of human genetic engineering projects. Of course, that film was not the only source of moral imagination about genetic engineering. Our societies have many varied art and literary forms, to engage the moral imagination. The point is that those imaginative elements magnify otherwise oblique elements of the nanotechnology quest. Imagery, fantasy, myth, science fiction, and other such forms are at the heart of Third Dimension Nanoethics and, in the final analysis, may prove to be a most powerful tool of reflection about ethics in a morally perplexing, technological emergence.

DIRECTED RATHER THAN DETERMINED

As the contributions individual citizens must make, and the consequences societies must bear in the evolution of nanotechnology are being negotiated, consideration of First and Second Dimension Nanoethics has already begun to be addressed through various efforts, including federally funded studies and projects. Conscientiously directing those changes requires that attention also be placed on Third Dimension Nanoethics, that is, the conscious recognition of the values embedded in nanotechnology development and the tacit meanings being ascribed to its quest. Nanoscale science and engineering, like all scientific revolutions in human history, involves multiple and multidimensional processes. It reflects, among other things, the internal, distinctively human processes of meaning making, which arise from the ambiguity of being alive and conscious of that aliveness in the world.

A U.S. report of the National Nanotechnology Initiative expressed that "nanotechnology has the potential to profoundly change our economy, to improve our standard of living, and to bring about the next industrial revolution" (NSET, 2004, p. 1). This is a rhetorical claim, subject to societal negotiations over values, ethics, and beliefs. Conceptualizations of the congruence of nanoscale science and engineering suggest that inevitable and radical material and economic changes are afoot as a result of the knowledge it will bring and the technological developments that will emerge as a result of the drive for this new knowledge. What might those changes mean for the emotional, psychological, and spiritual well-being of the human family, and care of Earth? As Hauerwaus and Jones (1989) reminded us, "The world is not simply waiting to be seen, but [that] language and institutions train us to regard it in certain ways" (p. 186).

BEAUTIFUL SCIENCE IN A SOCIAL CONTEXT

In *The Daedalus*, Haldane told the myth of the Minotaur. Dyson (1997) explained that Haldane did this to symbolize the shocks and horrors that well-meaning scientists are about to let loose on humanity, and to illustrate the historic role of scientists to do the unthinkable, to overturn cherished beliefs, and to kill gods. Haldane's reference was justified, according to Dyson, by the many ensuing events where science turned good into evil. So Dyson asked what we can do today to turn the evil consequences of technology into good. Among other things, he was concerned with the essential virtue of cheapness and with the use of science to provide toys for the rich rather than necessities for the poor. That, he clarified, is usually the result of market-driven applied science, the lap top and cellular phones being the most recent examples. Dyson worried about how the pure scientists have become detached from needs of humanity and applied scientists attached to immediate profitability. I worry about how nanoscience, which is directly linked to nanotechnology, can be motivated by addressing the fundamental needs of humanity, at least as much as by the allure of profit. Justin has thoughts about this as well:

ROSALYN: *What is it that you are working on?*

JUSTIN: *We are trying to understand transport phenomenon: How do electrons move in semiconductors, how does heat travel in the semiconductors? And then, electrons and heat—how do you put them altogether to get the highest performance of the device? Those are the kinds of things that we are dealing with right now. We are especially interested in using nanowires for energy conduction, novel energy conversion devices based on nanowire infrastructure.*

ROSALYN: *The applications are potentially very exciting. Perhaps you can replace old refrigeration technology, for example, then you would have fewer environmental problems.*

JUSTIN: *That's right, solid state, if there are no moving parts, there is no fluid.*

ROSALYN: *So what would be the risks of that kind of technology? What kind of problem might come from it? Do you think about that?*

JUSTIN: *To be honest, we haven't thought about it. I mean, certainly we are not, at least in the near future, going to replace refrigerators that are at home. Because of the amount of energy that it requires, we cannot do it with miniature devices. But if you can mass produce these things, maybe someday we should be able to do it. However, if we are thinking of portable devices which need power generators, instead of batteries, perhaps these are much more efficient. Such as the latest microprocessor, using chips and all that may need local cooling otherwise they just blow up. It gets so hot. And so those would require the devices we are trying to build. We haven't thought about the ethical and social consequences of these. We are more involved in the science and engineering right now. We haven't really spent time on that, to be honest. I mean there is a possibility, I'm not saying for sure, but there is a possibility that the environmental impact of this could be significant .... The way to actually have social impact is when you make something out of it and you commercialize it so that people can actually buy, and it goes out into the society at large. Otherwise it can stay in the lab, as beautiful science, but it may not have the technological impact that one could have if it is commercialized.*

ROSALYN: *Is that why you do what you do?*

JUSTIN: *Well, to some extent. Remember, I'm an engineer. I understand the science to some extent. And I convert that science. I exploit the science to make engineering progress. Part of my agenda is to make technology that can be useful to society. That's what I do as an engineer. If you are talking about engineering and technology there is clearly a social impact aspect to it. I look at this as a very positive thing. The venture people are investing, people are involved in start-ups, but at the same time I want us to be careful of all this. We need to understand how the nanoworld works. Because you know, you don't want this technology, this beautiful work, to fall into the wrong hands.*

Of course, many hands are already open, waiting to use the beautiful work of nanoscale science research.

Quoting Haldane, Dyson (1997) wrote: "In ethics, as in physics, there are so to speak fourth and fifth dimensions that show themselves by effects which, like perturbations of the planet Mercury, are hard to detect even in one generation, but yet perhaps in the course of ages are quite as important as the three dimensional phenomenon" (pp. 97–98).[8] Dyson interpreted Haldane's statement to be a warning about the progress of science and its destiny to bring confusion and misery unless accompanied by progress in ethics. Nanotechnology progress is hurtling forward unabashedly. Rhetorical claims made by various proponents of its power to bring renewed prosperity, to correct for environmental pollution, and to cure devastating diseases such as cancer keep all but the most courageous individuals from publicly challenging its momentum. Few research scientists can afford to do so, because the very agencies that are funding and supporting their careers are also the major proponents of nanotechnology development.

Horrible, unintended uses of nanotechnology may be irremediable. This is one reason why I would like to believe that nanoscience and nanotechnology research can and will be directed in humanitarian, Earth-

[8] As explained by Freeman Dyson, Einstein's theory of space time is described as a continuum with four dimensions. Haldane was referring to Kaluza's 5-dimensional vision relativity.

respecting, and spiritually affirming ways, now. But this is only possible if multidimensional ethical assessments of the possible cultural, social, environmental, and moral outcomes of nanotechnology are anchored in commitments to the conscientious control and guidance of nanotechnology development.

LUIS

ROSALYN: *So here we are. It seems to me that we're getting more and more powerful in what we are able to do with matter, with the Earth, with our material existence, and so it almost appears to be that there is another evolution occurring in our scientific abilities. At the same time, it's as if we're becoming more capable of controlling the course of human events, or so it appears. Some people look at that and they say, "oh, wow, wait a minute, wait a minute, are we trying to be God in terms of the ability to create life, change life," control life, and that's what I really want to talk about today, what that means, OK? Particularly with nano because when you hear some of the main spokespeople talking there's a lot of reference to controlling matter with precision. If, in fact, that's what we're trying to do, then that's really quite meaningful. It means controlling our destiny, controlling who we are, and maybe even recreating life. I just noticed on the elevator here there is a poster for a colloquium with a presentation on integration of the human being with the computer. Some people write about that as fantasy on where we're going. It will mean that human life will be almost unrecognizable. It won't be anything like the original creation. What does this mean?*

LUIS: *That's a very tough question to answer.*

ROSALYN: *I know, and yet this is really all I want to talk about today.*

LUIS: *That's fair enough. My feeling is that we underestimate the human mind and we underestimate the human emotions and I think one can control, one can talk about controlling matter and controlling, you know, the physical world. But controlling humans is an entirely different issue because of the complexity that arises. But when we talk about controlling matter, we talk about controlling these atoms here, there, you know. That is one element of one hierarchy that we're talking about and I think humans are several hierarchy levels above that. This is my perspective. Whenever we talk about putting things together (and we are matter put together), whenever you cross a level of hierarchy to the next level, we see complexity. We understand single atoms, we understand the periodic table, and we know how many electrons and neutrons and protons there are in an atom. Then you put two atoms together—molecules such as an oxygen molecule, nitrogen molecule—we still have a pretty good perspective on that. You put three atoms together, like a water molecule, and we don't understand.*

ROSALYN: *Don't understand?*

LUIS: *We don't understand completely how water behaves, which is just three molecules. Now to put together a thousand, a million molecules in a protein we don't understand how it behaves because the complexity of these many body effects are so interrelated, so complex. How does a protein fold? No one knows the answer to that. Given a sequence of amino acids, tell me what this final structure would be? No one can tell you that. And that's just one protein! Now to talk about humans, and that is billions of proteins put together, billions of nucleic acids in a hierarchy which is so many levels higher than just nanostructure, that to be even comprehending, to control, or to understand where emotions come from, where thoughts come from, I think, is a little premature. And so, I think in many respects one loses the humility. To say that we will control human beings, we will control the behavior, I think it is being a little too arrogant about our capabilities. It's very humbling to see what we don't know, what we don't know how to do, and even the simple things like water. We don't know how to explain water completely. So that we will have the power to do that may be a long, long way off. Nevertheless, what the scientific world is trying to do is to get to the point that we can take simple building blocks, atoms, for example, and try to put together nanostructures first and then maybe microstructures and try to predict its behavior. And that, of course, is a nice thing to do from the technological point of view, because it may have implications in improving human lives. But I think it is a very, very early stage toward what you were talking about to sort of impacting or controlling human behavior or interfacing humans with computers, etc. That seems to me a long, long way off.*

ROSALYN: *Why do people fantasize about this? Why do even some scientists fantasize about this?*

LUIS: *Well, I think you're right. The word is* fantasy. *Why do they want to do that? Well, that's a good question. I think there is an element of being able to have the ability to control things. I don't know, maybe it's the culture that we are in nowadays. We want to control our lives. And I think in some respects we think that we can control it. And that gives you some sense of, maybe a false sense, but some sense of security. That's one culture, one philosophy. There is another philosophy which says, why don't we accept the fact that we cannot control? There are some things beyond our control. Accept that fact and be happy with it, you know, be at peace with the fact that there are some things beyond our control, and that's OK.*

ROSALYN: *Can you have that philosophy and be a research engineer, a scientist?*

LUIS: *Absolutely, absolutely.*

ROSALYN: *Really? Do you carry that philosophy?*

LUIS: *I certainly carry that philosophy. You know, there are certain things beyond my control. What will happen to my kids tomorrow? I don't know that. They may get into an accident that's beyond my control. What I will focus on is the things I know, my limited engineering or science knowledge, and I'll attempt to control things that I think I can control, like build a bridge. Civil engineers do that. For me, I'll build a device that will probably test for cancer and, that's about all I can do. Beyond that there is a whole realm of things that I cannot control and I am at peace with that, absolutely.*

Well, I don't have a background in philosophy, but my perspective is that there is a certain limited amount of control we have and I think we should try to better someone's life using that control. But there are some things, and that's my humility, you could say, that tell me I do not have control of certain things.

ROSALYN: *Where is that control, or is that not the right word?*

LUIS: *It's ...*

ROSALYN: *Who has that control?*

LUIS: *I don't think anyone has.*

ROSALYN: *No one has?*

LUIS: *No one has. Can you predict an earthquake? I don't think so.*

ROSALYN: *We try.*

LUIS: *We try, and that's just one example. Predicting the future 20 years from now, boy, I'll fall flat on my face if I try to do that now. And anyone who says that 20 years from now this will be the way the world will run, I would look at with a very skeptical eye. I can tell you, a computer virus got into my computer and completely wiped out my hard drive, including the backup.*

ROSALYN: *Oh my, Oh no.*

LUIS: *Overnight, I lost everything I ever owned, data-wise. Everything. It's a life changing experience when you are frantically working and all that, trying to get this proposal out, and suddenly you have nothing. So what does one do then?*

ROSALYN: *Oh gosh.*

LUIS: *I actually went home and played with my kids and my wife said "Why are you home so early?" I said, "I have nothing to do." And it is a very liberating experience. I don't have e-mail, I don't have anything to work on, except meet with my students, chat with them, go to a lab, go out for coffee with them, go to my family, hang out with the kids, go take them out to soccer, wonderful.*

ROSALYN: *This is very important.*

LUIS: *A wonderful experience. Of course, things catch up, and that's reality.*

ROSALYN: *It's very important what you are talking about because part of what I'm watching is an increasing, an almost total dependency now on the technologies we have created. So you say I have nothing to do, except, and all the things you did do are the things that we used to value so much. It's the way we used to teach, it's the way we used to be, but now we build our day around the boxes on the desk.*

LUIS: *Right, right.*

ROSALYN: *I carry it with me on my travels because I can't be without it, all my intellectual work is in the box, and so then it goes away and who are we?*

LUIS: *Yeah, who are we, exactly. So I mean, I actually ask that question of myself. I mean, who I am without my computer.*

ROSALYN: *And who are you without your computer?*

LUIS: *I'm still the same guy.*

ROSALYN: *You're OK.*

LUIS: *I'm OK. I'm absolutely OK. Of course, you know, it catches up with you because you have to be connected to the world. You cannot be isolated.*

ROSALYN: *You say that's your connection to the world.*

LUIS: *That's what we've made it out to be. Often I think there is an issue about that and I think we are too dependent on it. There are better ways of connecting people (laughs).*

ROSALYN: *So one of the societal implications of nanotechnology is the increasing web of connections that we are creating through the technologies and our addiction to them increases. Might we then lose a grip on our sense of who we are on the Earth and in relation to other human beings? These are the kinds of things I think about. You've actually experienced it. It's really quite wonderful for me to hear you talk about it.*

LUIS: *It's a life changing experience. There's a song by Janis Joplin, "Bobby McGee." Have you heard it?*

ROSALYN: *Sure.*

LUIS: *"When freedom is another word for nothing left to lose ...."*

ROSALYN: *Nothing left to lose.*

LUIS: *In many ways, I am embedded in this world where everything, sooner or later, it catches up with you and I've got piles of work to do.*

ROSALYN: *Of course you do, things of this world.*

LUIS: *It's a funny situation and I hope no one has to go through that, of losing all your data in one shot.*

ROSALYN: *You responded to it well. I suspect some people would have gone into depression or panicked, or ...*

LUIS: *You can't do anything about it. It's beyond our control. The virus came into my computer and ate up my data. What can I do? I have no control over that. I might as well make the best of the situation and just move on, that's all. I thought about this issue about how dependent we are, and the only thing I could do was to change to a Mac, so that's what I've done.*

ROSALYN: *A solution! One of the things we believe in this culture is that the solution to our problems is better technology.*

LUIS: *Well, I mean, I can't get rid of my computer. As for the nanotechnology part, I really strongly feel that it has a tremendous opportunity to create technology that can really help the world, not just …*

ROSALYN: *The world?*

LUIS: *The world, and I may be naive about it, and it may be too premature. If you talk about technology as say, well, the icon of technology is a Pentium chip. That's the high tech that you talk about, right?*

ROSALYN: *Yes.*

LUIS: *So what is the price of a Pentium chip? It's about $200. And if you look around the world as to who all can afford a Pentium chip, you will find that it's a minuscule amount of people in the world and the number of people who can afford it are generally people who make more than $20,000 a year. I don't know if you have seen the world pyramid. Right at the top of these 100 million people who make more than $20,000 a year are the people who can enjoy the high technology that we talk about now. Right now in this world there are 6.4 billion people who are untouched by what we call high tech. So there are 2 billion people in the world who make $2,000–$20,000 and there are 4 billion people in the world who make less than $2,000 a year, OK. They are untouched by high tech. Nanotech is yet to fully flourish and is still in its infancy, but for the high tech that is already out there, and what we are hoping for is what I call in my own words: trickle down technology. We hope that our high tech products will someday become sufficiently cheap that it will trickle down, the cell phones will trickle down to the 4 billion. However, when the products were designed there were no plans to make it available for those 4 billion. They came as an afterthought.*

ROSALYN: *Why do you want them to have the cell phones?*

LUIS: *I don't want them to; I'm just saying that this is not a fit for those people.*

ROSALYN: *But why is that, you said it would help the world. How does that help the world?*

LUIS: *I'm not saying the cell phone will help the world, I'm not saying the Pentium chip will help the world, what I'm saying is that we are hoping that this will be useful for the rest of the world, but what does the world really need?*

ROSALYN: *That's my question.*

LUIS: *The 4 billion people or the 5 billion people, they need clean water.*

ROSALYN: *That was what I was going to say.*

LUIS: *They need energy, they need clean water, and they need the minimum of basic health care.*

ROSALYN: *That's right.*

LUIS: *And they need perhaps a little bit of, I don't know, entertainment, perhaps all human beings need a little, and so those are the clothes, textiles, some shelter, some transportation, and trade. Trade, that's been our system for ages now.*

ROSALYN: *Of course.*

LUIS: *So if that's what they need, is our high technology or nanotechnology providing that right now?*

ROSALYN: *This is the question.*

LUIS: *It's not.*

ROSALYN: *Will it? Can it ever?*

LUIS: *Well, it depends on us and how can we get this technology that we are developing right now to the world, to the 4 billion people in the world.*

ROSALYN: *Well, I'd like to know does nanotechnology become water, access to potable water? That's why I asked you what difference it makes if they have a cell phone or a Pentium chip.*

LUIS: *It doesn't make any difference right now, absolutely. So what if you give them a Pentium chip or a computer. I mean, people are saying that we are wiring up all the villages in India. So what?*

ROSALYN: *That's what I'm asking you. How does nano help the world?*

LUIS: *Well, it is possible. The question is can we direct nano.*

ROSALYN: *This is the question.*

LUIS: *Can we direct nano to make it to these grand challenges of the world, beyond the few people who earn more than $20,000 a year.*

ROSALYN: *I say we can but will we? Why would we bother?*

LUIS: *Because we cannot live in an isolated world, whether we like it or not, we cannot.*

ROSALYN: *OK, so when I look at the funding that's going to nano, and I see where the interests are behind the funding, so much of it is for military interests. It has nothing to do with getting potable water to people. So much of it is coming out of semiconductor interests, worries about Moore's law. We need to get into new markets, new products, and we're still now talking about that same little …*

LUIS: *Same little top of the pyramid.*

ROSALYN: *So if the funding agencies are not talking about access to water, who is? And how does it shift over?*

LUIS: *I have no idea.*

ROSALYN: *Yeah.*

LUIS: *Wherever I give a talk about nano, I talk about it. This is the pyramid. And you know, I tell them, and I'm not the only one.*

ROSALYN: *OK.*

LUIS: *OK, the administration in my college of engineering talks about it, because this is a place, I think, that is more socially aware than some of the other places that we know of. If we don't take this on, then who will? And so that's something that I certainly bring in when I talk to funding agencies. We had a gathering at NSF and that was my pitch—if you think you're isolated in this world, we are not.*

ROSALYN: *Our response to 9/11 is more military surveillance, more combat power. We reacted from fear to get more control and find the bad guys. You're talking about getting basic needs fulfilled.*

LUIS: *There is actual business to be made of this. I mean, if you're saying that the U.S. is very business oriented, there is nothing wrong with that. If you improve someone else's life and you make money out of it.*

ROSALYN: *It's good.*

LUIS: *It's good. But the multinationals have ignored that fact and they are trying to squeeze each other out in the top of the pyramid. But there is a tremendous market opportunity out there in this big bottom of the pyramid. There is a paper that came out of the* Harvard Business Review. *They take a look at this and they analyze the situation and so there is actual business potential. Unilever has a branch in India. They had been trying to get all these high tech goods out, like washing powder, and this and that. What they started was this little candy for the masses, and they thought, "Let's make it really cheap so that people can buy it." Now I hear it is the biggest growth product in that Unilever branch.*

ROSALYN: *How interesting.*

LUIS: *That is the biggest growth sector. It's a cent, so it's really cheap, there're a lot of people out there who can buy it.*

ROSALYN: *So you're thinking if nano can allow us to make products cheaper and more accessible ...*

LUIS: *Exactly. I mean, for example, why is a Pentium chip expensive? It is expensive because the processing is expensive.*

ROSALYN: *Sure.*

LUIS: *OK, the facility that goes into building the next generation Pentium chip is more than a billion dollars and that Moore's second law, which is the cost. It will continue to go up.*

ROSALYN: *But the nanofabrication labs are very expensive.*

LUIS: *Well, the question is, can we direct our nanofabrication toward ways that are really inexpensive. Can we use self-assembly to make things that are cheap, and then just put it together in a way that everyone can make. So the kind of change that we need now in our nanotechnology initiative, is how can we really impact the world, not just a few of us.*

ROSALYN: *We have to want to.*

LUIS: *We have to want to.*

ROSALYN: *I always tell my students, if we wanted everyone to have water, everyone would have water.*

LUIS: *That's the bigger challenge, convincing everyone that it's a good thing, that it's in our interest to make sure that people have water, that people have energy, that people have food. It's in our interest, and you may call it humanitarian, you may call it whatever, but it's only our selfish interest because otherwise it will haunt us, we cannot separate ourselves from the rest of the world.*

ROSALYN: *I was just reading a book where the author argues that not only do we have to be very clear about our connection to one another, but we have to be very clear about our connection to the Earth and its other inhabitants.*

LUIS: *Absolutely.*

ROSALYN: *And that our whole biological makeup is designed to be in direct relationship with the Earth and its inhabitants and that because we have cut ourselves off from that, humanity itself is disintegrating rapidly. So he talks even about our noses, they are designed to sense, not just, you know, that dinner smells good, but the winds and the energy fields and the animals and it's all connected. So, yes, I'm saying yes to you that what happens to a villager in Madras is important for me, even though it might not be apparent, it might not seem to affect my life, it's terribly important. But how do we teach our young engineers this? Or even, researchers who are less aware then yourself? I read the testimonies before Congress to appropriate more money for nano at the hearings of 2003 and most of the talk was about new market capabilities, a trillion dollars worth, but it is all about the top of the pyramid.*

LUIS: *The top of the pyramid.*

ROSALYN: *Health care changes yes, but again, it's all for the very, very elite of the planet and other than that I'm seeing dreams and fantasies about controlling the universe, and I'm searching for what you're talking about. Here we are again, we're on the brink of another major technological breakthrough, potentially, and what we can do, what are we going to do?*

LUIS: *So OK let me give you the reaction I've gotten.*

ROSALYN: *To your message?*

LUIS: *Yes, to my message. Some people love it: "Gosh, now we have seen the light." But it makes some people very uncomfortable and I'm still trying to figure out why is it, because it is there, if someone is looking at a Pentium chip and trying to improve it, and I'm using Pentium as an example, then their whole livelihood, their whole thing is being questioned, and that can be disturbing to people. So I may have made a lot of enemies by talking about this. But so be it. I mean, I have no problem with that because I think this is the right thing to do.*

ROSALYN: *Where does this come from inside of you?*

LUIS: *I don't know. I have no idea where it comes from.*

ROSALYN: *Is it spiritual orientation?*

LUIS: *We always think about technology as improving the human life and to ignore the majority of the humans is not even humane. It's inhumane to ignore them, if you're a human being and you want to improve the human beings life, let every human being be equal. That's my perspective on things.*

ROSALYN: *You just made a really big leap. You made a big jump from improving life to improving every life. How do you get from one to the other? The way we have things set up in our societies there are a lot of peoples lives who have been improved over the last few decades and a lot who were left behind.*

LUIS: *Oh absolutely, yes. Well, in my thinking there are very few absolute truths in this world. Everything else is relative.*

ROSALYN: *What are those absolute truths?*

LUIS: *Life, respect for life, and respect for the Earth and environment.*

ROSALYN: *Respect for life.*

LUIS: *Those are the two things I could find in my own search of what are the absolute truths. The others are all relative. I mean, say this is right, this is wrong, there are gray scales.*

ROSALYN: *Many gray scales.*

LUIS: *But there are few absolute truths.*

ROSALYN: *And respect for the Earth?*

LUIS: *Respect for the Earth, which is also life. You can think of it that way. I don't know what my basis for this philosophy is, because I haven't analyzed it any particular way.*

ROSALYN: *Well where does life come from?*

LUIS: *I have no idea. I don't know the answer where life comes from. I'm not deeply spiritual.*

ROSALYN: *So why respect it?*

LUIS: *Well, I don't know that. Why respect life? I don't know that answer. Why is life so precious? Is it precious because I think I'm precious?*

ROSALYN: *OK.*

LUIS: *I think my kids are precious, and if they are precious, everyone else is precious. And that, I guess, is where it comes from. I don't know, it's just respect for another human being. I don't have a good answer for you, that's my only answer that I can think of now. I have to think about this, why is life precious?*

ROSALYN: *OK, so related to that, when I hear research scientists and engineers talk about the manipulation of matter, I'm OK when it's carbon, but I get a little bit quiet inside when it's more complex molecules that involve living matter. Carbon, I know, is the basis for life.*

LUIS: *Right, right.*
ROSALYN: *But I don't see it as alive, something happens to bring it to life.*
LUIS: *Right.*
ROSALYN: *I don't know what that is, but something happens.*
LUIS: *So that's the question, what is life and what is not life?*
ROSALYN: *Yes. And what are we respecting, so to respect carbon we mess with it, I don't care what you do, you can make it into sheets at one molecule deep, wrap it up into tubes, and ...*
LUIS: *That's not life.*
ROSALYN: *That's fine, how fun. But if you then take some flagellum and you attach it to an electronic device and then you put it on cells and then you send them through the bloodstream into the veins, I want to be a little bit more careful to know what we're doing. If you somehow connect it to the root systems of the trees in an Aspen forest, I want to know a little bit more about what we think we're doing because then we have crossed over into another domain.*
LUIS: *Yes, you're influencing life and you're influencing the Earth, yes.*
ROSALYN: *Why does that matter?*
LUIS: *Well, that's a great question.*
ROSALYN: *Well, I want the answers.*
LUIS: *I think, my scientific view of it, is that life is anything that can spontaneously replicate itself.*
ROSALYN: *Is living?*
LUIS: *Is living.*
ROSALYN: *OK.*
LUIS: Spontaneously *is the key word. Spontaneously. Without us, without our input; it takes energy and matter from its environment and it replicates itself.*
ROSALYN: *Some researchers are trying to do that.*
LUIS: *I'm trying to do that in my lab.*
ROSALYN: *Yes?*
LUIS: *But it's not spontaneous, we are trying to coax it and do all kinds of things.*
ROSALYN: *But wouldn't you be successful if you got it so it became spontaneous, you initiated what ultimately became spontaneous, self-replication that begins a new life. Wouldn't that be successful?*
LUIS: *I mean, how do you define success, I mean, that's ...*
ROSALYN: *Would it be good?*
LUIS: *See, the way we are doing it, to be honest, we are not asking the question whether it's good or bad, because as I said, that's all relative. We are trying to just play around, essentially.*
ROSALYN: *That's what I think is the answer.*

LUIS: *Yeah, we are just trying to play around. Whether it's good or bad is a matter of debate, that's relative. Our definition of life is that it's self-replicating. As for our technical definition of life, I don't know what the answer is. I think there is a bigger aspect to it than just a self-replication aspect. I don't know what that is. I'm looking for answers myself. Can we possibly make something spontaneously, self-replicating 20 years down the line? I have no predictions on that but to be able to say that we have made life; that I think is a different matter. But your point is that if you now embed these things into living things, are we changing living things?*

ROSALYN: *That's one of my questions.*

LUIS: *That's one of your questions. I think that is certainly odd, whether that perturbation will have implications that are sort of nonlinear in the sense that it will have gross consequences by a little change, a little perturbation, that, I don't have an answer to. I think we should definitely put it on the table and ask these tough questions. It's a tough question to answer now because we don't know.*

ROSALYN: *We don't know and yet we're playing.*

LUIS: *Yeah, we are playing, but you're right, when you're playing with living things, you may be playing with fire at some point, and so I think it's fair enough, you should put it on the table and discuss and debate and hopefully collectively we can come up with some consensus and have some ground rules, a way to play and a way not to play. I mean, when we send our kids out to play we tell them don't play over there, play over here, and I think, I'm hoping that there would be something of that, some consensus.*

ROSALYN: *It's all relative. We have to decide what we care about.*

LUIS: *What we care about, exactly, that's the bottom line, what do we care about?*

ROSALYN: *I don't know of any ultimate rules out there that we can just go and try.*

LUIS: *Right, that's right. But if it takes someone's life, then we can say that this is a bad thing, that's obvious. As I said, there are a few absolute truths to me, that violate that truth and that's a bad thing, but more things are gray than black and white and I think that's where the debates come in.*

ROSALYN: *OK, I have one question left because I'm looking at the time and I want to respect your time.*

LUIS: *Oh, I can do another few minutes.*

ROSALYN: *I'm thinking back to the lab and back to the question of defining life technically around self-replication. What about the possibility that we as humans can learn to do that, to be the impetus, the energy, the cause of life forms coming into being and replicating themselves. Is that a fantasy or is that a dream? Or is that something that's part of the evolution of the human capacity as science itself evolves?*

LUIS: *It's a very hard problem to solve. Scientifically, there are people who are trying to make artificial cells.*

ROSALYN: *Why do you call it a problem to solve?*

LUIS: *It's a technical term we use.*

ROSALYN: *But you're using it as applied to life.*

LUIS: *So, for example, the work we are trying to do in our lab is not spontaneous replication, it's a very hard problem to make things replicate each other using self-assembly. That's why I call it a problem, but it may not be a problem for everyone.*

ROSALYN: *It sounds more like an intellectual challenge than a problem.*

LUIS: *It's a challenge. It's not a problem. Problem is a word that we use for having to solve this challenge.*

ROSALYN: *Yes, I hear that word in every one of these discussions with researchers. I'm thinking of a problem as the water, you know, the potable water to all the people.*

LUIS: *You're right. It's a misuse of English.*

ROSALYN: *No, no, no, it's used inside the language of science, and appropriately there. But my own work includes thinking about the meaning and use of words.*

LUIS: *It's a challenge, it's certainly a very, very challenging problem but there are some biologists that are trying to put the bare minimum stuff inside cells and make them replicate.*

ROSALYN: *Why are they doing that?*

LUIS: *I have no idea. I think it's just playing games, playing around with it. I guess you can connect it to saying that you can program the cells, you know, sort of like stem cells, program it and you can send it out there and cure diseases and there are all these health benefits from it.*

ROSALYN: *Yes. Does it tread on your respect for life principle at all?*

LUIS: *Does it what?*

ROSALYN: *Does it conflict with your respect for life principle?*

LUIS: *No, it does not actually. Well, all these things you can think of as neutral. If someone creates this artificial cell that can repair someone's kidney, wonderful, that's fantastic, and that certainly does not violate my respect for life.*

ROSALYN: *Yes.*

LUIS: *But you could also turn it in a very deviant way.*

ROSALYN: *Yes.*

LUIS: *OK, and that could influence living things in a very harmful way and so it is really, it's really up to us to decide how we take our technology.*

ROSALYN: *What I hear from a lot of scientists and engineers is that no matter what you do it can be used in deviant ways.*

LUIS: *It can, yes.*

ROSALYN: *And therefore there is no reason not to go forward because everything can be used for good or harm.*

LUIS: *Well, I mean, there is a point to that but what worries me sometimes is that, unless you have the debate, unless you have the discussion of what is deviant, what is not, and how one could be using it as deviant, then you could just use that argument to say, let's do everything and never put in any of the boundaries. It's incumbent upon us to be part of that debate and to make sure from our scientific knowledge, that at least within the scope of our understanding now, that you place some boundaries because otherwise, at least in my perspective, technology may be used to violate life as we know it. And that I think is disrespect for life.*

ROSALYN: *May I ask one more question?*

LUIS: *Sure.*

ROSALYN: *I want to get back to this concept of a higher being and whether or not there is any conflict between the evolution of scientific ability and the existence of something beyond the humans and higher forms.*

LUIS: *Well, I don't know whether there is a higher force or not, people talk about it, religions talk about it. But unless I experience it, I'm not going to say there is or there isn't. OK, could be, may not be.*

ROSALYN: *But you know there is life.*

LUIS: *The thing that I know is life and that is what I'll respect because I want people to respect my life and my kids' life, my wife's life, you know, my parents' life. So, that I will respect, and if that is the higher being, so be it, that's great, I don't have to search too far.*

ROSALYN: *So, is there a natural order to things, do you believe?*

LUIS: *I don't know, I can give you an answer but it may be totally wrong, I don't know.*

ROSALYN: *So the question of whether somehow our science is changing some kind of divine or intelligent order of things, is really not a relative question?*

LUIS: *What do you mean by order, I mean, how do you define that there is an order?*

ROSALYN: *I'm not sure there is. I think there's change.*

LUIS: *There is change I mean, if we call that order, that's fine.*

ROSALYN: *I don't know.*

LUIS: *But I don't know whether there is an order, whether there is a sequence of things to be done to make this whole Earth run. I don't know that. I don't know whether there is. I mean, I doubt it. Why should there be a grand scheme of things?*

ROSALYN: *So what you care about is doing what you can do, then.*

LUIS: *In my limited knowledge, in my limited scope of my intellect, I mean, I'm a professor out here. I guess the only way I can influence the world is by my technological skills, scientific skills, and through the students that I work with.*

ROSALYN: *That you have.*

LUIS: *Mother Theresa did it in her way, Gandhi did it in his way, you know, a nurse does it in her way or his way. I'm doing it my way and if it can help, that's great, that's my attempt to make people's life better, so that's all I can do.*

ROSALYN: *That's good, thank you.*

LUIS: *That's good?*

ROSALYN: *That's it.*

LUIS: *That's it?*

ROSALYN: *For this time.*

PART II

MEANING

I am interested in doing what technology doesn't today allow me to do.

—Ondrej

CHAPTER THREE

Conceptual Frameworks, Themes, and Values

PETE: *I hear myself sounding holier than thou and that's not really how I feel. I have got lots of friends who have been very successful, taking their work, commercializing it and starting a company and going on to another side of science.... And a lot of these are small companies that have sprouted up around nanoscience. Some of them are good and some of them are bad, and the good ones I think actually make strides toward functional devices. When I go out and talk about my research, I am very careful to say, "these are our goals, but remember we are doing fundamental research and maybe this particular molecule that we are using isn't good, because it might be toxic and actually kill people, but it serves as a good model, it's a classic physical chemistry model system that might allow us to understand how a functional device could be made." I try never to go out there and kind of puff out my chest and say, "this material is what we want," and claim it might be a model for all we want. That makes it easier for me to sleep at night.*

The current institution of Western science arose from its early European roots in the Royal Society, which determined that science would most successfully accomplish its goals of discovery if insulated from larger social concerns. The social contract held that science would be insulated from social concerns in exchange for the promise that its fruits would address social needs and bring benefit to the society at large. For many years, science functioned relatively free of public scrutiny and criticism, and was held in the highest cultural esteem, based on the implicit trust born of that initial promise. But the protective shroud of expertise has been increasingly difficult to maintain, and science is being asked by society to justify its practices. Given recent science and engineering history, various publics hesitate to assume that their social interests are heeded in technological development. The development of nuclear technology was one outcome of science that brought all of that into question. Today, "science" is not only being called on to be accountable, predictable, and responsible in its endeavors, but also to participate in the public discourse and to keep the public informed and aware.

Ellul (1990) traced five ideological periods of development in the history of scienced from 1850 until the present. In the first, which Ellul called scientism; science is the discoverer of all truth, which is associated with the finite, concrete material world. As such, science was thought never to be wrong. Its quest was toward the grasping of all that was knowable in order to solve all problems before men. The second period, 1900–1918, brought with it an educated public accustomed to the marvels of science, but without the understanding that big discoveries risked big changes. The main occupation of the people was the war, and although scientists' ideology persists as impregnable, it was now a minor consideration. In the third period, centrality of truth was given over to the centrality of happiness: Science was to assure us happiness. Spectacular progress had been made in medicine and surgery, and thanks to science, consumer goods increased and the standard of living rose. Happiness through consumption becomes well being. Comfort and freedom become the fruits of science. This fourth stage involves what Ellul (1990) identified as a long period of ideology of doubt and defiance regarding science: Truth is no longer the primary goal of science. There are more riddles to solve, and more numerous and difficult problems. Human phenomena seem too complicated. No one can any longer say what belongs to the category of science. Weapons multiply indefinitely and bacteriological warfare becomes possible. Science was everywhere and served every end. Scientists had become creatures of weapons and war, while also being creators of innumerable products whose effects could not be predicted.

The public, according to Ellul, remained astonished at the extraordinary discoveries of science, but there spread the conviction that everything depends on the way in which science is used, along with ambiguity about its value and positive nature. The fifth period, which Ellul dated back to 1975, represents a complete reversal of ideology respecting science. The view of science as independent and sovereign is no longer possible. Science ceases to be free. Its duty has become to save the national economy; its orientation is national greatness (pp. 172–176).

Science is no longer confined to unknown laboratories. In its social transformation, science has come to be about not just discovery of nature, but about a response to "everything that disquiets or troubles us." Ellul (1990) defined present-day science as a soteriology; that is, an ideology of salvation that holds the future up to society:

> This ideology of a divine, soteriological science in association with a dream world is reinforced by what we anticipate and by what is about to come seemingly with no human direction and in obedience to none of the existing classical laws. Science is becoming capable of absolute novelty and also of the regulation of a world, as is only proper for a deity. Like all deities, it has an oracular power. We ourselves can no longer will or decide. We leave this to the beneficent science in which we believe. (p. 185)

Scientists (and with them, research engineers) are held in modern, technological Western cultures as a voice of authority with intellectual and material access to that future. This position is reflected in the language used to talk about the work of scientists. "They say," "There's a study that proves," "According to science," and "Science says" are preambles to claims of proof, objectivity, knowledge, and expertise. Individuals casually base their beliefs and even personal decisions on the faith that "science" knows. "They" function as one entity, privy to knowledge and abilities no else has. This authoritarian ethos gives initial credibility and power to those individual researchers who are invested in the development of novel-appearing endeavors such as nanotechnology. It gives them a public trust that their work is leading us to a new era or realized dreams, a trust on which proponents and policymakers depend.

The "science" Ellul spoke of (which in nanotechnology would include the work of both research scientists and engineers) can also be understood as a cultural icon. In reality it is made up of many hundreds of thousands of individuals, with their own beliefs, dreams, hopes, and fears. As the so-called institution of science took on an independent identity beyond the individual and took claim of a relative independence from society, society (and the individuals that comprise it) endorsed and empowered that position. In turn, the individual researcher was diminished and relegated to an isolated and protected world. Even today, individual investigators, with their laboratories and graduate students, think and move and practice the profession as an individual within an exclusive institution of similarly trained people who use distinctive and exclusive languages. But they are not necessarily like-minded people. The voices of science are many and diverse, despite the publics inclination to unify them as one, or the professional associations intent to represent them as one as they lobby on behalf of the interests of "science."

In all respect to Ellul, whom I agree with in large measure, these individuals may be worker ants, but they are not one organism. "The beneficence of science" is a false premise, isolating and diminishing of the independent, morally responsible person who happens to be trained to do research science or engineering. It may be true that as a network of professionals they hold similar values, training, and goals. It may also be true that as such, they are subject to political and professional forces beyond their control, and on which they are largely dependent. This fact does not, however, erase the scientist as person. And the disparate, even sometimes inconsistent, voices they represent as persons cannot participate in public discourse as one, or serve to lead as one.

THE RESPONSIBILITY OF SOCIETY FOR ETHICAL NANOTECHNOLOGY

To some research scientists and engineers, nanoscale science and engineering is viewed purely as a process of observing, collecting data, and making experimentation toward the acquisition of new knowledge about novel phenomenon, from which will arise tools and devices for applications to addressing material problems. For others, it is also understood to be a more complex socio-cultural undertaking, which is fueled not only by scientific ingenuity, but also by political pressures, venture motivations, and conceptualizations of increased qualities of living. Most of the researchers I interview seem to appreciate that the emergence and development of nanoscale science and engineering raises a host of questions about ethics, and about the potential impact it may or may not have on various societies around the world. As Emily expresses, they tend to share the view that societal and ethical implications of nanotechnology are generally not pertinent to their own work:

> EMILY: *Social impact questions really should be a matter of government regulation. The people will deal with those things because we are a society and we talk about things and get concerned about different issues. If there is enough of a concern, there will be a large public pressure to change regulations and address those types of issues. Every time you invent something new and you develop it you might solve one problem, but you can create a bunch of others. And, we go through an interim process of dealing with these types of issues.*

ROSALYN: *Are you saying that potential social impact is really not a significant factor in terms of nanoscience research?*

EMILY: *I think it's actually almost insignificant, because I don't think we are in a position to guess and analyze all of the individual consequences of everything that we do, that we should think about some major consequences. I mean, if you are working on developing a bomb, I think you ought to be somewhat concerned about that. In the more gray areas, I think that we have the things that have popped up in respect to stem cell research, for example. They are doing stem cell research in controlled types of environments and limited settings. If major benefits are perceived from that research, then the scope of that will be expanded. And, desire from the masses will override today's policy. If there is a direct negative consequence of what you are doing, you have a moral obligation to think about what you are doing and decide how to respond and the government has an obligation to step in and play a role there. Remember the H. G. Wells' time machine? You go through and you pull a lever, you undoubtedly will change everything in the future. You can drive yourself absolutely crazy thinking about every little ripple that you create down the line from each individual discovery.*

In the simplest terms, nano research scientists apply their ideas and questions to understanding and solving problems at the nanoscale. They formulate hypotheses, observe the unique behavior of atoms and molecules, measure, image, model and manipulate matter at this length scale. Ultimately, they craft the blueprints for what is materially possible. Research engineers put their imaginations to the tasks of creating improved materials, novel working systems and devices, which exploit the new properties of the nanoscale of matter. On the whole, the researchers in this study view nanoscale science, engineering and technology as inherently good, as having the real potential for solving actual, pressing problems, and of addressing fundamental human needs. Some are focused entirely on the practicality of getting things to work in the lab, and believe that there is way too much hype in the rhetoric over the futuristic applications of nanoscaled science.[1] A few acknowledge that its potential impact on society is unknown and unpredictable, yet may be great. But most feel pow-

[1]Anthropologist Mikael Johansson found in his field work among nanoscientists that they have a "sober" attitude toward the new technology, and are much more concerned about getting things to work in the lab than about the hype of futuristic nanotechnology applications. See online issue of *Anthropology Today,* December 2003, http://www.blackwellsynergy.com/servlet/useragent?func=synergy&synergyAction=showTOC&journalCode=anth&volume=19&issue=6&year=2003&part=null

erless to affect that impact, to prevent possible harms that could come as a result of theirs or others' own research.

When I ask researchers about any particular ethics challenges or issues that would be of concern as a result of their work, most are unable to make direct connections between ethics and what they do. Interestingly, they point to stem cell research and other biomedical engineering as the fields of nanoscaled science that may have the most relevance for ethics and societal implications, because these areas of research deal with the human body. In them, the risks and potential harms are more apparent and dramatic. (At least, that is the way things currently appear.) But even the biomedical engineers, and those who are collaborating with them, tend to claim or affirm that any societal or ethical implications that emerge will best be taken up by the society at large and its government structures, rather than to be shouldered by the individual researchers. Researchers who are in fields such as mechanical and electrical engineering, where their interests are in electrical devices for such applications as information processing and surveillance, and chemical or materials scientists who are interested in novel properties for the design of new materials, are especially perplexed when I inquire about the social or ethical implications of their own work. They are interested and willing to discuss questions of ethics and society, but as for what they do in their labs, there just does not seem to be any real relevance. For example, Ryan expresses the sentiment that science is intended wholly for good, and is otherwise morally neutral. If it gets into the wrong hands and is misused, that is a matter for government concern:

ROSALYN: *Is there anything about what you do that you imagine could possibly bring harm to anyone?*

RYAN: *You know, you can talk about this in so many different contexts. Consider a ballpoint pen. You can write with it, but I have heard of people getting killed with pens.*

ROSALYN: *That's true.*

RYAN: *You can use a knife to make a beautiful flower arrangement, or, you can kill people with a knife. Humans use things they make for a positive or negative cause. And the same with those things they are endowed with at birth, like fingers. People create wonderful things with their fingers, and they kill with their fingers. So are my fingers inherently for good or for harmful use?*

ROSALYN: *You think that our creator put fingers on us to destroy one another?*

RYAN: *These fingers are lethal weapons.*

ROSALYN: *OK, well, alright.*

RYAN: *But some of the evils of today might be considered not so evil further down the road depending on how society figures it out.*

ROSALYN: *Is there any scientific inquiry that you would judge to be inherently bad?*

RYAN: *It's difficult to say because, it depends. Say somebody is trying to discover the most lethal toxin. That process alone would be considered evil. But how about the person who is studying chemistry and just trying to understand reactions? If the work were driven solely for destructive, morally questionable purposes, then certainly I would question those motives. But I don't question the motives of a pure research project, because that is for the purposes of discovery and creativity. Even if it just so happens that they are using certain molecules, which in this combination could make the human race extinct. To the scientist, that's not the desire.*

ROSALYN: *So you are suggesting that scientific inquiry, including nanoscience, is simply about curiosity—the discovery of how materials work, as a morally neutral endeavor?*

RYAN: *I think so. Actually, it's more than that. I think it's a very positive human development. Curiosity and inquiry are among the things that make us human and why we have evolved since the Stone Age. Monkeys, last time I checked, are pretty happy where they are. I think that all of the science work that we have done over the generations and centuries has been largely positive. Those who have evil or bureaucratic intentions have largely motivated the negative aspects. I mean, scientists may have developed nuclear technology, but it's the politicians who make the decisions about how to use it, right or wrong. Some scientists are oblivious to what it means. Ultimately, the decision to drop the bomb is made by a politician, not a scientist. I mean, you've got some people doing science over here, some engineer doing engineering work over there, an administrator over here, a lawmaker over there. And this person doesn't understand what that person is doing, but if everybody were taught that they had a responsibility to society, and had a basic understanding of what it means to be objective and to implement basic scientific method and to be unbiased, maybe everybody would come together on that common ground and be responsible people. As responsible people they would come to the right conclusions.*

ROSALYN: *Is there any point in the development of nanotechnology where you think it would become appropriate to stop?*

RYAN: *Well, there could be. Like with cloning. I have no idea what cloning is going to do to our society.*

ROSALYN: *What I am wondering is, is there any way to anticipate harm before the fact and redirect the development of nanotechnology?*

RYAN: *Well, it's hard because again, you have at times a very small group of people doing something that is outstanding. The science emerges and then all of a sudden, there is a new key to throw on the table, it opens up so many doors. Far from the legislators and policymakers, this is going on. That's why I come back to the fact that it's hard to construct a system. You need to construct a society to work on those issues. I don't think you will ever construct any kind of system that is going to be able to reach into that little world of science and say, "stop," unless we go to Big Brother to the* nth *degree. I just don't think it's realistic to even think about it. The only thing that you can do is to construct a society that responds effectively to science and technology. Then, when some researchers throw something really wild and potentially dangerous into the field, that we all don't just run for it to see what will happen, but that we as a society say, "Hmm, somebody has developed something that maybe we should have looked at more carefully. Maybe we should have redirected the research, but OK, it happened, now what do we do with it?"*

ROSALYN: *Interesting. You think about these things.*

RYAN: *Of course.*

ROSALYN: *Why?*

RYAN: *I think it's just interesting. If we look at history, to see where we are going to go and how things will change, we can see our society has a strong spiritual component to it. We thought that the Earth was in the center of the universe, because we believed we were the center of the universe. All of a sudden some scientist says, "Gee, I look through the telescope. That can't possibly be true. We are not in the center of the universe!" And the people asked, "What do you mean? It's impossible." At that point it wasn't a scientific thing; it was a real personal thing. It was about who I am in the universe. I'm special or our society is special and these kinds of beliefs gave comfort to the average individual. All of a sudden this scientific discovery emerged into the popular domain, and became something everybody talked about and then, "That can't be! It's not right!"*

ROSALYN: *Couldn't nanotechnology do that?*

RYAN: *Yes. I think that nanotechnology will do that, ultimately. It is going to challenge the thinking of average individuals. That will change things. I don't know how it's going to play out, but I think that if we educate our children and teach them to be constructive, positive, and not to be afraid; to be a type of an individual where you can look at these things and just say, "well gee, that's a very interesting thing" and you can process it, try to learn from it and see what insight it does give into your life or to society, and not have a negative reaction—I mean, whether there is a plan behind it all or not, there are little clues about how things work in life and I always think they are all meant to be positive. They are all meant to give us interesting insight, if they are*

> *meant to do anything at all. So, why not embrace them, try to use them as an opportunity for further development, rather than an opportunity to worry about destroying something.*

What does it mean to put trust in society? Does it mean that research done in laboratories doesn't require the deliberation of ethics by the individual researcher because society will address its social and ethical implications? What kind of relation does this so-called society actually have with so-called science? Here is the myth of neutrality, which holds science as an entity independent from society. The belief, which functions as a myth, may serve a very important purpose of self-protection against the unwarranted scrutiny and interference of outside interest groups. Unfortunately, it may inadvertently result in a conceptual roadblock to the formulation of sound ethical principles for the development of nanotechnology.

LAN

It might be argued that whether or not an individual researcher spends time and energy on reflection about ethical and social implications is a personal choice, not a moral obligation. One could even go so far as to say that to do so could become a distraction from the intensive focus needed for good basic science. A claim could also be made that scientists and engineers are not trained to think in terms of philosophical or sociological concerns, and therefore ought not try to do so. My own view is that individual researchers do have the power and knowledge to influence the development of nanotechnology toward humanitarian and Earth-respecting ends, but only if they are willing, like Lan and others, to commit themselves to conscientious pursuit of their own work and a willingness to step out as individuals to assume ethical leadership of nanotechnology development:

ROSALYN: *Is the precise and controlled manipulation of the atoms, by humans, a socially good thing?*

LAN: *That's a huge looming issue. How much are we supposed to meddle with what we are granted at birth? Perhaps we need to take a step back. Think about parents who give their kids human growth hormones so that they can become a professional athlete. Our genetic drive is supposedly to confer any advantage we can put on our children. What about the unintended consequences? It is hard to predict what eradicating all of these different diseases is going to leave us with.*

ROSALYN: *Perhaps some of those so-called genetic advantages are actually socially constructed ideas. The perfect body that runs really fast is not necessarily going to be the one that survives the longest or with the greatest health.*

LAN: *I asked my students, "within the next century, how many of you believe that there will exist highly augmented humanoid beings, so to speak, that would be quite different than what we are, and frankly, able to kick our ass in so many different ways that we would say that this thing is out-competing us?" And only two of them raised their hands, but I think that could actually happen in 20 or 30 years.*

ROSALYN *I do, too.*

LAN: *So, you know it's right here upon us and I am trying to suggest to my students that, like it or not, it's an unpleasant thing to have to wrestle with as a young person, when you just want to move on to your career and be happy and so on. And, all of a sudden your generation is the one that is going to have the opportunity to genetically modify your children. What are you going to do? Are you going to accept it, or are you going to fight against it? Or, are you going to just say, "well, that's progress." I don't want to be called a Luddite of the 21st century, but as an engineer, I am going to take a step back and say, suppose these things can happen? We won't worry necessarily about how they may happen, but what are the implications to society? One thing Eric Drexler said was that some people might be against these new technologies because they are coming too fast, and are going to do things that we don't like, so certain people refuse to be involved. It's sort of a pacifists approach, versus someone who is willing to go and fight, maybe war, etc. So, he said if you are not a player, then don't worry, someone else will be.*

But the problem is morally, how do you say at this point in this arena, "I want to put the brakes on" if you are involved as a player? That might influence your chance of getting funding. And there are personal, moral issues as well. The system calls for blind optimism. Everyone wants to hear a good story. Why put a downer on things by saying, "Whoa, this is not good, actually?" Perhaps that would be easier for engineers than for scientists. Scientists tend to just want to explore the unknown and where things end up being used is very difficult to predict, if it's really fundamental research.

You know, it could be 50 years before my own research [in nanoscience] ends up showing up in actual devices or processes, if ever. But even so, the human being needs to explore, to inquire. I think that is basic to the way we are and now it's likely to cause some really serious troubles. I mean, a lot of inquiry, even though it is fundamental research, is very likely to have fallout in things like biomedical advances that it leads to. Such as, augmented human beings, etc. It's just so transparent. But this is a process that cannot be stopped.

As the world gets richer, the capabilities of random, nonstate actors become greater and greater. You know it's not that hard to get a half billion dollars now in the world as a private person. The more people that have half a billion dollars, the greater likelihood that some small fraction of them are going to have very unusual worldviews which involve causing damage. As people in general become more capable, one thing they become more capable of is blowing things up, damaging things.

How long is it going to take before somebody, a private individual, actually makes a nuclear weapon, right? Eventually, it's not going to be that hard. On the bio end, it's even easier to hide and the chemistry end is easier to hide. Nanotechnology is going to make really nice sensors, really nice little machines, better processors, things that are more capable and have capabilities we hadn't thought of before. I don't see intrinsic danger in those capabilities. Any source of power is dangerous in the wrong hands. But, I don't see it bringing a change from any technology that we have developed in the past 20 or 30 years, as far as what we were able to do.

The social issues are a whole different story, but not as a danger to life. I see those as something that society has to sort out. What's not ethical is creating pain and hurting people, or creating a situation where that could happen in the future. It's just a matter of trying to understand the technologies and understanding what might cause that and what might not. I mean, trying to make a human clone nowadays would be a vastly unethical thing to do. It's quite likely that the baby would have health problems. And, it's not a very well-understood process. Another danger there is that if parents start getting a higher degree of genetic control over their offspring, it might change attitudes. One might say, "Oh, I didn't want my baby to turn out that way." You might think you have control over it, and then you can start becoming disappointed if the control doesn't work the way you want it to.

Eventually, people don't bat an eye at these social changes. Like with in-vitro fertilization. That was a big deal back in the 1970s. But people began to realize that fine people are using it, and they may have a kid through a test tube, and it's a fine kid. The kid grows up to be a regular adult, and what was the bad thing there? So I think the bad thing is, well there hasn't been a lot of longitude in the follow-up with in-vitro fertilization. The question to be answered still is whether or not test tube-type babies have a tendency toward health problems later in life. I don't think any one really knows. There is the potential to do some nasty things to some people along the way if you are trying. For exam-

ple, to make a clone without knowing what's going on and a baby gets born then they are dead by the time they are 5 for some genetic anomaly. There are definite problems with certain new technologies ....

New technologies that make people queasy are not necessarily a sign of something that is intrinsically bad. But it is a sign of something that intrinsically hasn't been thought of before and is not understood. I will caveat all of this by saying I really don't know what we are going to be in 100 years. The nanotech stuff is extremely difficult to visualize. The dangerous nanomachines have already been made by nature. People have gotten pretty good at playing around with the nanogears, so that's where I would worry. I will worry when somebody realizes "Hey, we can make Virus X a whole lot more lethal than it is right now." The machines that we are making are so crude compared to viruses that have had billions of years to figure out how to kill us. That's why I worry.

THE MORAL NEUTRALITY OF NANOSCALE SCIENCE AND ENGINEERING

In discerning what researchers work means to them in terms of its ethical implications, or how their work may affect society domestically or globally, three beliefs arise in these conversations. One is that ethical questions and social issues are pertinent to their research and warrant their consideration. Another theme is that while there are clearly socio-ethical issues in some fields of research, they are not in theirs, directly. A third is that basic scientific research is by nature an amoral pursuit, its good and evil being in the engineering applications of knowledge. In this view, it is the responsibility of the larger society and its legislating, policy-making bodies (not the individual investigator), to address ethics questions and to deal with stakeholder interests or social values that may be threatened, relative to that particular new technology. Those who adhere to this perspective describe their work as an endeavor that is inherently free of ethical challenges or moral considerations, as long as it is in their own hands as basic research. Once their findings becomes a matter of public knowledge and are out in the public domain for application, the ethical responsibility belongs to society. Those who do acknowledge a direct societal impact or ethical dimension to their work often qualify that acknowledgment with the contention that they are conscientious persons and ethical as professionals, but powerless to prevent what may be the adverse, unintended societal consequences of their work. Michael articulates that particular view:

ROSALYN: *I am being told by others that people who are working in biochemical engineering and biomedical applications of nano have more serious things to worry about, I suppose, so then could it be argued that your work is generally benign?*

MICHAEL: *I think so, but let me just go back. The standard argument, the reason that your field is fascinating, the reason that the ethics of this problem is a challenge is that you can do lots of things with knowledge. The classic case is Fritz Haber. Fritz Haber was a German scientist, and in 1911, I think, he invented something called the Haber process. The Haber process is a way to make artificial ammonia. Start with air and nitrogen and hydrogen and you recombine them over a catalyst to make ammonia. So if you drive through the farm country, you drive through Iowa in March, April and you'll see synthetic ammonia trucks injecting synthetic ammonia into the ground. It is the most important single fertilizer in the world.*

ROSALYN: *OK.*

MICHAEL: *That's how you fertilize crops. It is where all this productivity comes from. It's why we can have just a few percent of the people on the farm and they are growing stuff to feed all the rest of us. Take synthetic ammonia away, right, and your farm productivity drops, and it's no longer clear that there is capability to feed all the people on Earth. With synthetic ammonia it is clear that there is. You have to worry about distribution, but it is clear that there is the capability of feeding us. So now you take this synthetic ammonia, and one of the things you can do with it is react it with another catalyst with oxygen to form nitrates, and you can use those for various things too, they are also fertilizers. They are also in gunpowder. World War I almost ended 2 years earlier because the Germans were out of gunpowder, finished, they didn't have anymore. They were trying to get nitrates from Chile and they couldn't do that because basically the English, you know, were getting in the way of that. War would have ended. War would have ended because they were out of gunpowder. They couldn't shoot anything. And then they discovered that they could use the Haber process, which had been used to make synthetic ammonia to make gunpowder. Alright, so there is the knowledge, the knowledge is the Haber process; the knowledge is how to use the air to make ammonia. And you can use it for guns or butter, literally in this case, guns or butter. That's a societal issue.*

ROSALYN: *Sure.*

MICHAEL: *It's a huge societal issue. And everything you can think of that science and technology have contributed historically, you could argue have both a good side and a bad side.*

ROSALYN: *So you're doing pure research? You are a theoretician, you could argue that what you do is morally neutral, but what somebody does with it is where the moral questions come in?*

MICHAEL: *The work that I'm doing is, I think, even less probable a destructive source than other concerns. For instance, photolithography seems harmless.*

ROSALYN: *You can't imagine it being misused?*

MICHAEL: *It's pretty hard to imagine it.*

ROSALYN: *Have you thought about it?*

MICHAEL: *Oh yes. I have to think about it. You drive home everyday, look at your kids everyday. I mean just to give you an example, insect repellant. Not important stuff, right? Everybody who has ever gone camping knows insect repellant is really important. But you certainly would have to say that if first world countries invade third world countries the fact that they have insect repellant makes it much easier for them. So even insect repellant could be thought of as something that is socially not benign. But it is hard to think of anything, any object or process that can't be misused.*

ROSALYN: *When I was checking onto the airplane to get here I was given a random search and they pulled my glasses repair kit. My lenses pop out all the time because the screw was not designed well, so I carry a tiny eyeglass screwdriver. They removed it. I was furious, because I thought "what will I do if my glasses pop out?" I won't be able to function. This screwdriver was manufactured with the specific purpose of being used on glasses, but in the context of airport security, it could be used to hurt someone. So what you are suggesting is that if I did use it to hurt someone, the engineer who designed it, and the manufacturer who made it, would not be responsible for my actions.*

MICHAEL: *Well, it's a little worse. I mean, instead of the screws suppose that you had a piece of iron wood, which is a tree, and you take it and you polish it and you form it into a jabber, so that is responsible.*

ROSALYN: *Because the device was intended for harm from its very creation?*

MICHAEL: *Sure. But there is no science involved. I mean, are you going to blame the planter who planted the tree? After a while it gets silly. Because not only is iron wood not planted, it's wild.*

ROSALYN: *Some people ask me why I am talking to scientists and engineers who are just doing basic research and making useful devices. They ask "how can you expect to see socio-ethical implications in basic research?" It's just research, done for the sake of knowledge.*

MICHAEL: *There are different kinds of research projects that people can work on, like stem cells.*

ROSALYN: *So, microbiology and the like are a type of inquiry where it's easier to zero into the social-ethical challenges of such fields.*

MICHAEL: *Right. And the conflicts there are so much more obvious.*

ROSALYN: *They are so obvious and they slow us down. We have moratoriums on things because we don't know what to do. We are not going to have a moratorium on charge transfer.*

MICHAEL: *No, probably not.*

ROSALYN: *So why talk if there is nothing to talk about?*

MICHAEL: *I think there are things to talk about.*

ROSALYN: *Such as?*

MICHAEL: *I just mentioned two of them. The gun powder case is a classic, it's a century old.*

ROSALYN: *OK, so then what if we take the energy we've now learned to harness from the sun, and use it for say, weapons?*

MICHAEL: *Saddam Hussein could use it. Saddam Hussein could use it to run a factory in the desert, where he can't get energy any other way. But suppose he didn't have a portable power source. Suppose he had to run a power line. It is getting much more expensive, it's getting much more difficult, and you can trace power lines much more easily.*

ROSALYN: *Suppose your research allows us to provide potable water to soldiers in South Africa, but that also allows mercenary activities that are undermining to governments.*

MICHAEL: *Those are issues of policy.*

ROSALYN: *Does that mean that issues of policy have nothing to do with basic science?*

MICHAEL: *They have something to do with science. Science enables people to do things. What do you build swords for, to fight with people, right? But you don't just build ships to fight with people, you also build ships to take people on vacations, and you build ships to carry grain from one country to another. All you have to do is take the cargo ship and stick some guns on it and it becomes a warship.*

ROSALYN: *Are you suggesting that if you were asked to do research for specific weapons application, you wouldn't do it?*

MICHAEL: *That depends on who asked me and what they asked me to do. The fact is I have lots of money from the Department of Defense. I'm fine with providing to society understandings that permit society to move forward on lots of items on the social agenda .... Society is a terribly complex organism. And it's got control mechanisms and some of the control mechanisms are better and some of them are worse. I gave a commencement speech which asked, "what are your responsibilities?"*

ROSALYN: *What would you say you have sorted out as your responsibilities in terms of your research?*

MICHAEL: *Let me continue. In the speech, I quoted Francois Rabelais. He was a French writer who was born in 1448. You know the word* gargantuan*, referring to something huge? His great book is called* Gargantuan and

Pantagreuel, *and it's about two giants and their life. It's madness, an unbelievable book. So 1492—Columbus right? Spain, 1492—all kinds of trouble. People are getting burned at the stake, nasty stuff. Rabelais was born into this life, he becomes a monk, and then he becomes a writer and then he becomes a lawyer and then he becomes a doctor, and then he becomes a diplomat, and he writes these 5 books. At the end of the first book they build a monastery and the only rule in this monastery, the only rule is, do what you wish.*

ROSALYN: *Do what you wish?*

MICHAEL: *That's it, that's the only rule. But it's more than that. I mean it is do what you wish, and why. Do what you wish because people who have been well educated, who are responsible, who are at home in civilized company, who have good friends have an internal monitor that tells them when they are screwing up. Now he doesn't say that, but that's basically what it means ....*

Now the second quote—from Oliver Cromwell. Bad guy. He was a bad guy who killed a lot of people. But his famous quote is, "In the bowels of Christ consider that you may be mistaken." You have got to think about that. I mean, you can't just accept things on their face value, right? So Ashcroft tells us that all these guys in Iraq are bad guys. Maybe. You've really got to think about that. I think they are probably not all bad guys. Certainly some are bad guys, no doubt about that, but we just basically picked up everyone who was in a certain place and decided they were bad guys. Pack them off there, don't allow them to talk to anybody, don't allow anybody to talk to them. You have got to think about it. So, do as you will, put them all down there, but then consider that you may be mistaken. Those are some things you have got to balance as a citizen. I think most of us feel that we don't want the military to get out of control. On the other hand, you have got to have a military. Suppose there were no military at all in this country. Anybody can invade you, if you don't have any military. So you have got to protect yourself. So when does it stop being protecting yourself and start being stepping on the toes and ruining other peoples' lives? That's a societal issue. And that means very, very hard decisions to make. We arrested all of these Japanese people during World War II and stuck them in concentration camps.

ROSALYN: *And also innocent Chinese American citizens.*

MICHAEL: *It was terrible what we did. "Consider that you may be mistaken."*

ROSALYN: *Do you, in your science?*

MICHAEL: *You have to in science. You absolutely have to. People put forward ideas and they publish their ideas in the open literature, and other people around the world will look at those ideas and see if they work. If they are experiments, they will try to reproduce them. If they are theories they'll try to apply them. I've thought about it really hard, the question of certainty. There is a question of certainty right now. I am pretty certain that because they have been making noise outside for the last hour, they are going to make noise for the next 3 minutes. That doesn't mean they are going to. But I'm pretty certain.*

ROSALYN: *And that's where you have to be comfortable, is with your certainty?*

MICHAEL: *How else can you do it? I mean, otherwise it's like you write a sentence and every time you write a sentence, you put words in it and there are conventions on spelling and you think you have spelled them right. You think y-e-s is yes. It is not y-e-s-s, and it's not y-e-e-s, you are pretty sure, so you go forward with your sentence that you wrote. Now sometimes there are words you are not quite sure about. But you go forward anyway. But then if you really think it's wrong or you are worried you look in the dictionary and figure it out. You eventually need to have trust in yourself, and I think you need to have trust in your society. You have got to trust society. You have got to. You drive down the street (and this is my favorite example), you are driving down the street, you are going 46 miles an hour in a 50 mile zone, you're fine, right. It's a beautiful day, no problem. There is a guy coming toward you going 46 miles an hour on his side of the street, if he swerves 3 feet out you're dead. At 46 miles an hour, you are not going to make it. You have implicit faith that the guy is going to abide by the social contract. The social contract says you stay on your side of the road. You have to trust. And sometimes you have to trust for emotional and you know organizational reasons, because you can't prove everything logically.*

As nanoscale science emerges as fully appropriated technologies, it will shape and mold society to its possibilities, just as society is currently shaping and molding the science. In order to assure nanotechnology develops ethically, toward humanitarian aims, it will be necessary for researchers to ask questions about their own work, like, "Does my own work in nanotechnology represent a progressive technological force?" "What harm might it cause?" "How can what I am doing be controlled and contained?" Could what I am doing in the laboratory possibly adversely or positively impact personal spiritual or psychological well-being—my own or that of others? It is no longer a matter of academic dispute whether technology in general has a direct influence in shaping the social, cultural, and material elements of the modern world. This is an accepted and recognized premise inside of the various humanities and social science studies of science and technology. The question of whether researchers' rejections of, or disinterest in, that premise is problematic for the ethical leadership of nanotechnology.

THE INEVITABILITY OF NANOTECHNOLOGY DEVELOPMENT (DETERMINISM LIVES ON)

Researchers are very much a part of the social, political, economic, and institutional environments in which nanotechnology is forming, and from which it emerges, thereby influencing the development of nanotechnologies themselves. Its very existence as an international undertaking is due largely because of the intellectual contributions and hands-on work of individual scientists and engineers and their graduate students. But is it their responsibility to consider or in any way direct nanotechnologys' potential impact on the human communities? Does such consideration and leadership rightly belong with its researchers, that is, those experts whose work it is to study, innovate, and develop the workings and products of nanotechnology? Some researchers, like Stan and Carl, are quite clear about the societal context of their work:

Stan

ROSALYN: *As an engineer, to what end are you trying to understand the material world so that you can change it?*

STAN: *It isn't that the process is centered on being able to effect a change. There is a different test for whether the scientific learning was really of interest or not. And that is, not, will it change the world? The question is, in what direction does one change the world?*

ROSALYN: *I see.*

STAN: *It's not an easy call. Look at science over the last 100 years or so. It has brought great advances in a number of areas. People often point to medicine and longevity, but really, it's the more common-placed things that have most changed our lives. Look at the changes that computers have brought and how low cost communications have really transformed our lives. I can remember when I was a college student. I would call my future wife who was 150 miles away and it cost $.85 for 2 minutes. That's all we could afford at that point, so that's how long we would talk. Well today, you can call anywhere you want, even across the world, for pennies a minute or a couple of cents a minute. That electronic revolution will continue. However, I would be careful about characterizing them necessarily as good. They are changes and some of them are going to be good and some of them will be less obviously good. For example, the same thing that brings low cost computation and low cost communication, which has a great impact on productivity, allows the average work environment to be arguably quite a lot better than say, 50 years ago. But for most people, it also has brought the dominance of* *media, and sort of uncontrolled media culture to society, which at least from my perspective is not a good thing. People tend to be controlled often by the lowest common denominator, so there is almost no limit to the degree to which you can use something for ill if that's what you choose to do. Look at the Internet. It's a great convenience, for work. If I need information about something, it's often the first place I turn, and yet today Americans and others spend a lot of the Internet capacity pushing pornography back and forth. People also use it for online gambling. (I'm surprised that there is even an interest in that.)*

ROSALYN: *So, you really think about these things. Do you also factor these issues into your own work, as an engineer?*

STAN: *Well, yes and no. The example of that that people often turn to when thinking about the ethics of science is nuclear weapons. The attitude is probably not very useful. The perspective that people have is that, if these people at Los Alamos had not made the bomb, the world would be a better place. The world would be a different place, but we would probably not be a better place, and the reason for that is the technology that is required to build an atomic bomb. In the context of the time when it was developed, the technology was really heroic in scope and scale.*

ROSALYN: *Because it ended the war?*

STAN: *There was a Danish scientist who predicted that it wouldn't be possible to make an atomic bomb. He believed you would have to turn the whole nation into a bomb factory in order to do so. Then much later, when the World War ended, he was smuggled out of Europe and brought to the United States and actually was shown the facilities that were being used to build the bombs, and told that his prediction was wrong. His reaction was that the prediction was exactly right. They had done exactly what he had said and if you looked at the scale of the production facilities at Tennessee, for example, you would understand his argument. So the lesson there is that it was really a heroic undertaking. The most powerful economic and scientific communities in the world could barely do it. But the lesson in history is that today, because technology inexorably advances, misapplication of its resources can present a significant risk. Consider Iraq. So the reality is that once the scientific developments, which were not done with the goal of producing energy or building bombs, but were done because they were interesting scientific questions; once those had been done, then the bomb became one of the necessary results of the research. So you could choose not to build the bomb here, but then somebody else builds it and that doesn't necessarily make the world a better place, it just makes it a different place. You can't just make the bomb go away. It doesn't work that way.*

ROSALYN: *So, we can't blame Einstein for the knowledge that* $E = mc^2$.

STAN: *You could blame God I suppose, but you can't blame Einstein.*

* * *

CARL: *It's all for the good. I really think nanoscience has brought a new level of creativity to science understanding and discovery. We are discovering things that we didn't even think we could look into 50 years ago, things that we couldn't even possibly consider. And then there are the applications that are available as a result of that knowledge.*

ROSALYN: *Do you think these new approaches and instruments for working at the nanoscale will really mean we are more able to control and manipulate matter?*

CARL: *Probably to some extent, the extent to which we think we can. Human arrogance gets in the way. I never thought more stupidly than the day I got my PhD. The realization has come to me how much I don't know. It hit me hard. And at the same time, it's very enlightening that there is still so much to learn about—great job security for us scientists. I think nanoscience has shown us how much we still don't know ....*

And if we think that we really know much more, then due to arrogance you start to fall into traps. Like how we developed antibiotics then all of a sudden we find that we weren't smarter than the bugs and they have developed antibiotic resistance and now we have an even more serious problem because we can't figure out how to treat the infections. So, to some extent, yes we can better manipulate and control matter, as long as we understand that we still don't have a very good understanding of biology, of life, of medicine and how organisms work and interact with one another and the best lesson we can learn is balance, to take things carefully, step by step. The problem is that we just keep pushing to go further and further, and make policy or come up with treatments that take certain things into consideration, but don't take into consideration what we don't know ....

It's a balance. You have to come up with medical advances, knowing that there are things that you will learn 50 years down the road. You just have to be humbled by this. Then there have been real examples of times when scientists and the public have taken science too far. We have decided that science was somehow absolute and it's never absolute. We need it to be so, because we want to come up with medical developments, so we take things as is and we forget that nothing in science can be proven, that we can only work on certain theories and certain theories can survive longer than others, but they will never go beyond theories. You can't have a proof, yet at the same time we can have medical advancements that you can prove make the difference in things. But consider the antibiotic; this was the wonder drug and, my God, absolutely made a huge difference in humanity in our ability to treat disease and our ability to survive and now, in fact it's come back and it bites us ....

I think we would be better if we were a little bit more cautious and had that balance. I still think we would progress, I still think medicine would move ahead and technology would advance ....

I am always concerned about the balance. For example, I think we always have to put people first. I think that is important except that putting people first means that sometimes you can damage the environment. And maybe it might be good now, but 20 years from now it's going to hurt. We have to take people into consideration but I also think we have to worry about other life forms. So, for instance, there is an organization that funds researchers to come up with alternatives to using animals in research. They are animal rights activists in the sense that they have acknowledged the importance of animal research, but they have also acknowledged that as we get more technologically advanced, and as we start to understand more, we probably can come up with alternatives. I actively support animal use in research, but I would absolutely put my money behind their organization because I think they help develop a balance. They don't let research continue at the expense of something else.

To some extent I think that you have to have animal research because we need to advance medicine for people and that to me is more important. And technology is to advance ourselves in this humanity, to be able to progress and to contribute more and for us to live longer.

ROSALYN: *Is there an outer limit on how long we should live?*

CARL: *I don't know. I have been thinking a lot about that lately, because I have a feeling that barring anything extraordinary, I am going to manage to live a very long life.*

People once thought that living to 50 was a good long life. Dying at age 50 just doesn't cut it for me at all. So I have been wondering about that and I don't have the answer. I don't know if there is a limit. I think we would like to say there is a limit now, but trust me 20 years from now when we are probably living to be 120, we won't think the limit is the same ....

At the same time if we are managing to live longer, then society has to come up with social change, that means that there is going to have to be more medical support for older people and for diseases that we didn't see. We will have to work hard and put more research toward Alzheimers and Hutchinsons disease and things like that.

So it's a balance. If science is going to help us to live longer, then we have to balance what the consequences will be, which doesn't mean any of them are wrong, but we have to take those things into consideration. I am pretty liberal when it comes to ethics and where I think science should and shouldn't progress. At the same point, I don't ever think that you should discount what the more conservatives are saying or the considerations that they would like for you to have .... Don't just progress without thinking about what the consequences might be. Now maybe you should think of the consequences and figure out how we are going to approach them and then you can proceed, because you have thought about what the problems might be or what new findings we might come up with and how that is going to change our thinking.

ROSALYN: *Do we have to control the process of scientific discovery?*

CARL: *Not the discovery process. But we should certainly control its applications. The initial discovery always to me has a lot of serendipity in it. But once a discovery is made, you then decide to pursue that line of investigation or what types of applications come out of that line of investigation. I was fascinated by the fact that we had DNA cloning. This whole new technology was going to be available and scientists stopped and they paused to ask what could be some of the implications. Now, it's not possible for them to have thought of everything, even up through now, and we are not even talking 50 years, but they did stop and think, "it's going to have an effect on more than just our research? It's going to have an effect on technology, and society and public policy and medicine and humanity, so let's at least acknowledge that we are going to have an impact, not do the science in the void of a vacuum."*

They had this self-imposed moratorium as a time for them to educate themselves, to educate the public, to educate policymakers, to educate government. We didn't have people trained to make genetic monsters, but they acknowledged the fact that somebody might want to use that technology in such a way.

And even if we can't prevent them, we should acknowledge that and then we need to push toward ways to avoid it, and then when we do recognize that perhaps somebody is going beyond what society and science thinks is reasonable at the time, the time is always going to matter, because it does change the frame of reference. Science interplays with everything else. Sometimes science is missing that, scientists get a little bit on the high horse and over trust things.

I take issue with anybody who takes any discipline and says that it shouldn't be dealt with in the context of the rest of society. I think anybody who chooses a political agenda without concern for constituents is doing unethical politics. I think anybody who pursues science without an understanding of what it is, anybody who receives federal support, has no business in science, as far as I am concerned. They

> *are taking taxpayers' money, they are taking public money, and they are taking some of my money, too. As long as you are taking peoples' money, societies' money at large, you do have a responsibility ....*
>
> *Then I think there is an obligation to society in terms of the science. If you really are doing pure discovery-type science, and publishing the results, then maybe that argument could be a little bit different, but I think most nanoscience researchers enjoy the idea that there are some medical implications or useful technological implications to the work. If you are going to take your science and put it out into any aspect of society, then you have to take into consideration society as a whole. You have to take into consideration that you are in a university doing research and you have to balance the responsibilities of training new scientists and teaching students and of doing your own research. You can't just pursue any one line of those responsibilities without considering the others. So, even as a university researcher, they usually have at least three responsibilities: training graduate students, undergraduate teaching, and doing their own research. You have to be able to balance those and you have to ask "where does my research fit into the rest?" ...*
>
> *Scientists really do believe that there are all kinds of possible advancements that could result from their research. So they get very gung-ho and forward thinking and think about the medical implications and the good things that can come and then those are so overwhelming that even though there might be bad things they may have considered, the good is so overwhelming to them that they just push ahead. To some extent, I think they are right to do so. They are simply good people who know they have worked a lot of years and they are successful and they get to a certain point and they are experts. But you have got to be careful, because the public fears those things they don't understand—it's human nature. If you really can't make somebody understand, they are just probably going to stay afraid. Scientists need to be saying "we are going to be cautious and we want to work together with everybody.*

Some nanoscale science and engineering research projects are specifically intended to solve problems that are of a purely scientific nature, such as measuring, calculating, and predicting particular kinds of behaviors and reactions. Funding for these projects tends to come almost exclusively from federal government agencies such as the NSF, which is interested in and committed to basic research. Other projects are specifically aimed toward particular technological applications related to specific industries such as health care, semiconductor, or the military and are more readily funded privately by corporations and venture capitalists, as well as by agencies of the government (e.g., the Department of Defense or the Department of Energy). These are often collaborative projects involving multiple investigators from private laboratories and / or different universities. Some projects are undertaken inside of National Laboratories with very specific federal mandates. Regardless of the sponsor, there is a sense of helplessness for many researchers regarding their ability to take control and direct the outcomes of their own research. (Some feel bound to the interests of their sponsors, or motivated by the economic opportunities that a patent would bring.) Despite that feeling, most express in their words a conscientious moral commitment to contributing in good ways to the betterment of human life, although most are unwilling to assume responsibility for the effects of something they cannot control—even their own research. What happens in the laboratory of individual researchers and their group will matter in the larger society. What happens beyond the laboratory may not be controllable, but the most control will come from doing it one's self. This is a common assertion by the researchers. If they don't do it, then someone else will; one of the persistent themes of these narratives, as Ermias declares:

ERMIAS: *We are not randomly moving through this process. That doesn't mean that serendipity can't play a big role, that you shouldn't rely somewhat on serendipity, but I think you should always have some sort of rationale for what you are doing, what you want to learn, not necessarily what you want to invent, but what you want to learn from the particular project you are working on. Then, if everything works the way you anticipate it, what are you going to get out of it? Was it worth doing? You should ask those basic questions.*

ROSALYN: *Have you asked them for yourself?*

ERMIAS: *Oh yes, every time.*

ROSALYN: *Do you have the answers?*

ERMIAS: *Sure. Part of the purpose is to recognize problems that need technical solutions, like the anthrax problem. They brought in five or six different technologies to determine whether or not the letters had anthrax on them. They went through 2.5 days of testing. I believe one of the patients was dead by the time they got the results. They still had a level of uncertainty after that. To me that tells us that we have a big hole to fill in terms of the technology needed to address these problems.*

ROSALYN: *Ideally, would you like to see us be able to detect every possible disease?*

ERMIAS: *I think that's the goal, yes.*

ROSALYN: *And how about for every possible genetic abnormality?*

ERMIAS: *Sure. But there are all sorts of ethical issues. Such as, how do insurance companies use these diagnoses? Ultimately, of course we would like to have a home test so that the patient can decide what they want to do without disclosing that information to other people. Because if somebody screens you for all of these different possibilities and you look like a high-risk cancer person, then obviously your insurance company is going to learn about it.*

ROSALYN: *Yes, and it may affect your employment opportunities.*

ERMIAS: *Absolutely. If somebody can screen you and determine you have AIDS then, there are all sorts of issues. I don't think this world has ever shied away from knowledge. I think given a choice, people always want knowledge. But we can always deal with those types of issues after we have that capability.*

ROSALYN: *Do you have faith that we really will?*

ERMIAS: *Absolutely. If we can't we are all doomed.*

ROSALYN: *Well, we put a moratorium on stem cell research because we couldn't deal with it.*

ERMIAS: *I think that will change. It wasn't because we couldn't deal with it. That was a political decision. Don't confuse the two. I think a political decision doesn't determine what we can and cannot feel. I think you know with a different administration those decisions might be different and with time those decisions might be different. Once something is open, you can't just sweep it under the rug. It's better for you to understand its potential positive and negative implications and deal with those rather than letting somebody else decide how to deal with it. Because if you develop it, you actually have more control over it versus the other way around. I don't know if you read the Bill Joy article.*

ROSALYN: *Sure, I use it in my classroom.*

ERMIAS: *To me a lot of his arguments were somewhat silly. When you develop new technology, you develop a hammer. You can either use it to pound nails in the wood or you can use it to pound somebody's head in.*

ALL KNOWLEDGE IS GOOD AND TECHNOLOGY IS NEUTRAL

Bruce and Bryan explored with me (among other ideas) the moral nature of scientific discovery:

BRUCE: *In fundamental research, there are a lot of things that are unpredictable. Those targets of research may not be the ones that generate the most profits.*

ROSALYN: *What do you mean?*

BRUCE: *A lot of discoveries are accidental. Some directions cannot be really planned and then if you are working on something else you may find something that was unexpected and that turns out to be very useful.*

ROSALYN: *So you keep going back to fundamentally just simply learning about the physical universe?*

BRUCE: *Um hum.*

ROSALYN: *For the sake of learning. Not necessarily because there is something that you want to do with what you learn?*

BRUCE: *There are two different ways to do science. There is the very planned situation like development of the atomic bomb. There is the other; phenomena that you are not looking for, but you find it, and it turns out to be quite useful. So some by-products become quite more important sometimes. We may have some incremental goals or maybe long-term goals, too. There are certain things that you want to do eventually. OK, but you are just moving toward that direction and don't know whether the advances in your lifetime will reach that goal or not.*

ROSALYN: *It sounds like you are on a journey and one day you will be through the other side. There is a place that you are investigating, but as you are going along, it's like you take a hike through the woods, and you notice, "Oh look at the moss here, look at the birds here, look at the crystals there" even though you are trying to get to lichens on the trees. You are discovering things along the way.*

BRUCE: *You are studying the probabilities. Along the way you understand that a little more. That's also true with technological advances.*

ROSALYN: *Can you imagine at any time in your research thinking, "Oh, this particular observation would not lead to good knowledge."*

BRUCE: *Not good knowledge?*

ROSALYN: *Right. Is there any knowledge that we shouldn't have?*

BRUCE: *No. But if it is not an important problem now, I don't spend my time on it. You do have to make decisions on whether it is promising or not, in terms that that would have some potential to make some impacts.*

ROSALYN: *Yes. But what I am pressing you about is whether all knowledge is good knowledge in terms of scientific discovery and understanding?*

BRUCE: *In basic science yes, because we are building up the framework of a certain field. Every understanding is built on top of others. As you understand more, you can move toward other directions.*

ROSALYN: *When will humans be finished learning?*

BRUCE: *Never.*

ROSALYN: *Why is that?*

BRUCE: *Well, I cannot see that we will solve all of the problems. I don't even know what new problems will come out.*

ROSALYN: *Are the problems infinite?*

BRUCE: *I don't know. I don't really know that. Nobody does.*

* * *

ROSALYN: *What happens if nanoscience makes it so that humans no longer die from cancer? Is that a social, ethical good?*

BRIAN: *I think you are being really philosophical about this.*

ROSALYN: *Well that's my job.*

BRIAN: *I think if we can do that, it will be an ultimate triumph of science and an amazing triumph of the human race, which always has been about adaptation. And one of the reasons people argue that we get cancer is that because we are not very good to ourselves. We are not very good to our environment. The human race, in its pursuit of domination of everything around it, has turned its own achievements against itself, because a lot of the things that we put into the environment have increased cancer rates. So if then we can overcome our own success, by defeating cancer, then to a certain degree you have staved off what may have been the planned obsolescence of the human race. That is, you know at some point every species becomes extinct. And if cancer and a lot of other diseases are those things that were destined to make the human race extinct and we overcome those things then that just shows that the power of adaptation that the human race has is a lot stronger than the diseases that we have brought upon ourselves.*

So whether it's a good thing or a bad thing, I don't know. The smarter and smarter we become about how we can understand our own natural world, our bodies being one of the more complex things in the natural world, maybe that degree of adaptation is something that can be applied outside of ourselves. You know, nanoscience is being used in many applications, not just in biotechnology.

ROSALYN: *Yes, absolutely, such as for new materials.*

BRIAN: *Right, and to address environmental issues, wastewater remediation, things like that. In fact, people have developed materials that catalyze and break down things that we put into the environment a hundred years ago.*

I strongly believe in freedom of choice and I think that almost all of the quests that we have about our environment, etc., today are from social/economic pressures. Maybe that wasn't the case when a hundred years ago there were some very bright minds who dominated the scientific landscape, and were really out for pure fundamental understanding of the world around us, but I think now so much of it is dominated by social economic pressures ....

I am not entirely sure that a complete and utter understanding of the world around us is predestined in our makeup. I mean, there are still groups of people living in remote parts of the world who are living as they did 500 years ago.

What about technology? Is it also morally neutral? One answer some researchers give is that what's good or bad is the uses people make of it. Such an argument is almost convincing. Yes, there are people in the world with horrible intentions who misuse technologies that were otherwise aimed for good. But that fact does not make technology neutral. Nor does it allow individual researchers, policymakers, consumers, or whoever has a stake in it to abdicate their responsibility. Is artifact really meaningless in and of itself? The thin black box that sits in front of the computer screen on the desk is only a keyboard because of the intentions built into its very design. The distances between each key, the lettering, even the order of the lettering all reflect values, economics, politics, and beliefs. Its intended use was very much a part of each and every element of its development, and even the research that made its development possible, came with intentionality. Once that box was produced, distributed, and brought into use, whether or not someone picks it up and hits a colleague over the head with it is irrelevant to the fact that some values and intentions were embedded in its actual design.

The continuing quest for precise material control that is connoted in the nanotechnology initiative is far from neutral. It is saturated with human dreams, ambitions, fears, and fantasies. The well-being of humanity and other species in the face of this quest, and certainly the health of the planet Earth, will depend on the evolving searches for meaning that a conscientious nanotechnology quest requires. If left to an unexamined, independent evolution, nanotechnology could lead to anywhere. The claim that the nanotechnology future is out of anyone's hands is highly objectionable. Despite the unpredictable outcomes and seemingly uncontrollable factors of influence, the creators of nanotechnology are still building the future of nanotechnology. Three clear factors of determination already at work are the individual researchers own aims, their sponsors' hopes, and our leaders' ambitions.

Material objects are not in reality separable from the social institutions, social practices, and social relationships necessary to create and sustain them. The way we typically think about technology—as material objects—is an abstraction. Johnson argued that the abstraction is contrived, "somewhat like a thought experiment in that it involves the mental act of separating the artifact from its context" (Johnson, 2004).[2] She made the point that when that abstraction is rejected in exchange for the broader

[2]Cited from an internal, unpublished work in progress as written by Deborah Johnson in 2004.

view of technology, then the ethical issues can be seen "in" the technology that could not be seen before. Johnson's work concerns primarily the ethical issues in the design phase of technological development, but she saw values at work, and correspondingly, ethical issues to be identified at each stage of a technology's development from design, to manufacture, to marketing and distribution, to adoption and use:

> At the design phase there are issues about what values and whose interests are being facilitated or constrained with a particular design. For example, what assumptions are being made about who will use the technology? Are the designers designing for disabled individuals, for men or women only? In the manufacture phase there are issues about the labor force needed, risks to individual workers, risks to the community in which the manufacturing will take place, and so on. In the marketing and distribution phase there are issues about how the technology is being sold and who is getting access to it (who gets access first and who doesn't get access at all). In the use phase there are issues regarding what the technology will be used to do and what social arrangements will be facilitated or constrained, and what anticipated or unanticipated effects and side effects its use will have.[3]

What about in the research phase? Are there values, and therefore ethical considerations, to be made there? Some of the researchers say yes, absolutely. Others suggest there is no such responsibility because the basic science and engineering research stage of technological development is value free and morally neutral. That, at least, is the presumption of the argument so consistently being made that an object, artifact, device, process, chemical reaction, material, and so on created, developed, and tested in the laboratory is only as "good" or "bad" as its actual user. Assuming Johnson to be correct, then that premise is false. Novelty and intrigue are very much a part of what motivates nanoscience/nanotechnology researchers. And so is candidacy of the research for usefulness in particular applications, which can determine the ability to secure funding to pay for laboratories and graduate students who will work in those labs. It also plays a role in the personal satisfaction that comes from knowing that ones work is being used in the world for some good purpose. Researchers themselves become of value when their work is useful, and being held in high esteem as successful and creative is in and of itself a value to the researchers, which has immediate effects on how the research is done. There are

[3]Ibid.

values, assumptions, beliefs, and intentions at work in the very experimental questions researchers pose, which may be shaping the research itself.

VALUES IN NANOTECHNOLOGY RESEARCH AND DESIGN: THE CASE OF THE AEROGEL

The Aerogel Research Laboratory at the University of Virginia, directed by Pamela Norris (2004), was founded with the mission to investigate both the fundamental properties and the cutting-edge applications of aerogel:

> Aerogels are the lightest solids ever produced. Highly porous and almost wispy in appearance, aerogels produced from such materials as silica, alumina, or zirconia can have densities as low as just three times that of air. These microporous materials have other unusual and desirable properties as well, making them candidates for a wide range of applications. In addition to being the best thermal insulators ever discovered, aerogels have the lowest dielectric constant, and the lowest sound velocity of any known solid material .... Although discovered in 1931, it was not until fairly recently that scientists could take advantage of this impressive set of qualities, because aerogels were extremely difficult and dangerous to synthesize.

Multiple values are expressed in the language used to explain aerogels. Consider, for example, use of the words "unusual" and "desirability," and the statement that "scientists could take advantage of this impressive set of qualities." What is a "quality" in material science, and what makes that quality impressive? Quality is generally defined as a distinctive attribute or characteristic, and the word impressive as something that invokes responses of admiration. Both of these words are value laden in their very meaning, which suggests that the aerogel is being deemed worthy of attention, significant financial and space resources, and time. Why? The web page continued:

> Aerogel is an extremely adaptable material. The sol-gel production process offers the ability to tailor the material properties for specific applications. Properties such as pore size distribution, density, and surface chemistry can be controlled during the chemical preparation process, and substances can be added during production to impart a desired functionality. Applications include superinsulation, substrates for chemical catalysis, acoustic delay lines, sea water desalinization, subatomic particle detectors, micrometeoroid collectors, and supercapacitors. An example of a novel application is in

> space exploration. Aerogels were used to insulate the Mars Rover, a mission where its lightness and strength proved ideal.

One answer to the question "why" seems to be in its usefulness to industry and to the interests of the government, which are funding these projects. What does that mean in terms of the ethical issues of the research phase? It means there are important questions to be asked about the intentional use of these materials: Who stands to gain what, who may be harmed, what kinds of impacts will their prospective uses have on employment, labor, education, warfare, semiconductor products, and in the home ...? Who are their intended users? What kinds of purposes do those users have in mind? If new markets are yet to be developed for those products, then what kinds of assumptions are being made about those who might one day seek to purchase aerogels for their own companies, schools, shops, or homes? Is the material disposable, biodegradable, toxic? What amount of energy is required to produce it? At what point in the research and development do such questions become pertinent? Consider the following:

> Research at the Aerogel Research Laboratory is currently concentrated on three projects. With funding from DARPA and in conjunction with Veridian-PSR, the laboratory is working to develop new sensor technology for the detection of biological warfare agents. Laboratory researchers are studying the fundamental flow and collection properties of aerogel materials in their efforts to develop a multifunctional bioaerogel to collect, concentrate, and detect biological warfare agents. This research could also be applied to hospitals to detect harmful viruses and bacteria, or used in manufacturing plants to detect toxic airborne chemicals.
>
> The laboratory is also developing an aerogel/polymer composite thin-film material with electrical insulating properties superior to those currently in use. The purpose of this research, sponsored by IBM, is to create a novel composite interlayer dielectric material to sustain the miniaturization of microelectronics.
>
> Finally, the Aerogel Research Laboratory is conducting fundamental research, sponsored by the National Science Foundation, on how thermal energy diffuses on a fractal-length scale. This research has the potential to enhance the understanding of the thermal energy transport processes in amorphous materials used in the microelectronics industry. (Norris 2004)

Language reflects values, as is apparent in this passage from the laboratory's Web site. Use of the word *detect* speaks to the investigative nature of scientific research, and the value of discovery in finding what is otherwise hidden from view. Citing the source of funding reveals a relationship of dependency, and a willingness to participate in the mission and aim of particular organizations (in this case, a division of the U.S. Army, which is particularly interested in warfare). Reference to bacteria and viruses as harmful reinforces at least two commonly held beliefs; first, that scientific research itself is of value and essential in defending against certain elements of matter; and second, that there are elements of living matter and of human-made chemicals that are a threat to human life and therefore must be discovered. The word sustain in reference to miniaturization of microelectronics suggests that increasingly small electronic units is of importance. Each of these values points to questions of ethics. In the case of aerogel research, ethics considerations would include the relation between basic research and government interest in military capacity. Although on one level the answers seem to be obvious and without any moral difficulty, there are questions to be asked about the microelectronics industry, and its accelerating push for new products of exceedingly small dimensions.

Ethics questions abound in nanotechnology research. Maybe one day some terrible person will get hold of a large amount of aerogel and somehow use it to block the flow of a central air duct system in a government building. Pamela Norris and her colleagues certainly cannot be blamed for that kind of atrocity. But do principal investigators have the obligation to bring to their research conscientiousness about their own embedded values, and those of their sponsors? The problem is not that the work is value neutral. Perhaps the problem lies in the organizational structures or in the values of the institutions in which nanotechnology research occurs, which provide few opportunities for researchers to ask such questions. What is an individual researcher to do?

TIMOTHY

ROSALYN: *Have you been interested in science as long as you can remember?*

TIMOTHY: *Yes, even as a little child. I didn't know exactly what form it would take. As far as I can remember, I was always intrigued by how things worked—chemistry and so forth.*

ROSALYN: *So in high school did you take all the science and math courses?*

TIMOTHY: *Yes, I certainly did.*

ROSALYN: *What is the focus of your current research?*

TIMOTHY: *It's in two pieces right now. The first piece and probably the biggest and most established piece are technologies related to gene therapy and gene delivery.*

ROSALYN: *So did your work in any way hinge on the completion of the Human Genome Project?*

TIMOTHY: *I think it did, indirectly. There will be more opportunities now that the human genome is known and is accessible. There will be more targets of intervention that are known. But we look more at the interaction between the chemistry of the drug delivery agent if you will, and the cells and physiology. So we are really looking to see how the whole process can be more effective without really concentrating on any particular disease.*

ROSALYN: *You are working on drug delivery for genetic problems?*

TIMOTHY: *Yes, that's right. It's not just genetic problems though; it can be things that we don't really think about as genetic problems. For example, one of the big targets for us is cancer, because there are gene therapies that can be specifically directed toward a particular kind of cancer.*

ROSALYN: *Tell me what that means to have a gene therapy.*

TIMOTHY: *So, there are a number of diseases and disorders that really are derived from an error in the genetic coding of some cells.*

ROSALYN: *Yes.*

TIMOTHY: *So in cancer there are many possible errors, the result is a cell that grows unboundedly.*

ROSALYN: *Right.*

TIMOTHY: *And so if you can somehow identify that gene and do a variety of potential manipulations to turn that gene off, silence it someway, or take advantage of something unique about that cell, presumably it had some unique characteristics now that it is growing unboundedly, and you take some advantage of that on a genetic basis to target it for traditional anti-cancer pharmaceutical. You can focus all of these drugs just at the tumor rather than everywhere.*

ROSALYN: *OK. So you were talking about the genetic structure of an individual cell.*

TIMOTHY: *Yes.*

ROSALYN: *So working with that particular structure.*

TIMOTHY: *Yes.*

ROSALYN: *To sort of reprogram the cells response to the message of the genes.*

TIMOTHY: *Yeah.*

ROSALYN: *This is in laypersons terms. I am just trying to get it.*

TIMOTHY: *Yes.*

ROSALYN: *So, there are genes in the cell which are possible targets for you?*

TIMOTHY: *Yes, exactly. In many cases it is because there is a particular protein that is being made that shouldn't be made, or one that should be made that's not being made and so that's what the gene therapy would seek to do. Gene therapy would seek to turn off the expression of this protein that should be made. Many times what happens is there are some proteins, these tumor repressors, which prevent the cell from growing unboundedly and for genetic reasons, they are suppressed. So that allows the cell to grow unboundedly. It's kind of like taking the brake off.*

ROSALYN: *And the brake is inside the gene, not just generalized in the cell.*

TIMOTHY: *Right, exactly.*

ROSALYN: *You get to the true source of it.*

TIMOTHY: *Exactly, and so for example in that particular case you can just reintroduce that genetic material to make that protein again, without having to really modify the cell in any way. Just add that genetic construct, make it, get it expressed so that the protein is then expressed, reapplying brake.*

ROSALYN: *This is really interesting.*

TIMOTHY: *Um hum.*

ROSALYN: *So basically you tell the gene to add this piece to your construction.*

TIMOTHY: *Um hum.*

ROSALYN: *Because it's missing or we don't like what it does.*

TIMOTHY: *Um hum.*

ROSALYN: *Will you take out some element of the gene?*

TIMOTHY: *Um hum.*

ROSALYN: *Because of the message it's sending to the cell?*

TIMOTHY: *Exactly.*

ROSALYN: *Gees.*

TIMOTHY: *It's pretty wild, but our interest in it turns out to be the introduction of genetic information into the cell. For over millions of years our cells have been programmed to resist this kind of activity.*

ROSALYN: *I suspect so.*

TIMOTHY: *Exactly, and so in fact it looks something like a viral infection. That's what we are trying to do, that's what viruses do, and they enter the cell and reproduce. So what we would like to be able to do is get our material in the cell as efficiently as a virus does. In fact, lots of gene therapies are based on viral strategies, but they have had a host of problems in the clinic. In the laboratories, many of these therapies seem to be very efficient and effective, but when they get to the clinic, there have been all sorts of problems associated with them. So, we do nonviral gene therapy, which is more difficult, but safer.*

ROSALYN: *OK, now hold on.*

TIMOTHY: *Yes?*

ROSALYN: *You are going to have to work with me.*

TIMOTHY: *OK.*

ROSALYN: *What material do you use to attach to the gene or the genetic material itself? The code itself is made up of DNA?*

TIMOTHY: *Yes, DNA. Four base pairs structured together in a very specific sequence.*

ROSALYN: *OK, so you have to change the sequence or add DNA?*

TIMOTHY: *Generally with most of these strategies we are adding something rather than changing it, although there are folks who try to do the change, but we actually just try to add something.*

ROSALYN: *What do you add?*

TIMOTHY: *So what we do is we identify, let's say a protein that's not being produced.*

ROSALYN: *OK.*

TIMOTHY: *So, what you do is you find the gene for that protein which is just a sequence of DNA that can be translated into that protein and we manufacture that.*

ROSALYN: *So you do some kind of chemical reaction?*

TIMOTHY: *Actually, we employ the most efficient way to make DNA we know of, which is to use bacteria. There are some bacteria that, if we provide them with the template for making this DNA that we would like, will make many copies very efficiently, and then we can extract that and purify it.*

ROSALYN: *Do you have to teach the bacteria what you want, somehow?*

TIMOTHY: *It's very easy. It's basically a template, so we just provide a relatively few number of these DNA strands.*

ROSALYN: *Of what you like.*

TIMOTHY: *Yes.*

ROSALYN: *It's from the source.*

TIMOTHY: *Yes. Or there are ways to chemically produce it. But generally we make small amounts of this. There are a variety of ways to do that. But once you have the small amounts and you would like big amounts, you would use bacteria to make that amplification plus.*

ROSALYN: *Wow.*

TIMOTHY: *So we grow the bacteria here and make lots of this DNA that codes for the gene of life.*

ROSALYN: *Then you extract the gene and purify the DNA, and then you insert it and that's the trick.*

TIMOTHY: *Then we have to formulate it into some structure that can efficiently get into cells and then once it gets into the cells, do what we want it to do. That's the hard part, because the DNA by and large is reasonably fragile. It's a rather large molecule, so it doesn't get into a cell very effectively by itself.*

ROSALYN: *But it exists in there on its own.*

TIMOTHY: *So in the cell it actually exists in the nucleus.*

ROSALYN: *In the nucleus.*

TIMOTHY: *The nucleus is a specialized structure which is protected and that's where we would like our pieces to go. But it's very difficult to get it there. There are a variety of obstacles that keep DNA from being translated there efficiently and that's the real trick of nonviral gene therapy; how do you take this bit of DNA and wind up in the nucleus of that cell?*

ROSALYN: *That's the delivery that you were referring to?*

TIMOTHY: *Yes. That's what we do, so our real interest then is to try and understand the mechanism of how that happens, understand the rates, which step is the slow step, so that we can focus on improving that and then we also test a variety of proposed strategies to improve it. For example, there are some methodologies that are proposed to target delivery to the nucleus once it gets inside the cell. But we haven't found that, so we tend to be sort of gadget makers both in terms of equipment as well as in terms of methodology. We have developed a few methodological gadgets that will allow us to study this very carefully and we haven't found any evidence that any of these proposed strategies actually deliver more of this to the nucleus. They might improve things for other reasons, but they don't improve it by delivering it to the nucleus. It's very interesting; it has been the focus of some of our work. There continue to be industrial and academic efforts to try and identify these nucleus-targeting strategies and so far, you know everything is in control.*

ROSALYN: *Is there any known method for getting the gene into the nucleus?*

TIMOTHY: *Yeah, viruses.*

ROSALYN: *That's the dangerous one.*

TIMOTHY: *Yeah and the viruses have been programmed over evolution.*

ROSALYN: *Right.*

TIMOTHY: *They are pretty efficient at doing it, at least some of them are. But they have had problems in the clinics, so they may not be the best choice. But, it's still an active area of research. Right now it's just two different kinds of approaches.*

ROSALYN: *OK, so nonviral, transport chemicals.*

TIMOTHY: *Primarily it tends to be a lipid, so we try to mimic the cell membrane. The lipids have some special characteristics. Cells are negatively charged and so is DNA and so we have to overcome those charged propulsions, so the life zones are partially catatonic. We provide a global net positive charge. That is reasonably effective at getting those particles inside the cell, but then there are subsequent steps inside the cell that we are trying to overcome.*

ROSALYN: *To get into the nucleus.*

TIMOTHY: *Yes. So that's where most of the tweaking is going on right now. These particles initially wind up in the endosome of the cell, which is the compartment that is initially brought in from the outside. The cell tries to inactivate anything it brings in from the outside, so for example, the PH is very low in the endosome. So, that tends to degrade the DNA. So we have some strategies to promote endosome release to have that endosome break early so that you can avoid that really low PH. Strategies like that. And then there were strategies to try to take that structure and deliver it more efficiently to the nucleus. So there are lots of different proposed strategies to make the whole cellular thing work more efficiently. Some work better than others.*

ROSALYN: *And because the nucleus is so darn tiny we are talking about working with the nanoscale?*

TIMOTHY: *Yes, the nucleus is actually a few microns, at most.*

ROSALYN: *Relatively speaking, is that large?*

TIMOTHY: *Most cells are about 10 or 12 microns total and the nucleus is about 3 or 4. So most of our nanoscale work is in modulating this particle. The particles themselves, these DNA particles are on the order of—again the particle size is a variable that people like to play with—but the most effective ones seem to be reasonably small, somewhere between 15 and 100 nanometers. So the particles themselves are reasonably small. One of the things that we would like to do is use some nanotechnology to improve our ability to measure what happens, to use fluorescent tracers like some of these quantum dots, for example. So you could employ a quantum dot strategy to better track where the material goes maybe in time, get better kinetics faster, that sort of thing. So that's one of the aspects of our work now.*

ROSALYN: *I don't know much about these quantum dots, but can they be developed to have kinetic properties?*

TIMOTHY: *I think that is an open question. Right now they have interesting fluorescent properties, which are controlled by their size, primarily.*

ROSALYN: *Yes.*

TIMOTHY: *But other than that, I think they have been used only for tracking and tracing and monitoring things, kind of a tagging system. I think there are some opportunities to do other things. There was a speaker on campus just very recently who said they essentially make quantum dots rather than solid particles, as sort of hollow shells.*

ROSALYN: *Um hum.*

TIMOTHY: *Which opens the door to putting something inside.*

ROSALYN: *Inside?*

TIMOTHY: *Inside.*

ROSALYN: *What's curious to me about your work is if these are really flaws—we consider cancer a flaw, why doesn't the body willingly accept treatment, because it doesn't recognize it to be a flaw? Because the individual cells have their own determination irrespective of the whole unit of the body and what's good for it? Is that what's going on?*

TIMOTHY: *I guess that's certainly one way to look at it. The current thought about how the body treats cells that are not behaving properly is that this probably happens on a reasonably continuous basis in everyone and many times, 9 times out of 10, 99 out of 100, I don't know, but a large percentage of the time when a cell becomes transformed and is misbehaving, the body recognizes it. Most of the time the immune system in our bodies, the white cells primarily, will recognize it as misbehaving and kill it.*

ROSALYN: *OK.*

TIMOTHY: *But by chance, there will be one in a hundred or a thousand of these events where a cell misbehaves but externally it can't be seen.*

ROSALYN: *Oh.*

TIMOTHY: *Can't see this misbehavior.*

ROSALYN: *OK.*

TIMOTHY: *It's a pretty efficient process, but every once in a while you wind up with one of these transformations that to the rest of the body looks like a normal cell and it just keeps growing. So, the only thing our immune system can do is recognizing things on the outside of the cell. If that doesn't change, we don't know that it's foreign or that it is misbehaving. Most cancers, most misbehaving cells presume we do have those changes on the surface and we deal with those effectively. But, it's that one in a hundred that ...*

ROSALYN: *Can't be seen.*

TIMOTHY: *Can't be seen that wind up being the problem.*

ROSALYN: *So then, we come along with our scientific abilities and say we don't like this part of the human body, what can be done about it?*

TIMOTHY: *That's right.*

ROSALYN: *What would it mean to be successful in your work?*

TIMOTHY: *There are many different kinds of disorders I think that could be conceivably treated by this method. We have been focusing mostly on cancer right now, partly because we have really strong collaborators here who are interested in that work. One of the collaborators is the Chairman of Radiation Oncology. Not only is he a great physician, clinician, but he is an excellent scientist as well. He has a big research laboratory. So it's really a pleasure to work with him. He is one of the physicians that you can work with that understands the science very deeply as well. They have many patients who have tumors, that despite the best traditional efforts continue to re-emerge and generally speaking it's probably because the surgical interventions don't get all of the cells or because the chemical or radiation treatments again don't kill all of the cells. You can reduce the volume of the tumor but then eventually it regrows. So what we would like to be able to do is seek out those sorts of remaining cells and we are hoping that our (gene therapy) strategy will allow us to do that. It will be in some ways a supplement to traditional strategies, which are reasonably effective in terms of say a radiation therapy or a chemical therapy. But, then following that, maybe simultaneously we could deliver some of these gene therapies to try to catch the remaining cells that were resistant to these other treatments.*

ROSALYN: *Oh, that's interesting.*

TIMOTHY: *Then hopefully have a longer term modification to the progressive ones.*

ROSALYN: *If you could actually do that, would it be a success for you?*

TIMOTHY: *Yes, absolutely, absolutely. That's one approach. There are lots of other approaches, some of which are just purely the traditional paradigm of gene therapy. That is to fix genetic disorders, rather than cancer, and that would be another wonderful strategy. For example, something like hemophilia, a genetic disorder in which a molecular clouding protein is not effectively made. If you could somehow reintroduce that gene and have that protein expressed, the symptoms of the disease would be abolished. That would be good.*

ROSALYN: *And conversely, what does failure look like to you?*

TIMOTHY: *Oh boy failure. I think it would be hard to identify failure, because there is always, even when a particular approach is unsatisfactory, for one of many reasons, there will always be some new idea. "Yes, but what if we do this ..."*

ROSALYN: *Yes.*

TIMOTHY: *And so, it's hard to imagine what complete failure would be in the sense of throw up your hands and walk away and say I'm done. That's hard to imagine. I think there is always a creative new potential solution to whatever problems arise. Most of these problems arise from the biology, rather than the engineering and the chemistry. They are biological responses, or failure of biological responses to our interventions.*

ROSALYN: *If you were successful and you found a transport system that the nucleus responded to and then you could broaden that to many, many different genetic bases, theoretically we could combat or cure many diseases known to humanity, right, because, so many of them are genetically based?*

TIMOTHY: *Right.*

ROSALYN: *What would that mean if we don't have to any longer worry about illness?*

TIMOTHY: *Would that be wild?*

ROSALYN: *Yes.*

TIMOTHY: *Oh gosh. I have to say that that seems like it's so far off, I guess I don't really think about it very much.*

ROSALYN: *Do you really think it's not possible to get to that state?*

TIMOTHY: *I think everything is possible, but I think it won't happen in my lifetime, maybe not my children's lifetime. I think that is a long, long-term process. I think what we are going to find is this naive view that you modify one gene or one protein and fixing things is going to turn out to be much more complicated. I think what you will find is that modifying one protein will have six effects downstream and upstream and to cure a disease like some kind of heart failure, you are going to wind up having to intervene in not just one gene, but in ten, or a hundred, or a thousand. It's hard for us to modify one right now, trying to modify let's say 10 and do it in such a way that all 10 are controlled in concert, because generally there is some kind of a balance and most of these diseases represent some imbalance. It's not just expressing the gene, but expressing the right amount. So it may be easy to think about going from no gene expression to some, but what happens when you get to these more sophisticated levels of diseases where you want to double this one and half that one. Well, how do you control all of this? That's a whole new layer of investigation for which the very beginnings are just emerging. I think it's going to be a much more complicated system. I can imagine that there will be particular disorders that will effectively be treated by gene therapy over the next few years and that might in some small way overall, extend life span tremendously for the people with these diseases. But for the general population, only a tiny blip in the overall average life span. Over time, that will probably grow and so I doubt we will see the sort of step change in the duration of life. We will probably just see a progression in the increase that we are seeing now.*

ROSALYN: *Is it when you do your work, you are focused on this immediate time frame; "what can I do now to make this?"*

TIMOTHY: *Right.*

ROSALYN: *Somehow every contribution goes into the whole?*

TIMOTHY: *Absolutely.*

ROSALYN: *So the contributions made by Pasteur a hundred years ago, make it possible for you to do what you are doing now?*

TIMOTHY: *Absolutely, like standing on the shoulders of giants.*

ROSALYN: *Or somebody we don't know, whose name we never heard of made a contribution somehow.*

TIMOTHY: *That's right.*

ROSALYN: *People with only five papers to their name …*

TIMOTHY: *Absolutely.*

ROSALYN: *So what I am learning from these conversations is that there seems to be this body of knowledge that everybody contributes to and pulls from to grow in their own knowledge. Where is that body of knowledge taking us?*

TIMOTHY: *We rustle with that often since graduate students do most of our work. So there is this relatively limited time horizon for a student to come in, take an extensive amount of courses, learn something about their desired research area, mostly from the literature, and then conduct their own experiments and do all of the other things we expect each candidate to do. But that middle step—trying to understand where this literature is now and how it is propelling us—is one of the more difficult things we have to do. The line here is "a month in the laboratory can save you a day in the library," so we wind up with students who spend a lot of time doing experiments that, had they evaluated the literature a little bit more carefully, it would have saved a tremendous amount of time, because there is a big body of knowledge out there and sometimes it's hard to window through it.*

ROSALYN: *Yes.*

TIMOTHY: *One of the things we find with our multidisciplinary approach is that there will be information out there in the literature somewhere that we don't look at. So for example, Medlines are a big search tool, but just the other day, this is a good example. There is a journal that I get; it's an engineering journal;* Annals of Biomedical Engineering. *It's generally not categorized in Medline. So it's kind of a funny story. I was in here looking for some student work that was lost and I was in this big box here; there is all sorts of stuff … I am tearing my office up looking for this and one of my, one of my targets here, I had stacks of journals here on the table. I said, OK, these are just getting in the way; we are going to look through them, see if there is anything important, and then throw them out. So there is this one journal on biomedical engineering, just from a month or two ago, and what I generally do is just scan the table of contents and 9 times out of 10 I throw it away. So in this particular one there was an article in there about molecular transporting tumors, which is right up our alley. This is exactly the kind of thing we are interested in. So I looked at the abstract. It was like, boom! I ripped it out, threw the rest of the journal away. We would have never found that, probably. So, it was just there. One of the things I worry about is that can we spend more time with our students trying to get them to evaluate the literature which is growing every day. And so as that body of literature grows, it gets harder and increasingly more difficult to have stu-*

dents do that and do the things in the laboratory we expect them to do in the amount of time they have. I am not sure how to make that process more efficient, but it seems to me over the last 10 years that it is becoming less efficient for us.

ROSALYN: *Maybe as the technology changes and information processing changes there will be a more efficient system for search.*

TIMOTHY: *Wouldn't that be great if every journal had opportunity for us to search them electronically?*

ROSALYN: *Yes.*

TIMOTHY: *That would be fabulous. Heck, I guess the traditional paradigm of the researcher was somebody with the filing cabinets bulging with papers and now it's hard to bulge the PDF files. Actually we have a difficult time trying to sort and characterize them and figure out where they are. How do you archive them in some way that is searchable and findable? It's a big challenge.*

ROSALYN: *Long range, what is this process of contributing to the literature all about? And where is it going? If it's sort of like this entity, the literature entity, and it moves through time. Alright, where is it going, where is it taking us?*

TIMOTHY: *I think that's really determined by the students we produce.*

ROSALYN: *Oh, you do?*

TIMOTHY: *Yes. I do, because they will be the future architects of where the literature goes. It's interesting I think because it's difficult for us to get students excited about the literature. But ultimately that is a primary source of their progress. It's how they will be judged, evaluated. It's their contribution to literature and certainly later on depending on where they go, academia or industry, they will continue to rely on either contributions to or borrowing from the literature to help them in whatever it is they are doing. So, where is it going, I don't know, I think my graduate students will hopefully determine that, because they will be the ones 10 and 20 and 30 years from now who will be contributing to it and extracting information from it so they will have some feeling about where it ought to go.*

ROSALYN: *Do you actually think we can direct the focus of our scientific inquiry?*

TIMOTHY: *To some limited degree. I think a lot of that is controlled by money.*

ROSALYN: *Yes?*

TIMOTHY: *And so there are some scientific inquiries that you can put a spin on that makes them fundable that otherwise might not be. So you can perhaps do things that might not be otherwise possible, but in large measure I think the big picture is really dictated by where the resources are. So, for example, as we were talking about nanotechnology, I think the presumption right now is there is this enormous pot of money earmarked for nanotechnology. Maybe even more spe-*

cifically from my point of view, a section of nanotechnology and biotechnology, I have colleagues who come up and say, "You are the luckiest guy in the whole world. An enormous vat of resources out there and all you have got to do is ask for it and it comes raining down on you." Well, I don't really see it that way. It's still a very difficult, competitive field.

ROSALYN: *Yes.*

TIMOTHY: *But what we do is at least partially controlled by what we can get funded.*

ROSALYN: *So the National Nanotechnology Initiative makes your work more possible than it would have been without it?*

TIMOTHY: *Absolutely.*

ROSALYN: *And what do you think is the impetus of the federal government to put so many resources in this area?*

TIMOTHY: *Isn't that a great question. I don't really know. We think about that all the time. I guess we think about it from the context of how decisions are made, who decides, or how it is decided that this area will be a focus of a particular funding agency. I don't know. That's an interesting question. I think it's a very complicated feedback loop, which is probably not perfect, so there are some disconnects. But I think senators have to be accountable to the constituents so they would like to be able to show some high technology and say, "this is your federal government at work" kind of thing. I think they are at least in some ways looking for things that can be made easy, but sexy to understand to the general public, you know cover of* Time *magazine kind of things: "God, isn't this great for sequencing the genome."*

ROSALYN: *Right.*

TIMOTHY: *So I think those things play a role if you are doing something very esoteric, super difficult to understand, but perhaps scientifically and medically very important, I think there is a little marginalization of that. It's a little less sellable. But there are sources for that kind of money, like DARPA.*

ROSALYN: *Right.*

TIMOTHY: *Those guys do all sorts of crazy things.*

ROSALYN: *I understand.*

TIMOTHY: *Maybe two thirds of which never really pan out, but the one third that do are outstanding; paradigms are shifting. There is this kind of spectrum from say DARPA, which might be the most hysterical kinds of ideas that may not have any direct foreseeable application that are more fundamental, all the way down to very mundane, where we did A, B, and C in the previous years. I know it's clear that we should do B next. OK, why don't we do D?*

ROSALYN: *Um.*

TIMOTHY: *But it is interesting to think about how those decisions get made in terms of how programs are emphasized.*

ROSALYN: *It almost seems as if, if money were no object, science might take a very different turn.*

TIMOTHY: *I think that might be true. Although, I think there is some fairly good agreement between national means and technology people who are interested in doing it. I don't think there is a tremendous disconnect there. I think if there were, there would be another feedback loop. For example, I think initially when AIDS became a public health problem, the NIH [National Institutes of Health] fairly rapidly mobilized large resources to try to understand that disease and after a reasonably short while, a few years, scientists began pointing out, I think this is right, I am not sure, that at least at one point the money spent on AIDS was equivalent to the amount of money spent on something like heart attacks and strokes. You show the numbers of how many folks are inflicted with these two diseases, there is this big disconnect. We were spending as much money on these two different afflictions, but there is a hundred or a thousand times more people afflicted with this particular disease, maybe we needed to focus more money there. There was this gradual kind of redistribution of resources. Again I think AIDS, HIV money got put in that pot because of …*

ROSALYN: *Politics?*

TIMOTHY: *Public demand mostly. Call it political demand and then eventually there was this other feedback that put everything into balance. So I don't think things can get too out of whack for too long, but they do occasionally.*

ROSALYN: *OK.*

TIMOTHY: *So it will be interesting to see if nanotechnology investment is perceived to be out of whack with what society is interested in. I don't think so; I think people are, as you mentioned, the lay public is fascinated by small things now. I think they are easy to sell from the point of view of some concepts, some possibilities.*

ROSALYN: *Well, some of the security issues are being attached to nanoscience.*

TIMOTHY: *Um hum.*

ROSALYN: *And that's easily sellable right now.*

TIMOTHY: *Right.*

ROSALYN: *Some of the futuristic visions though …*

TIMOTHY: *May be harder.*

ROSALYN: *Still, people don't want to think about them as far as I can tell. They are frightened. They haven't made it from the science fiction domain into the popular press very well. I do wonder what you make of some of that material.*

TIMOTHY: *Specifically?*

ROSALYN: *Such as the scientists who are convinced that our inevitable evolutionary path is to a state where the body is no longer DNA based, but in fact becomes a combination of our mechanical creations and circuits, and biology. Some say it is our only way to survive as a species.*

TIMOTHY: *Um, well that's interesting.*

ROSALYN: *Ray Kurzweil believes that we won't survive otherwise because machines will soon be able to outpace our computational powers and become way more intelligent than we; that in order to sort of manage that we will have to combine circuits and disks and chips into our bodies so that we can keep pace.*

TIMOTHY: *It's interesting.*

ROSALYN: *To the extent that individual scientists contribute toward that end, he argues that even in a way, what you do contributes toward the integration of machine and human. If your transport system becomes a creation of the laboratory rather than a creation of the body, then it does move us one step closer to a state where the body is no longer just biological. Do you think about it? Do you worry about it? Does it excite you? What does it mean?*

TIMOTHY: *I guess I see that there will be some clear applications for that kind of intersection of silicon and biology.*

ROSALYN: *Right.*

TIMOTHY: *But otherwise I would disagree. I think that we are always going to be smarter than the machine. Maybe that's because I come from a biological perspective rather than an information technology kind of perspective. I think we will always be smarter than the machine. We will always be a little more creative, being able to keep technology working for us, not against us.*

ROSALYN: *Do you want that or do you believe that?*

TIMOTHY: *I believe that.*

ROSALYN: *So, he and others like him feel that the human body is essentially about information processing. And if we could create a machine that processes information, faster and on a more sophisticated level, it actually will eventually develop cognition and self-awareness, because he believes that with heightened intelligence comes self-awareness. Would you classify the human as essentially an information-processing machine?*

TIMOTHY: *I think there is something a little more to it than that, but it's a very, a very difficult concept to try to get your head around, isn't it?*

ROSALYN: *Yes.*

TIMOTHY: *If we were just information processors, I imagine that a particular input would always relate to a particular output. There wouldn't be the human capacity for change. So for example, it's different to know a book. You can know every word on a page easier than it is for me to know you, because you might have a Starbuck's coffee every morning for 20 years and then one day we will do something different. How does that happen? I guess I still see machines, maybe my narrow-minded view as having a relatively predictable output for a predictable, for any given input. I see humans in a very different way.*

ROSALYN: *How interesting.*

TIMOTHY: *Yeah.*

ROSALYN: *So when Kurzweil's computer writes poetry, which he says it does ...*

TIMOTHY: *Right.*

ROSALYN: *He says it's because he has taught the machine all of the variations of poetry known to humankind. He has put every single possible combination of words from every tradition over time so that the computer says, "Oh this is what poetry is" and then it outputs poetry, and you read these poems and you think, I can't tell whether a person wrote those or a machine wrote those.*

TIMOTHY: *Right, so then I guess I would say that tomorrow there will be a human who develops a new kind of poetry, that the machine won't know about, since it only knows the history of poetry, and it might be able to mimic that, and it might be able to produce something new in that same form, but could it create a new form? Does it know what's pleasing? Why do we write poetry, because it's somehow pleasing to us.*

ROSALYN: *Or, we write because we suffer, or because we are afraid, or we are enraptured.*

TIMOTHY: *Sure, exactly. So can a machine do that, maybe some day they will. I don't know, but I see them at some level primarily as mimics of humans, and they might be able again to write new poetry of the same form, but what happens when somebody produces a new form? I don't know that a machine will be able to do that, because it might not be able to understand the genesis of that.*

ROSALYN: *You are distinguishing the machine's ability to mimic from the human's ability to truly create.*

TIMOTHY: *Really, I don't know.*

ROSALYN: *You don't know?*

TIMOTHY: *No, I don't know. I'm sorry. I don't have all of the answers. But, I really think if that's one of the differentiating characteristics then maybe, someday there will be a way for the machine to mimic that process. I don't know that it will intrinsically be able to do it, but maybe we can give it some rules that mimic say, the creative process and maybe they could, for example if we were asking them, write a great new form of poetry. It might look through an entire dictionary of every word in the human language in all languages and think about all of the characteristics—the way they rhyme, their pace, other characteris-*

tics, words, sentences—and then create every possible new iteration of every possible word and decide somehow this is very pleasing or this is very descriptive or this is a new form of poetry. But I don't think that is the way we do it.

ROSALYN: *Right.*

TIMOTHY: *I think that even if you could mimic that process of inventing something new, it would happen by a completely different methodology presumably than we would do it.*

ROSALYN: *So we may end up with machines that have the information-processing capacity to mimic us, but that wouldn't be human.*

TIMOTHY: *Right.*

ROSALYN: *And we won't be machines.*

TIMOTHY: *Right. I think.*

ROSALYN: *Unless we come together with them, somehow.*

TIMOTHY: *I think there will be some forms of that, but I don't know. I don't think that will be the prevailing desirable form. But, I do think there will be some folks who will try it and combine it at some very fundamental intimate level, machine processing, biology, there is no doubt.*

ROSALYN: *There are scientists who are aiming to do exactly that.*

TIMOTHY: *Sure, right.*

ROSALYN: *One of the motivations seems to be to have perfect and indefinite health of the physical body.*

TIMOTHY: *Sure, biology always seems to run down or run amuck in a way that maybe some people believe machines don't. My car breaks though, my computer breaks, too. So I don't know.*

ROSALYN: *Would that be a goal for you?*

TIMOTHY: *To have a?*

ROSALYN: *If your contribution could take us to that place.*

TIMOTHY: *To have an indefinite life span, is that what you are saying?*

ROSALYN: *Um hum and unlimited health and vitality.*

TIMOTHY: *Well I don't know. I don't know that I would call that a goal. I think that some deeper consideration of that goal, that outcome would reveal lots of problems that would have to be resolved before we got there.*

ROSALYN: *You mean biological problems?*

TIMOTHY: *Other kinds of problems, sociological problems, all sorts of problems.*

ROSALYN: *Huge problems.*

TIMOTHY: *I think it might be possible in some future world that life span could be tremendously extended hundreds of years, maybe thousands. Is that what you really want? I do think there will be other pressures that modify the way we behave.*

ROSALYN: *Access to potable water?*

TIMOTHY: *Right, exactly.*

ROSALYN: *Well, not too many folks seem to be concerned about that.*

TIMOTHY: *Those continue to be problems of today. I think it's funny we have a friend whose wife is a physician. She just went away for 4 weeks on a trip to Honduras, on a medical kind of mission. Her husband was talking to me, he said, "I don't know really what she is going to be doing down there, she is a pediatrician. Gosh I guess she will be taking care of all sorts of baby health problems." I think down there the best thing you can do for public health is get clean water and get the animals out of the streets and provide some sanitation. Those are the things that really contribute to public health there and we just take them for granted. But if you think about longer life spans and populating the planet by a factor of five, ten, hundred, hundreds of thousands, all of these things come back to haunt you even in a modern way.*

ROSALYN: *Yes.*

TIMOTHY: *So I think there are big issues of that sort.*

ROSALYN: *So for now, making cancer go away would be good enough?*

TIMOTHY: *Boy wouldn't that be good. That would be outstanding.*

ROSALYN: *It's a horrible disease. I don't see social complications to curing it.*

TIMOTHY: *Probably not.*

ROSALYN: *Some?*

TIMOTHY: *Some maybe. Yes.*

ROSALYN: *Everything has an effect.*

TIMOTHY: *But not big things, not things that wouldn't be easily overcome I think.*

ROSALYN: *Yes.*

TIMOTHY: *I think it would be a very big change when you think about a world where you could just throw a light switch and say, that's it, nobody will ever get cancer again!*

ROSALYN: *What does that mean?*

TIMOTHY: *It's a very interesting concept.*

ROSALYN: *Well, we did it; didn't we do that with small pox?*

TIMOTHY: *Sure.*

ROSALYN: *Except it came back, didn't it?*

TIMOTHY: *There are some diseases that are no longer problems for us. But I wonder if that would change people's behaviors, for example, a trivial case if you could just take a pill like an aspirin every day and guarantee that I will never get cancer.*

ROSALYN: *Um hum.*

TIMOTHY: *OK, so everybody starts smoking again.*

ROSALYN: *Right.*

TIMOTHY: *I will agree, yeah, bring on the bacon and the fatty foods, who cares, tumors, laugh at them, I am taking the magic pill.*

ROSALYN: *That's true.*

TIMOTHY: *So how does that change human behavior and does it cause other problems? Probably it would. But I don't think we are going to get to the magic pill that prevents you from ever getting cancer. I think we will probably have a purchase that tends to provide longer term moderation of the existing disease so you wind up being afflicted with cancer, but it's a more treatable kind of thing. So you have a longer life span and a better quality of life rather than what we just described which is some utopia, which you never get, regardless of what your behaviors are.*

ROSALYN: *Interesting, because what you actually are talking about is treating the symptoms of something, right?*

TIMOTHY: *Right.*

ROSALYN: *So we want to say well "that's genetic," that's not my fault.*

TIMOTHY: *Right. And some are. There are certainly some cancers, that regardless of your behaviors ...*

ROSALYN: *But they are arbitrary.*

TIMOTHY: *Right.*

ROSALYN: *But it seems to me that there are an awful lot that we can influence.*

TIMOTHY: *Self-inflicted, yes. So you change people's behavior. A perfect example is my parents, who quit smoking about 2 years ago. They smoked for 50 years and then they quit very recently. They had a friend who died of lung cancer and that affected them deeply and they decided, even if it might not modify their health very much, they thought that was the right thing to do. Now in a world where nobody got cancer, that wouldn't have happened, for example.*

ROSALYN: *So it takes away the individual responsibility.*

TIMOTHY: *Responsibility, yeah maybe so, at least in some ways.*

ROSALYN: *Well, I do think if we didn't have AIDS and STD's ...*

TIMOTHY: *Well there you go.*

ROSALYN: *We might have very different behavior in the population.*

TIMOTHY: *Well.*

ROSALYN: *Because of biology.*

TIMOTHY: *Well that's another good point. But that's exactly right, I think that's right. I think there are some things like that, that would really modify human behavior.*

ROSALYN: *Interesting.*

TIMOTHY: *Yeah.*

ROSALYN: *Well this has been great.*

TIMOTHY: *This is fun.*

ROSALYN: *Thank you very much.*

TIMOTHY: *You are more than welcome.*

CHAPTER FOUR

Meaning Making

Technology, or the making and using of artifacts, is a largely unthinking activity. It emerges from unattended to ideas and motives, while it produces and engages unreflected-upon objects. The need to think about technology is, nevertheless increasingly manifest.

—Mitcham, 1994, p. 1

The unattended to ideas and motives referred to in Mitcham's opening statement have roots inside of a largely symbolic process. Lakoff and Johnson spoke about the cognitive unconscious, or a system that functions like a "hidden hand" that shapes how we conceptualize all aspects of our experience, and how we automatically and unconsciously comprehend what we experience. This hidden hand shapes everyday commonsense reasoning, as well as philosophical concepts (including time, events, causation, essence, the mind and morality). Metaphor is the primary means of operation for this "hidden hand" (Lakoff, 1980).

THE MAKING OF MEANING

The primary human responsibility is to actualize the potential meaning of life. True meaning, according to Frankel (1992), is discovered in the world rather than within man or his own psyche. Take, for example, something as seemingly simple and apparent as the hand attached to one's wrist. Using the eyes to observe the hand, the brain registers its various qualities: shape, size, color, texture, dexterity, and so on. The nerve endings below the skin on that hand stimulate signals to the brain, providing information about the sensations it receives when it performs certain functions in certain ways, or when it is in certain conditions, such as wet, cold, hot, or sticky environments. But what does that 5-digit thing mean? Making meaning of that part of the human body is essential to understanding the significance and value of being alive in the body. Making meaning about "hand" (as humans have agreed to identify it) is essential to mental well-being in that feeling a sense of self and place in the body, in the world, and in contact with other sensual beings is a source of comfort and reassurance about one's own existence. When that thing named "hand" is used for the creation of beauty—or food, or essential work—certain types of meaning are placed on it. When used for destruction, or to otherwise do harm such as when it pulls a trigger or forcefully hits the face of another sensual body, then other types of meanings are ascribed to hand. With the determination of meaning, shared or personal values and beliefs are evoked, such as whether the person using the hand is "good" or "bad." If tragedy leads to the loss of hand, then new meanings must be made around being in the body without it.

The psychological survival of humans requires that meaning be made of both perception and experience. Where there is no meaning, there is an incomplete, pathological sense of self. But meaning is fluid. It is subject to continual change. It is being made and remade as consistently as air flows in and out of a living body. It is subject to negotiation, competition, illusion, and fantasy. In the case of scientific inquiry, conceptual understanding is formed as meaning is ascribed. As new technologies are introduced into human life, humans simultaneously change and adapt the sense of what those technologies mean to living, and the understandings of the world in which we live with those technologies. The nanotechnology initiative is more intricate and complex than the simple search for new knowledge and the development of new tools, because it implicitly involves the renegotiation of selfhood in the search for meaning.

Where will humanity go with nanotechnology? As with any new technological development, that will loosely be determined, in part, by random social forces and conceptual negotiations over competing interests, both tacit and explicit. Individual and commonly shared beliefs about who we are as humans, and what it means to be alive in the body and in the community, are among those forces. Ongoing negotiations are being made within society over the meanings ascribed by various individuals and communities, about where they wish to go and how it is they wish to live with themselves, with others, and with nanotechnologies in that future. Beliefs and meanings are also constructed over perceptions of winning and losing opportunities, resources, power, and control. Personal meaning making and the sociocultural negotiation of those meanings are as much a part of the development of nanotechnology as is its laboratory research and development. They arise from the fundamental and uniquely human quests to place self in the context of one's sensing and perceiving, to assert and establish a sense of purpose, significance, and control over one's life.

THE NEGOTIATION OF SELF IN TECHNOLOGY

Social orders and individuals arise in and through processes of ongoing negotiation. Selves are constructed through a process of cooperation, in specific negotiating contexts. The question of what there is to be negotiated is answered in consideration of contributions participants make, and the consequences they face in participating in the social order. There are a number of values yet to be negotiated in the development of nanotechnology, each of which will have their consequences in the reconstruction of self. Consider the appropriation of the computer as an example of the reconstruction of self in a social negotiation process.

Computers rapidly shifted down in scale from room size monstrosities that individual citizens had little access to, to relatively small, very fast, convenient processors and purveyors of vital information. As such, these devices now make it possible for many of us to work with mobility: in an airport, on a train, in a hotel room, or at our own home. In its inception, this newfound capacity to work and communicate from nearly anywhere was spoken of in futuristic terms with promises of improvement to the quality of life. The act of using it promised to give more free time, to make everything more efficient. As such, it was widely held to be a social and economic good, and accepted with its associated costs to society. Every radically new technological change requires a negotiation of changes to our social and cultural norms and expectations. In the case of the miniaturization of computers, those changes include the rapid and perhaps irretrievable erosion of established boundaries between work life, family life, social life, and personal life with their independent responsibilities and even their own sanctities. For many, a once highly valued activity; time at the family dinner table, has been supplanted by competing home computer time, telephone time, and other demands of technological living. Many young people now use the computer (and cell phone) as their primary source of information, and as their preferred medium of communication with other humans. Instant Messenger (IM) replaces face-to-face communication. To large measure, the World Wide Web (WWW) supersedes three-dimensional, tactile sources of knowing and experiencing human connection, such as musical instruments, books, and the living face of another person.

One of the early allures of the compacting and personalization of computers was the efficiency and speed offered for the processing of military and business systems of information. Another was the incredible profit potential from the prospective new global markets associated with producing and distributing computer hardware and software. The revolution in computing happened not simply because of the determination of business enterprise, or because of the government support for the development of these technologies. It was also driven by a cooperative process that took place simultaneously, a negotiation in the public domain involving beliefs about quality of life gains and perceptions of possible changes to human conditions.

Some values that will surely arise with the miniaturization and hybridization of commonly used electronic devices regard the assumption that faster and cheaper is equal to better. Among others, new nanoscale devices may demand examination of how market imperatives supersede other social goods and respected human values. Nanotechnology represents a collection of new tools and the development of new devices, many of which may be used in ways that have profound implications on the way we construe our lives, including its meaning and significance to various human, technological societies. This is what happens with the development of new technologies. The transference of computing technologies into the broader society is one example. Computing technologies changed the pace and tenor of living for many, and especially, the rate of communicating. The effect has been to increase the volume of information individuals and institutions are expected to exchange and process, without increasing the amount of time devoted to those exchanges. One consequence is a dramatic increase in energy consumption. Another is a personal feeling of exhaustion and being overwhelmed with daunting tasks under the pressure of rapidly moving time. The way technologically socialized humans perceive time has also changed, and the amount of work some expect to accomplish in that time has increased, simply because so many individuals have incorporated into daily living a technology that works many times faster than the human being is constitutionally designed to respond.

In the United States, the National Nanotechnology Initiative draws from American taxpayers multiple billions of public dollars for its research and development. Many thousands of research scientists, engineers, and graduate students devote their intellectual resources to the pursuit of knowledge about the nanoscale of phenomenon. Educational institutions from K–16 will soon be asked to make significant adaptations of teacher training, curriculum, and pedagogy toward building a future workforce that is trained in the languages and techniques of this "technological revolution." Job losses, environmental effects, social restructuring, unanticipated financial costs, and revisions of public policy are just a few of the constraining factors that must be dealt with in the negotiation process. Resources and opportunities otherwise not available will be offered on the table of societal negotiations to assure public support, which is required if the "revolution" is to go forward. Thus, a federal funding proposal claims that the discovery of the novel phenomena and material structures that appear at the nanoscale will affect the entire range of applications that the grand challenges identify (Roco, 2003). Those grand challenges include the following:

- Chemical-biological-radiological explosive detection and protection (homeland defense)
- Nanoelectronics, -photonics, and -magnetics (next generation of information technology devices)
- Health care, therapeutics, and diagnostics (better disease detection and treatment)
- Energy conversion and storage
- Environmental improvements

These identified areas point to great possibilities for profound changes to the conditions of living for collectives, as well as for individuals. Each reflects values generally upheld in our society regarding healthy life and longevity, physical security and freedom from aggression, and rapid and free access to information, cleanliness, and safety in the environment. These values have evolved in tandem with new technological developments, without which they could not be assured.

INTENTIONALITY AND RESPONSIBILITY

Popper (1992) delineated and distinguished the reality of three, interconnected worlds:

1. The physical world of bodies and physical states, events, and forces

2. The psychological world of experiences and unconscious mental events
3. The world of mental products, which includes our technological creations.

World 1, of solid and material things, gives us our central and most basic sense of reality. Within that world, humans seek to extend their freedom. They are continually problem solving, in search of better living conditions, greater freedom, and a better world. From Popper's perspective, World 1 is an environment in which "a tiny little living creature has succeeded in surviving for billions of years and in conquering and improving its world .... We inhabit a world that has become more and more agreeable and more and more favorable to life, thanks to the activity of life, and its search for a better world" (p. 15). Popper reconstructed the Darwinian ideology from life in a hostile environment that is changed by evolution through cruel eliminations, to say that "the first cell is still living after billions of years, and now even in many trillions of copies. Wherever we look, it is there. It has made a garden of our Earth and transformed our atmosphere with green plants. And it created our eyes and opened them to the blue sky and the stars. It is doing well" (p. 17).

"It," also referred to by Popper as "the first cell," seems to him to be completely responsible for our creation, our evolution, and survival. His argument continues by attaching to the evolution of human materiality, the evolution of human consciousness:

> My basic assumption regarding world 2 is that this problem solving activity of the animate part of World 1 resulted in the emergence of world 2, of the world of consciousness .... My hypothesis is that the original task of consciousness was to anticipate success and failure in problem-solving and to signal to the organism in the form of pleasure and pain whether it was on the right or wrong path to the solution of the problem. (p. 17)

And from there our species arrives at the creation of World 3, the world of language and of the other material products of our consciousness. But because World 3 is a world of our own inventions, it creates problems that depend on us. These problems are unintentional and unexpected. And as such, they react on us.

Popper summarized this particular paper by putting forth that the shaping of our reality is the result of interaction between Worlds 1, 2, and 3. The products created from our technology come from human mind, dreams, and objectives. In turn, those very products shape us. Popper (1992) indicated, "This is in fact the creative element in mankind: that we are in the act of creating at the same time transforming ourselves through our work. The shaping of realty is therefore our doing ...." (p. 26). As for the recreating of our world atom by atom, emerging as a result of our knowledge and of our desire and dreams to reshape and improve on our material condition, we once again face the unforeseeable consequences of our activities. Popper recognized this fallible condition of our humanity, and warned of the potential danger of belief in a political utopia, as connected to our search for a better world. "We are right to believe that we can and should contribute to the improvement of our world," he wrote. "But we must not imagine that we can foresee the consequences of our plans and actions. Above all, we must not sacrifice any human life (except perhaps our own if the worst comes to the worst). Nor do we have the right to persuade or even encourage others to sacrifice themselves—not even for an idea, for a theory that has completely convinced us (probably unreasonably, because of our ignorance)" (p. 28). Popper believed that scientific knowledge arises from our conscious engagement with our material condition. That it is expressed through the material realities we create, and can become an objective aim for peace and nonviolence. To that I will add that the increasing power and control that will come to us as a result of knowledge of nanoscale phenomenon requires of us the additional objectives of providing for basic human needs for all in our species, and the stewardships, care, and maintenance of our earthly home. This is no more of a political utopia than Poppers' ideology of peace and nonviolence. And, only after those fundamental objectives are fulfilled, might we, with full moral conscience, begin to play with the dreams and fantasies of the material reshaping of our reality.

Through the will, those who make the commitment do have the power to guide and determine the future of nanotechnology: to make explicit how it will be used, to what ends, for what purposes and by whom, as well as which developments should be avoided. In Chapter 1 the question was posed, "Where is nanotechnology development leading?" One answer to that question is: Nanotechnology is not a force to be followed. Its future is neither predetermined nor independent of human will. It is being established in the present, by human action, intentions, desires, and beliefs about who we are, how we want to live, and what we aim to achieve. Neither is the nanotechnology future simply a matter of time. It is a matter of intention. Whether explicit or not, those intentions are there nevertheless, at work in the development of nanotechnology.

Although it may be obtuse, there is intentionality in the development of any technology, and with that comes responsibility. The pursuit of new knowledge is commonly held to be a moral good. That value is not being questioned here. Neither is the worthiness of finding good engineering solutions to material challenges and technical problems. What is being questioned here is humanity's emotional, intellectual, and spiritual capacity to use it well, in light of the prospects for the increased ability of humans to manipulate and control matter with precision and to specification. Although many wonderful new products and processes may be made possible to address myriad material needs, the unintended consequences of nanotechnology development could potentially be more disruptive than society is equipped to handle. As Richter (1972) explained, "Science has made possible immensely potent new technologies, which have created critical new problems of social organization and control" (p. 93). Richter used weather control as a compelling hypothetical example of the critical social and organizational problems that surface when a potent technology is newly appropriated, and society is not ready to make the needed adjustments to it. Richter (1972) pointed out:

> Given such a new technical capacity, problems of the following sorts would be likely to emerge:
>
> a) Different people would want different kinds of weather and diverging interests in this respect would presumably create new political cleavages.
>
> b) Long-range interests of society might differ from dominant short-range preferences. Thus most people might want "sunny and mild" weather every day, but this could be disastrous in the long run. Thus issues of societal discipline vs. immediate gratification would be superimposed upon issues arising from the clash between opposing groups with incompatible meteorological interests.
>
> c) Expansion of the capacity to control natural events often outstrips the expansion of knowledge concerning long-range effects of any initiated changes. Perhaps, for example, hurricanes, tornadoes, and other meteorological phenomenon which men generally would be inclined to dispense with, may have important functions of which no one is aware.
>
> d) Boundaries of political jurisdiction commonly fail to coincide with boundaries of natural phenomenon that may become subject to artificial control. Perhaps, for example, weather in the United States could

not be controlled without thereby also influencing weather in other countries. (p. 94)

Current understandings of the kinds of societal or ethical challenges that might arise with nanotechnology development are nascent. But there are quite a number of scholars and institutions whose research interests are in this area.[1] Consider the following question: What *should be* the nanotechnology future? Might it be developed for purposes and ends that are respectful of humanity and the Earth?

RECREATING THE WORLD, OURSELVES, AND OUR SENSES

As Winner (1986) made clear, "As technologies are being built and put to use, significant alterations in patterns of human activity and human institutions are already taking place. New worlds are being made" (pp. 10–11). Consider again the example of information technologies. Human engagement with these technologies has qualifiedly altered the tenor and rhythm of human communication in the industrial world, and in turn, has changed perceptions and experiences of being in that world. One effect has been an exponential increase in the volume of information that is exchanged. There is an assumption that with phenomenal increases in processing speed there would be a correlating increase in free time available for other activities. In fact, the amount of information exchanged has increased exponentially, whereas the amount of time devoted to those exchanges remains relatively static. The result is that users of information technology are increasingly pressed to sometimes overwhelming and exhausting tasks under the pressure of accelerated time. The way time is perceived has changed, and typically the amount of work one is expected to accomplish in that time has increased.

Technological humans have come to rely on computing and processing technologies that work many times faster than the human being is constitutionally designed to respond. As such, new meanings and modes of human-to-human communication have evolved. In incorporating various information and communication technologies into daily work and personal living (from the standard telephone, to faxes, to the Internet and e-mail, and cell phones and pagers, satellites, two way radios, etc.),

[1]Such as the work on the societal implication of nanotechnology done by Davis Baird and his colleagues at the University of South Carolina, or by Jacqueline Isaacs of Northeastern University, and her Center's collaboration with other New England area universities.

the world has been recreated once again. The world has become smaller and in some ways more dynamic with the intensity of information exchange. Through technology, most people who inhabit the world are now accessible to one another. Many are electronically interconnected through high speed transmissions. Technological changes have led also to physical alterations, especially to shifts in sensory acumen. Smell, touch, sound, and sight senses that were once critical to human communication (and to human survival) are now auxiliary to electronic and other forms of sensing and perceiving (Abram, 1996). Messages are now exchanged beyond the human body in ways that no longer depend solely on that body to engage itself for signaling and interpretation. The meaning of words and the means by which feelings are expressed, along with the nature of ideas and the means to understandings, have also taken on new forms. What it means to speak and write, to sense and communicate with other living humans has been reconstructed as new communication technologies are assimilated.

Another example of significant alterations in patterns of activity is how humans living in technologically dependent societies have come to engage music. Music was once experienced directly, through multiple senses, in live performance. Musicians were visible, performing in the immediate vicinity, playing actual physical instruments with their bodies that are audible to the listener's ears, in real time. Instruments each had their distinctive scent. Watching the musician was as much a part of the experience as hearing the musician. Being in physical contact with other listeners was also part of imbibing music. In contemporary, Western technological society, music listening is increasingly done alone. Access to it is largely through digital recording; much of it is from synthesized electronic processes, rather than individual human performance on individual instruments. In many ways, MP3s[2] have become the icon of music listening. With this technology, headsets assure that external auditory sensing is minimized, and the listener is moved from awareness of being in body, in community, to an experience that is almost entirely inside the human skull. To experience music is coming to mean something entirely different from what it meant, say, 100 years ago. Abram (1996) revealed,

[2](MPEG Audio Player 3) An audio compression technology that is part of the MPEG-1 and MPEG-2 specifications. Developed in Germany in 1991 by the Fraunhofer Institute, MP3 uses perceptual audio coding to compress CD-quality sound by a factor up to12, while providing almost the same fidelity. MP3 music files are played via software or a physical player that cables to the PC for transfer.

> In a society that accords priority to that which is predictable, and places a premium on certainty, our spontaneous preconception experience, when acknowledged at all, is referred to as "merely subjective." The fluid realm of direct experience has come to be seen as a secondary, derivative dimension, a mere consequence of events unfolding in the "realer" world of quantifiable and measurable scientific "facts." (p. 34)

Of course, this orientation away from the senses began well before the development of the information technologies of the past century. Its origins are ancient. It is a phenomenon, which Abram traced back to a very early technology development: the alphabet. He explained that the process of learning to read and write with the alphabet engendered a profoundly reflexive sense of self:

> The capacity to view and even to dialogue with one's own words after writing them down enables a new sense of autonomy and independence from others, and even from the sensuous surroundings that had earlier been one's constant interlocutor .... The literate self cannot help but feel its own transcendence and timelessness relative to the fleeting world of corporeal experience. (p. 112)

From Abram's assessment, the societal implications of this trend are profound.

Recent human evolution (as externalized in technologies we develop) is deeply rooted in scientific inquiry. Toulmin's (1962) account of the history of science suggests that once we were able to conceive of atoms, we became determined to find sources of knowing that were beyond the senses. Atoms being invisibly small, senses could not penetrate far enough to observe them directly. This frustrated the concern of scientists to construct a plausible system of nature. Stymied by this problem, the Greek scientist/philosopher Demokritos claimed that "we have no accurate knowledge of anything in reality, but can be aware only of the changes which correspond to it in the conditions of our bodies, and of those things that flow on to the body and collide with it" (Toulmin, p. 57). From this belief he asserted the scientific imperative to use only intellect as the guide in exploring the world of the invisibly small. According to Demokritos,

> There are two kinds of understanding, one authentic, the other bastard. Sight, hearing, smell, taste and touch all belong to the latter: but reality is distinct from this. When the bastard kind can help us no further—when we can no longer see, nor hear, nor smell, nor taste, nor feel more minutely—

> and a higher degree of discrimination is required, then the authentic variety of understanding comes in, giving us a tool for discriminating more finely. (p. 58)

As Toulmin explained, this doctrine and the belief it reflects has played a key role in the development of physics and chemistry. Its affects are apparent in belief and meaning in the larger society. Due, in part, to the collective faith and trust in science as the determination of material reality, and in our saturation in technological living, modern, technological humans struggle to negotiate the role of the senses in knowing and being. At one turn, we haphazardly dismiss the senses as an inferior means of knowing reality, and shun the use of those senses in connection to others. At another, we invest tremendous resources in devices to awaken and stimulate our senses (e.g., those that enhance sexual pleasure, intensify entertainment, make outdoor sporting "extreme," reconfigure, enlarge and/or replace body parts, etc.). For example, in most engineering and science (or business) conferences, the use of computer-generated slide shows is the expectation for presentations. The art of reading carefully prepared text or of oration without visual stimulation is devalued under the demand for the visual and auditory stimulation of the colorful, moving, electronically transmitted images. An academic paper read aloud word for word becomes a bore to an audience of today's students and scholars who crave more titillation.

Having numbed our senses, we technological humans have come to hope and believe that technology will enliven them again. It cannot. It is therefore worth considering whether the opacity of sensing consciousness is likely to be amplified when nanotechnology is transferred into the broader society, which in turn could lead to further crisis in meaning. Abram's (1996) thesis suggests that it may:

> Caught up in a mass of abstractions, our attention hypnotized by a host of human-made technologies that only reflect us back to ourselves, it is all too easy for us to forget our carnal inherence in a more-than-human matrix of sensations and sensibilities. Our bodies have formed themselves in delicate reciprocity with the manifold textures, sounds, and shapes of an animate Earth—our eyes have evolved in subtle interaction with other eyes, as our ears are attuned by their very structure to the howling of wolves and the honking of geese. To shut ourselves off from these other voices, to continue by our lifestyle to condemn these other sensibilities to the oblivion of extinction, is to rob our own senses of their integrity, and

> to rob our minds of their coherence. We are human only in contact, and in conviviality, with what is not human. (p. 22)

Nanotechnology is touted as opening the way for myriad new devices and processes that will touch on nearly every aspect of human life. If this is true, then it will likely also change our senses of what it means to be human. For example, if nanoscale engineers are successful in shrinking the size of transistors say 100-fold (one of the goals of some semiconductor researchers), then there will be the incredible capacity to place computing devices in and around the body, in order to have access to information currently unavailable. In order to adjust to the flow of that information, a radical, perhaps subconscious, reconstruction of cognitive processing about self within the body will have to happen. There are many millions of processes occurring inside of and around the body about which the individual has no conscious knowledge. Some researchers are working toward the development of miniaturized, internal sensors for gaining information about particular physiological functions. If and when individuals gain access to certain types of information to even a few of their bodily processes, such as particular biochemical changes, temperature changes, exposure to viruses and bacteria, breathing rate and heart rate changes, and so on, it will require that meaning is made of that information. (Again, information individuals normally have no access to, processes about which one is totally oblivious.) New meanings will need to be made of that information. To do so will involve a social renegotiation of what is believed about the self, and what it means to be alive in the body. It is not clear if individuals now adapted to highly technological societies are still capable of reshaping the meaning of human bodily awareness in a way that affirms and reinstates the sense of self in nature.

Stated differently, as the externalized, material world changes, the inner world of the person does as well, along with relationships to self and other, and the meanings made of those worlds. This externalized technological transformation is reflective of the internalized processes of the searches for meaning, the quests for survival, and the human need to establish a sense of self and purpose in life. To that end, humans pursue visions and dreams of control, progress, improvement, and well-being in life. The knowledge that science creates, and the technology arising from that knowledge, are mechanisms on which industrialized humans have come to utterly and completely depend.

MEANING MAKING AND CHANGE

Natural history points to the proclivity of all living entities, especially humans, to pursue change as a means toward self-improvement, not simply for survival. What is called into question here is not the desire or even the worthiness of change, but rather the direction and intention of it. As nanoscale science matures into nanotechnology, many profound changes to human capabilities and experiences are likely to follow. (Radical new technological development always brings profound changes.) Research scientists and engineers are relied on to find new knowledge and understandings, to be creative and imaginative, and then to share those capacities with the larger society, in the form of ingenuity and application. They are asked to refine and develop the human capacity to manipulate and respond to the material universe. They are also asked to bring relief to some of the discomforts and sufferings of human life, and to provide humanity with increasing levels of material comfort. I would posit that ethics problems with the development and appropriation of nanotechnology will come, not so much because of lacking of moral commitment on the part of research scientists and engineers, but more so from the voracious consumption, childlike faith, and unexamined acceptance with which we, in technological societies, imbibe, consume, and even devour the idealism of self-improvement through technological development.

In large measure, technological society depends on scientists and engineers to anticipate, predict, and even buffer change. For example, communities expect warning of hurricanes, and further, not to have to suffer extreme losses or even inconvenience as a result. We are threatened by the loss of electricity and hold fast to the conveniences brought to life by refrigeration, computers, televisions, telephones, heat and air conditioning, lighting, transportation access, and so forth. Emotional stability is challenged by radical changes to material conditions. Engineers and scientists are expected to protect these states of mind as well. When, in fact, nature's wrath overwhelms our technological capabilities, individual members of the society quickly blame those who we expect to protect us—scientists and engineers. Scientists and engineers are integral to the creation of meaning. Changes to the senses of self—the meanings ascribed to human relationships to technology and one another, to mythologies and beliefs—are one predictable result of the development of nanotechnology.

Research scientists and engineers work directly with matter, and are particularly aware of its elusive and changing nature. They believe it behaves according to fixed laws, but they now have learned that behaviors exhibited at macro- and microscales do not necessarily occur at the nanoscale. Many researchers are awe inspired and deeply curious by nature, while also thinking of it in terms of a force with which to be reckoned (see chap. 5, this volume). With that awareness, they position themselves (as agreed on by society) at the "frontline" of human engagement, to "battle" with nature. As such, researchers most directly contribute a significant yet often unspoken, unacknowledged role in the searches for meaning through enhancement of fundamental, rational understandings of the world and humanity's place within it. They contribute not just to the material world that human beings experience, but to the symbolic world of change that humans collectively and fundamentally fear.

Perhaps that fear is one reason why most individuals in technologically dependent societies generally fail to pursue a conscious and conscientious relationship with the technologies developed and consumed. We tend to ascribe to the false notion that technology evolves independent of us. That assumption breeds unwise and irresponsible consumption of new technology. Our failure encourages the dominance of commodity-based scientific explorations, as well as passive, complicit uses of new technologies. As individuals, we become eager consumers, unconsciously driven by, changed by, and dependent on that technology to live. We participate in the collective perception that rapid, new technological development is both essential and inevitable in the evolution of our species. Meanwhile, we leave to the research scientists and engineers, the evolution of our material lives.

PSEUDOVALUES DISGUISING TECHNOLOGY'S BLACK BOXES

The public rhetoric about nanotechnology is profuse with pseudovalues, such as universal access to good health care, when in fact other deeper and more obfuscated values may be at the core of the nanotechnology quest. The rhetorical effect can be the numbing of consciousness. The undertaking of nanotechnology research and development comes to be construed as another simple objective new reality, rather than a socially constructed, culturally negotiated process. As long as the dominant understanding remains obtuse in this manner, nanotechnology development can and will evolve in quiet and unbridled ways, behind what Marcuse (1964) called the technological veil, and what Latour (1987) referred to as the black box of technological development:

> When scholars open the black box of technological innovation, they find social, cultural and political choices through and through. If one looks closely enough, the creation of hardware, software and large-scale technical systems is never simply a matter of invention and application, but of complex negotiations and sometimes fierce conflicts among competing groups. Choices that affect the distribution of wealth and power in society are intricately woven into the very substance of technical design, right down to the last pipe fitting, circuit breaker and computer chip. (Winner, 1997, p. B06)

If, on the one hand, nanoscience is "pure," which is to say, pursued solely and entirely for the sake of new knowledge and understanding, then the existence of the symbolic black box may serve some important functions, such as creating a safe, socially and politically unencumbered space for scientists to do their science. There already exists a broad and well-established social agreement to give science freedom from interference and scrutiny by parties outside of its disciplines. Only when humans and animal subjects or hazardous wastes or materials are involved does society require of its scientists a formal process of external review and regulation. Otherwise, scientific inquiry is sanctioned to proceed on peer review alone. Another valuable function of a black box for nanotechnology development may be to provide needed isolation for intensive and undisturbed focus in the pursuit of scientific understanding.

One hint of nanotechnology's placement inside a cell (in a "black box" or behind the veil) is the prevalent claim made that because nanoscale science comes from existing science, there is really nothing special or distinctive about it. The fallacy of this claim is that although that may be true, there is great significance in the increasing human ability to precisely manipulate matter and in social and cultural changes that may be the result of that ability. Nanoscale science and engineering represent an exponential increase in our ability to control matter. This, in turn, points toward our increasingly powerful ability to alter the fundamental constitution of human material experience. And yet, the potential qualitative changes to come as a result of nanotechnology are, at this stage, quite difficult to conceive and anticipate. To categorize it as an ordinary scientific "revolution" indirectly asserts that it warrants no special moral consideration or preemptive study.

Another indication of its opaqueness is the language barrier. Whether or not intentional, these serve to keep the general public at bay in the development of nanotechnology. Until there are more efforts to explain in layperson's terms what scientists and engineers are doing in their labs, nanoscale research and development will stay sealed from public understanding. It is interesting that many national laboratories in the U.S. currently are doing nanoscale research are in fact inaccessible to the public, not simply in terms of the expert technical language used in them, but literally so. It is exceedingly difficult to gain entry into some national laboratories without special provisions being made. These spaces are physically and intellectually inaccessible and, therefore, constitute another factor of assuring that nanotechnology's development will be off limits. Another indication that nanotechnology is evolving and developing inside a relatively impermeable cell of knowledge creation is its exponential speed, irrespective of any ethical, legal, health and environmental, economic, or social implications to be considered. Research in these areas has not yet kept pace with the speed of laboratory experimentation with nanoscale particles, or with the progress of development (Mnyusiwalla, Daar, & Singer, 2003). Another reason for the public access problem may be the mysterious nature of nanoscale phenomena, and their complete invisibility from the unaided human eye. Also daunting is the sheer difficulty of trying to anticipate where this is all leading coupled with the common belief that there is no possible accuracy in predicting how nanotechnology may materialize. Another factor may be the prevalent belief that there is no apparent way to control its development. The seemingly overwhelming task of addressing those challenges and the societal disinclination to do so are among the factors that keep the development of nanotechnology cell walls impermeable from public intrusion.

There are multiple and definitive values asserted by researchers in these conversations about nanotechnology. Those include values of material prosperity, physical health, and enhancements to the body; freedom of choice, access to basic consumer goods, and material pleasures; speed in the transmission of information; open access to information; military power; and safety from harm. One wonders whether the values inherent in those claims are authentic to the human searches for meaning, or if they are masking desires for a meaningful life. Frankel (1992) warned that the human will to meaning is sometimes met with existential frustration, which entails problems of existence itself, its meaning, and the searches for meaning within personal existence. He purported that it is not the meaningless of life that man must endure, but rather his incapacity to grasp its unconditional meaningfulness in rational terms. In Frankel's language, the only real transitory aspects of life are the potentialities. And, therefore, everything hinges on our realizing the essentially transitory possibilities.

Frankel's research led him to believe that striving to find meaning in one's life is the primary motivational force of human life. And thus, the will to make meaning invokes an individual process, which focuses on the future and on the meanings to be fulfilled by the individual in that future. Frankel's thesis is interesting with regard to the question of whether and why nanotechnology is developing inside of a black box. The nanoscale science and engineering quest is for acquisition of new knowledge and the development of new technologies. It is also about the creation of new meanings. Frankel recognized the ability of human beings to live and die for the sake of ideals and values, but he also acknowledged that there are times when individual concern for human values is a camouflage for hidden inner conflicts. In these cases, pseudovalues need to be unmasked so that one can confront the genuine and authentic desire for a life that is meaningful. On one level, those meanings are made apparent in the explicit, rhetorical expression of values and beliefs, which are accessible through discourse. At another level, those meanings are being made tacitly. Some of the values and beliefs that are most apparent, and presented with the greatest conviction, are not necessarily the only or most deeply rooted in the development of nanotechnology. It may very well be that some of the values and beliefs that appear to be fueling and driving the nanotechnology quest are "pseudovalues," those that serve to mask hidden inner conflicts of the human unconscious. Metaphorically speaking, nanotechnology probably is developing inside of a black box, but maybe not because of any benign neglect or social conspiracy to keep the public at bay. Perhaps it is there simply because it is entangled in an inaccessible web of pseudovalues. One way to remove it from that box then, is to make those values less opaque.

Scientific and technological undertakings that take place in a democracy should never happen without some provisions for conscientious, public involvement. There is no excuse for black box developments, even in the event of a "war" on terrorism that leads to the creation of projects involving technological security such as "homeland security." Even in a case such as this, provision ought to be made for honest, open, conscientious, public discourse. Changes in the law over privacy rights and the like do not preclude the deliberation over ethics that technology appropriation requires. Sheltered inside of an impermeable cell of knowledge creation, and obscured by a blind belief in progress, nanotechnology development could proceed without discourse that engages all stakeholders, including the general public. At risk is the possibility that not until appropriation, when actual nanotechnology products come fully into consumer markets, might it become obvious that nanotechnology has come with profound social and ethical complexities, some of which societies may be ill-equipped to confront. In order for a nanotechnology using-consuming society to stake moral claims over the evolution of nanotechnology, there must be some open exchange of knowledge, intentions, ideals, and beliefs for the honest appraisal, recognition, and negotiation of its core values. Otherwise, "unintended consequences" may be far more unmanageable than we might expect. And, the determination of what futures are possible will be left to the influence of an elite and powerful few.

NATHAN

ROSALYN: *First, is there anything about nanoscience and its possibilities that make it a distinct inquiry and that would beg us to question more carefully where we are going with it?*

NATHAN: *The answer is: "I don't think so."*

ROSALYN: *Yes, that's what I keep hearing.*

NATHAN: *There are people who for whatever reason, seem to think that nano is qualitatively different from chemistry. I don't agree with that. Nano is chemistry. Some people would say: "No, no, its physics." But the fact of the matter is that I don't even see it as a significant departure from the domain of chemistry, except that, at least on the scale where I operate, its easier. Real chemists, real live chemists, who make new compounds, they do hard things. They have to figure out how to get two atoms to come together one way or another. They are actually involving what are, on my scale, large energies. They make certain kinds of bonds, to give new properties to molecules. I can't make any new properties in the molecules I use. I use old molecules and put them together in new ways. For example, I could take a piece of steel and I could make a chair like the metal-framed one I am sitting in, or I could make a key out of it. It's the same molecular structure in both types of steel. The only difference is when the chair is in this fashion, it makes something slightly rocky and when I make a key its reasonably rigid on its scale. They aren't really different materials, but they are put to different uses. A more classic example is a brick. I could take a brick to make a house, I could take a brick to make a laboratory, I could make many bricks, and I could take a brick to make a pyramid, or a tomb, or whatever for a pharaoh. They are all bricks, although pharaoh bricks are bigger bricks. It's the same concept. The utility that you are getting out of these things is based on the way in which you are organizing them, but you are not changing the properties of the material itself. Maybe the key is a better example of that. That's the scale where we operate. The material we work with obviously is DNA and we are making devices or objects or periodic arrays. Now, we're even starting to make arrangements that will eventually lead to what I call nanorobotics, which isn't little men running around, but more structurally variable and controllable states of the physical system. Its all just chemistry; its all just playing around with things on a nanometer scale. I don't know if we talked about this before, but the nanometer scale is the structural scale on which biology also works. To some extent many of us in nano started off looking at biological structures in one way or another, often from the investigative/analytical point of view, and certainly I was inspired by such things to do that [go into what is now called nanotechnology]. The only potentially qualitative difference is that if we are actually able to make things on that scale, we can interact with molecular biology. Ultimately we will, although not easily and not soon I think. In the same way that millimeter scale technology now works with larger parts of biology, you can get angioplasty because we have the materials and we have the ability to make things small, flexible, and blunt enough to actually run something up into your heart. There are other kinds of microsurgical things that we can do. Its clearly a very effective way of solving the plumbing problems associated with cardiovascular diseases. The molecular problems of some other diseases may ultimately be amenable to a smaller scale of operation as well. But again, its not a departure, its just that once upon a time they treated people with cups and trepanning and bleeding, and now we are a little more sophisticated, just a little, still not terribly.*

ROSALYN: *So, why is nanotechnology being described as the next industrial revolution?*

NATHAN: *I don't know. The industries that are involved in nano things right now, the companies out there actually making nano products, insofar as I can tell, make paint. I mean, nano coatings.*

ROSALYN: *Right.*

NATHAN:	*Coatings with nanosized particles are more effective than coating with micron size particles.*
ROSALYN:	*Yes, in fabrics. I think some are working on fabrics.*
NATHAN:	*I don't know about that.*
ROSALYN:	*One of the flyers from one of those Nano Inc. things you talked about says that Gap, or someone, has a product that is made with nanofibers. We've done microfibers and I think we are now doing nanofibers.*
NATHAN:	*My guess is they are calling microfibers nanofibers. I don't really know. The only thing that I know where there are different physical properties on the nanoscale isn't from my end, from the chemistry end of building up to the nanoscale. Its from the other end, of working down from the top. I heard a lecture not long ago where somebody was talking about magnetic nanoparticles. What he pointed out was that if you had a bulk material that's magnetic—I don't know if this is true because I was nodding during the talk—but if you have a bulk material that is magnetic, you actually have a sort of majority rule of little domain. Where microdomains in there are mostly pointing this way if you magnify your magnet, but some are still pointing that way because you didn't switch them all around in your materials. By contrast, a nanoparticle is so small that its of the size of a domain. Every atom in there will polarize in the same direction and that's qualitatively different from the majority scenario, and they are trying to take advantage of those properties. The part where I fell asleep was when they started telling me about how to use them, take advantage of them. I don't remember the answer.*
ROSALYN:	*Other researchers are talking about how, at that scale, what we observe are different kinds of properties.*
NATHAN:	*Right, that's the advantage.*
ROSALYN:	*They are unfamiliar and we haven't really …*
NATHAN:	*I see. But you are talking about examples where people are saying let's compare the nanoscale to the micro.*
ROSALYN:	*Yes, I am. That's what I am hearing.*
NATHAN:	*Or the macroscale. I started off on the other end.*
ROSALYN:	*I know, years ago. You have been there right?*
NATHAN:	*No, no, I was never there. I have been on the chemical end. I am building my way up to the nanoscale.*
ROSALYN:	*So there is no novelty in the nanoscale for you?*
NATHAN:	*No, for me there is nothing novel. It's still very hard to do what we want to do. Again, the energies are small, there are reversible reactions, and there are all sorts of things that are not so easy to do on the macroscale, on the chemical scale. We don't go down much. The chemical scale is not much smaller than the so-called nanoscale. It's*

about one order of magnitude. The word I use is, Ångstrom, *you are familiar?*

ROSALYN: *Right.*

NATHAN: *Alright, so that is actually one tenth of a nanometer, 10-10 meters ... So a nanometer is 10Å. If somebody asks you how long is a chemical bond, you are not very far off if you say an Ångstrom and a half.*

ROSALYN: *OK.*

NATHAN: *Even though I think about nanoscale things, this model isn't such a bad example of how I think about DNA. But the point is: What have we actually got here? The way that I grew up thinking of DNA was like this. Or like this. You see here is the bond, and this is what a base pair looks like when I think of a base pair. So the issue is on what scale is your mental universe ruled? How far apart are the gratings before you say two things are in the same place? For me the answer is about a tenth or a quarter of an Ångstrom, 10–25 Pico meters. If two things are closer than 25 Pico meters, or maybe 10 Pico meters, I don't care, I will never care. When I think about these things, plus or minus an Ångstrom or two is actually good enough. Because this [the diameter of DNA] is 20 Ångstroms, two nanometers in width. What happens on the chemical scale is you are creating new molecules with new properties. All the drug companies are making new molecules, and they are making new molecules that have different electronic properties. For whatever reason, you put two nitrogens next to one another, and that's different from having a nitrogen separated from another nitrogen by a carbon, and so forth and so on. When we do build our way up to nanoscale, we ignore all that and its hard to make covalent bonds like that. What we do is we ignore all of that, we just say, "hey its all DNA." Its all the same stuff except maybe for the sequence, and we might or might not want to take advantage of those differences. We just slap things together without much regard to what the new chemical properties are going to be. We are interested in new functional features. So, that's how we work. For me the nanoscale was just an easier place to work than the Ångstrom scale. When I started off 20 years ago and I said I am doing nanoscale chemistry—it took me years to figure out what to call what I do— because I wasn't coming out of any tradition. I really wasn't coming out of a chemistry tradition. I wasn't coming out of a biology tradition really. It's interesting actually. I looked in the literature to find out, am I in any tradition at all here? I spent a lot of time reading the older books. I read D'Arcy Thompson's* On Growth and Form; *I don't know if you are familiar with that.*

ROSALYN: *No.*

NATHAN: *It's a classic. It's a classic from .... I think the first edition was published shortly after World War I. I got it out of the library and read it. You could tell the era because when a footnote was to something in German, it was in German, in Greek and so on, and an educated person knew all of those languages. There was one reference in there to a guy in the 1870s by the name of Harting and he looked to Wöhler, who had started organic chemistry by synthesizing urea, much earlier, maybe in the 1830s, I don't know. He said what he wanted to do was synthetic morphology. He sort of played around with various calcium containing things and got some shapes out of it and so maybe this is my tradition.*

ROSALYN: *Where you belong?*

NATHAN: *This is the tradition that I came out of. I am a long lost prodigal son. It's funny, about 7 years ago I went to a meeting and there was some guy talking about doing exactly that kind of stuff that Harting had been talking about doing in 1871 or 1872. I actually made a slide of the first page of Harting's paper. I was in a biology department and these guys there didn't know what the hell I was. So, for my tenure seminar I showed a slide of the abstract of this guy Harding just to say: "Hey listen you know, I am kind of a biologist guy here." Anyway, so one of these guys from Canada gets up there and starts talking about making these sort of shapes out of calcium compounds, and I asked how this compares with the stuff that Harting was doing in 1870, and he had never heard of him.*

ROSALYN: *That's great. That's exciting that you have stayed connected to the original field ...*

NATHAN: *If there was a field, it wasn't as much a field as it was an intellectual notion. Let's see if we can do synthetic morphology, that's the term he used.*

ROSALYN: *Morphology.*

NATHAN: *Synthetic morphology.*

ROSALYN: *Morphology. That makes sense. Well, then what is so special about nano?*

NATHAN: *Nothing. It's just the next step.*

ROSALYN: *The next step toward where?*

NATHAN: *The next step in progress.*

ROSALYN: *What's that, where are we going?*

NATHAN: *Chemical progress I would say.*

ROSALYN: *Which means the ability to ...?*

NATHAN: *Chemistry is about the ability to control the properties and structure of matter.*

ROSALYN: *Right.*

NATHAN: *It is presumably never ending.*

ROSALYN: *So refinement of that control.*

NATHAN: *Refinement, yes.*

ROSALYN: *To what end?*

NATHAN: *Well, I think to the end of simply being able to make largely new chemicals. Not so much new chemicals in the sense of new chemical properties, new drugs, that sort of thing, but more new materials. I was just thinking about this, this morning in the shower, I don't know why. We are really extraordinary in the ways in which we make materials. I know what I was thinking about. I noticed that my sock drawer got stuck on a Lycra pair of socks. It said Lycra on it. I said to myself, how did these guys know what to put into Lycra to give it the particular properties of those socks and I realized they didn't know. It's all trial and error. We now know enough that at least with the kind of stuff that we do, we can't predict what the microscopic properties are, but we can predict at least what the microstructure is going to be and that is a step in the right direction toward making the things that we want to make. Again, it's sort of fabrication. The history of our species could be described as the history of the materials that we use. We've been working our way from animal skins to polyester. I am not sure that is a step in the right direction. From whatever it was, bronze to iron, to whatever the Hell we have now, steel I guess. Likewise for other things, to more efficiently use the resources that are available in the planet and maybe someday in the universe in which we live.*

ROSALYN: *I think that's where people like me are saying: "Oh wait, wait, is that a good thing to become more efficient in exploiting the resources of our world?"*

NATHAN: *Well, to become more efficient in exploiting the resources means that you are getting more bangs for the buck, right? Everything derives from solar energy, except maybe a few things that are derived from nuclear energy. Everything comes out of it, either from the biosphere or whatever else you have got here. If you can get something that is more effective with less energy, you exploit the resources more efficiently, right. It would be ideal if we could heat our homes with solar energy. We are not doing that yet. A friend of mine in the nano business is in fact working on nanocrystals and they are working on solar collector paints. That would actually be useful. I don't know how you turn that product into energy.*

ROSALYN: *Through a collection of paints.*

NATHAN: *Yeah paints, so take nanocrystals, paint them on a surface.*

ROSALYN: *And they absorb the energies.*

NATHAN: *They absorb the energies, sunlight. How do you transmit the usefulness? I don't know, but …*

ROSALYN: *It's a beginning.*

NATHAN: *It's a beginning, yeah. So, new materials, better uses of energy and other natural resources .... It's relative to a large extent, but I think it differentiates us from our ancestors of maybe 500 years ago. From 500 years ago, to 5,000 years ago, I don't feel there is much difference.*

ROSALYN: *No, no the Bronze Age was when we really started to be able to work with metals.*

NATHAN: *Yeah.*

ROSALYN: *Are there worries over the implications of nano? Are there risks?*

NATHAN: *Well again, it's a question of which implications of nano?*

ROSALYN: *The social, cultural, ethical implications.*

NATHAN: *I don't see those as being any different from any other endeavor.*

ROSALYN: *You mean any other scientific endeavor?*

NATHAN: *Yeah.*

ROSALYN: *Then you think it's generally the same no matter what we are doing in science, generally speaking?*

NATHAN: *Yes. Again it's one of the differences between us today and us 50 years ago. It's a role in which I am starting to gain consciousness. Vehicles are about the same as they were then. Planes are a little faster. Jets existed but they weren't passenger vehicles then. If I had to say what the major difference between now and then is, I would say the computer.*

ROSALYN: *Right, and then you get to Moore's law, right.*

NATHAN: *Right. You are getting to the ending of Moore's law.*

ROSALYN: *You think so?*

NATHAN: *Oh yes, inevitably. I mean things get faster, smaller, and cheaper. At some point it's got to stop.*

ROSALYN: *I would think so.*

NATHAN: *It can't get smaller. It may continue to get somewhat faster and somewhat cheaper for a while. Although cheaper ... I don't know.*

ROSALYN: *Well isn't that one of the so-called promises of nano, that it continues Moore's law?*

NATHAN: *In a while, Moore's law has taken another step. A chemical bond is an Ångstrom and a half. There is no way you are going to be doing anything smaller than that scale. We can talk about 2020 or we can talk about 2070 or whatever, but pick a safe number like 2200. Forget it. Moore's law will have run out by then, in terms of computers as we think of them. Possibly, quantum computation may change things. And perhaps DNA computation or some other form of molecular computation will change things as well. I am a member of the molecular computation community, but I am a member of it for other reasons.*

ROSALYN: *Interesting people.*

NATHAN: *They are great people, I love working with them. It gives me an excuse to talk all the time with mathematicians and computer scientists. It's a lot of fun. But, I would be surprised if a lot comes out of molecular computation.*

ROSALYN: *OK, so I am just trying to sort out where the concerns are coming from.*

NATHAN: *Well, you have talked to 35 people.*

ROSALYN: *Oh, I am telling you about the concerns that are being expressed outside of the nanoscience and engineering research communities.*

NATHAN: *None of the 35 people are concerned?*

ROSALYN: *No, most of the 35 people are saying: "Well look at anything. Anything that has one intention behind its development can be misused and redirected."*

NATHAN: *Yes, we are all saying that.*

ROSALYN: *That's uniform, right. Are we supposed to put the brakes on research and development?*

NATHAN: *No.*

ROSALYN: *Then my question is certainly not on research, because that's about knowledge and understanding.*

NATHAN: *Um hum.*

ROSALYN: *And that's a good in and of itself. But what about on development?*

NATHAN: *Well, you know it turns out there is a well-known line from Popper, or maybe a not so well-known line from Popper. He said that we talk about science driving technology, but in fact, it's the other way around.*

ROSALYN: *Absolutely.*

NATHAN: *If you don't develop things you are putting the brakes on. You learn through the experiment.*

ROSALYN: *The new microscopes, for example, that allow you to see these materials. You couldn't do without the technology.*

NATHAN: *Without the development, you are right. What we talk about when we talk about development is really about taking something that works under very complicated circumstances, and making something that will work more robustly under less complicated circumstances. The very first computers not only took up a room, but you actually had to walk in the room.*

ROSALYN: *Yes, you did.*

NATHAN: *You know where the term* bug *comes from?*

ROSALYN: *I don't. I know my little car was called a bug, but you mean the computer bug?*

NATHAN: *The computer bug, right.*

ROSALYN: *Where does it come from?*

NATHAN: *The story I heard was that in the very early years, where every bit was the size of an old fashion tube radio, somebody went in one day—at least as I heard the story. It hasn't been verified for me—somebody went in, somebody was at work and they went in there one day and they found a moth in it.*

ROSALYN: *A moth.*

NATHAN: *A moth sitting there in the thing screwing up one of the lines.*

ROSALYN: *OK. I had no idea. It's a good story.*

NATHAN: *Compare that with this thing [laptop computer] that is sitting behind me or behind you that is so robust that I carry it with me. I abuse it, I carry it with me all over the world and it's so much more effective than the earliest computers. The very first computer that I used—I spent 3 years programming—had 32K, for the whole computer.*

ROSALYN: *Wow.*

NATHAN: *The operating system took up about two of those, maybe four of those.*

ROSALYN: *Wow.*

NATHAN: *The thing I did before I came to this university, writing codes and things like that, it all fits into about 20M, which is a very small piece of that. I get files today that are more than 20M.*

ROSALYN: *My earliest memory of computers is through my father using punch cards.*

NATHAN: *Well, I used punch cards. When I moved from graduate school, my whole backseat was punch cards.*

ROSALYN: *Same with his. One day he put them all up on the roof of the car, opened the door, got in the car, and drove away forgetting them. Thousands of punch cards were flying into the city streets and I think it is the first time he cried. Because that was years of work.*

NATHAN: *Yes, yes.*

ROSALYN: *There was no disk. That was it. So, that's a good example of how the development of technology makes us more efficient.*

NATHAN: *Yeah.*

ROSALYN: *It makes us more efficient at collecting and storing information, right.*

NATHAN: *Right. But see, that's really the difference between now and 50 years ago. Furthermore, the first computers were developed to solve differential equations, for the war department of course. But, the first time I bought my first computer was about 20 years ago, maybe 21. I remember, I bought it because I wanted to do some color graphics and it was an old Apple II Plus. I got a special card with it to do the graphics and I got a bunch of programming software for it. In total, I was spending like $6,000. It was a lot of money to spend on a computer, but I had to spend out a grant. There was a word processor program available, so I said, what the Hell, and using it became 90% of what I did with the computer.*

ROSALYN: *Yes, isn't that funny?*

NATHAN: *Yes. That was a II Plus. It wasn't until I got a Macintosh a few years later that I started drawing with the computer. That's all I do, I draw and I write. I don't really do much else. I have so much calculation power in there.*

ROSALYN: *Do you think you are a better scientist for it than you might have been when you didn't have that capacity?*

NATHAN: *I am not a better scientist; I am a more effective scientist.*

ROSALYN: *Meaning?*

NATHAN: *Meaning I can do things that I couldn't otherwise do. Let me just go back to my own field of my scientific childhood, crystallography. The first time I saw a crystallographer was in a movie in 1961, it was an old black and white movie. It wasn't an entertainment movie; it was a movie for our chemistry class. This guy was talking about crystals and their cracks, "You cleave it this way and the thing will suddenly split apart." He was wearing a short sleeve shirt and I noticed that his right arm had a welt under it. I know now where the welt came from, because what you did with crystallography was to add. In the old adding machines, you would go bang, bang, bang, bang, pull, bang, bang, bang, bang, pull, like that, and that had developed this huge muscle on his arm.*

ROSALYN: *Wow.*

NATHAN: *The guy went into crystallography in the generation maybe 1.5 before mine, before computers. The only thing you were doing was basically adding up numbers. When computers came in, it was a whole other bag. Now you could just solve a few problems in logic and you would get your calculation done like that. A whole different kind of person went into the area because you weren't going to be spending your life adding up column after column, after column, after column. You were going to be doing something more interesting. And actually solving the problems rather than overcoming technical hurdles that stood between you and the solution to your problems.*

ROSALYN: *So you can be less of a technician and more of a thinker, because you don't have to spend time.*

NATHAN: *Right.*

ROSALYN: *So, it's a more efficient approach to the science. We would just assume that this is a good, if we are going to qualify it as such. I am sitting here and I am thinking, "OK, huge change in our ability to manage and store information, huge change." Huge implications for what we understand and can do, major changes to the society really, because of the flow of information and access to it. I don't know if that's a relevant question.*

NATHAN: *Yes. I don't know what's good and bad.*

ROSALYN: *Me neither, but, it's change right?*

NATHAN: *It's certainly a change.*

ROSALYN: *Change is inevitable, right?*

NATHAN: *Yes.*

ROSALYN: *Can it be directed, or is it random?*

NATHAN: *Um.*

ROSALYN: *That's where I am.*

NATHAN: *That's an interesting question and I don't know the answer to that. I guess the first question is "can it be directed?" The second question is "should it be?"*

ROSALYN: *Well, maybe. As for nano, there will be changes as a result of what we can do, and as we refine our ability to work with matter. It's inevitable?*

NATHAN: *Right.*

ROSALYN: *There will be bodily changes, other material changes potentially, which bubble out into social cultural changes. There could be changes in the way we perceive the world and one another and all kinds of things happen when we refine our abilities. It's inevitable.*

NATHAN: *Correct.*

ROSALYN: *Can't stop it right? That's what research scientists and engineers are telling me.*

NATHAN: *I don't think I disagree with you, with the others. You can't say change is inevitable, maybe you can, it's hard to say, because not much changed for maybe 5,000 years, and I am not quite sure what catalyzed the changes that have taken place in the last, say, you know three to four hundred years.*

ROSALYN: *Well, there is that theory about paradigm shifts. So, we hit one and then we go boom and then we are static for a while and then we hit another one and we go boom.*

NATHAN: *Well, what was the paradigm shift? Was evolution a paradigm shift?*

ROSALYN: *That's actually where that phrase came from. It's inside of scientific revolutions that we have these shifts to change our perceptions and our abilities.*

NATHAN: *But, what changed in anybody's life because of Copernicus?*

ROSALYN: *That was more political, it changed political relationships. When Galileo placed the sun in the center, right, it changed everything and threatened the status quo!*

NATHAN: *That was actually Copernicus.*

ROSALYN: *Oops.*

NATHAN: *Whose life changed because they suddenly knew that the Earth went around the sun? Even the theologians, I think didn't get really picky about that after a while.*

ROSALYN: *They were really pissed in the beginning.*

NATHAN: *In the beginning, they kind of put up with that. But in Conan Doyle, for instance, as a joke Watson makes some comment to Holmes saying that "Its' as sure as the Earth goes around the sun." Holmes says, "The Earth goes around the sun?" That's fascinating, it doesn't affect my life.*

NATHAN: *I would say the major shift was probably the shift from an age of faith to an age of science. With a shift I should say from scholastism to Baconian science. Namely, instead of debating about what would be the answer to actually see what would be the answer.*

ROSALYN: *Yes and looking in a different place for the answer?*

NATHAN: *Right, just looking. Just looking at all. I became aware that there, that this even was a scientific revolution, somewhere in my life. One of the problems about the education of scientists is that it actually teaches us the history of philosophy of science at a technical level.*

ROSALYN: *That's fascinating.*

NATHAN: *Yes, I know. If you want to know whether a feather and a ball are going to fall at the same rate or a plum and a peach are going to fall at the same rate from the Tower of Pisa, should you actually go through the experiment or should you just talk about it? Apparently, this was a really major thing when Galileo did that. In part, because we asked for answers to our questions through faith. Or even legend. He would say, "Well, they didn't say God has told me that the peach is heavier than the grapefruit or whatever."*

ROSALYN: *Sure.*

NATHAN: *They would say the grapefruit is heavier, therefore it should fall faster. It was logic. It might have been flawed logic, but it was logic.*

ROSALYN: *Well OK, but there continue to be shifts that change our understanding and our perceptions and our worldviews.*

NATHAN: *Of course, that's what some of this is. That's why we keep doing it.*

ROSALYN: *But you are suggesting none of them really matter?*

NATHAN: *I didn't say none of them really matter.*

ROSALYN: *Which ones matter? You said that "the sun is in the center" doesn't matter.*

NATHAN: *No, the fact that the sun is in the center doesn't affect anybody's life. The shift from scholasticism into science was a major change. That affected everybody, ultimately.*

ROSALYN: *Then science, particularly when Descartes decided, it's basically all a machine. That's still the operative paradigm.*

NATHAN: *Who came first, Descartes or Newton?*

ROSALYN: *I don't know.*

NATHAN: *That's an interesting question because I think of Newton as this sort of clockwork universe guy.*

ROSALYN: *I think Descartes decided that the body and the universe were like machines and could be treated as such. That mechanistic view of the world changed science again.*

NATHAN: *It was an important way of thinking about things. I just don't know whether that was Newton or Descartes?*

ROSALYN: *I thought it was Descartes. Maybe it was a combination of both thinkers that mechanized the universe.*

NATHAN: *Or, at least recognizing the person.*

ROSALYN: *Particularly the person. Alright, so we are going to continue to have these shifts and nano is going to accelerate this in some way?*

NATHAN: *I don't know, I have no idea. I don't see nano as a .... I hate to use the term* paradigm *....*

ROSALYN: *How about in terms of the devices that come out of our science and what we are capable of doing? You are back to the nuclear age. The two German scientists who figured out the fusion is a very good example of "hey look what we did." Then, because of the times ....*

NATHAN: *Yeah, they buried it.*

ROSALYN: *But other people picked it up and said "let's get going on this." This is pretty powerful. Then the science got driven by the war.*

NATHAN: *Sure.*

ROSALYN: *Something is going on.*

NATHAN: *Well, I would back off on that.*

ROSALYN: *You wouldn't say science can be driven by an external force?*

NATHAN: *No, no I am just talking about historically. Science was not driven by the war. The development of the nuclear weapon was driven by the war. That's a different matter.*

ROSALYN: *You are right.*

NATHAN: *We didn't know any more about nuclear science between the start and the end of the war, not that I know of.*

ROSALYN: *But we knew what to do with our ability to split the atom.*

NATHAN: *We knew how to do it, but I don't know when it was recognized that there were nuclear forces. I think anybody with half a brain would have had to recognize it at the time of the Rutherford experiment. You know the one I mean.*

ROSALYN: *No.*

NATHAN: *Rutherford discovered the nucleus. Previously people thought the atom was this sort of electron pudding in this cloud of positivity. They realized from Rutherford that all of the protons were in this thing about 10^–5 of an Ångstrom. Something had to be holding them there, and it certainly was not electromagnetic forces. So, from that time on it was known that there were forces stronger than electromagnetic forces. Over the course of the next 20 or so years, it was obvious that there were neutrons in there and you might be able to smash up nuclei with neutrons and so on. Then Szillard came up with the idea of the chain reaction. Once all of that was known, it was just a matter of who was going to build the bomb first.*

ROSALYN: *Is there pretension inside of nanoscience to come into a new level of understanding that we are just waiting for?*

NATHAN:	*Again, nanoscience is chemistry, OK?*
ROSALYN:	*OK.*
NATHAN:	*Chemistry is not there to discover new forces.*
ROSALYN:	*They have already been discovered?*
NATHAN:	*If there are new forces out there that are not known, then eventually we will talk about a fifth force. But that's physics.*
ROSALYN:	*Right.*
NATHAN:	*And it's not nano.*
ROSALYN:	*It's physics.*
NATHAN:	*It may be Zepto rather than …*
ROSALYN:	*Zepto?*
NATHAN:	*Zepto. Nano means 10 to the minus 9, Pico means 10 to the minus 12, Femto is 10 to the minus 15, Atto is 10 to the minus 18, and I think Zepto is 10 to the minus 21.*
ROSALYN:	*Oh gosh. OK.*
NATHAN:	*In chemistry, things that happen on a much smaller scale than 100 picometers, I think don't affect us, as human entities. We are sitting here being bombarded by cosmic rays. They are going through us. We don't absorb them because they are so fast and so whatever. None of the things in us responds to cosmic rays, thank goodness. So, it's a matter of absorption. You only respond to things you can absorb. I remember I had an argument with my botany professor when he claimed that infrared light had more energy than the ultra violet light, or visible light. I said, "Wait a minute. Energy is h? and IR has got a longer wavelength, how can that possibly be true?" He actually used X number of foot-candles, which is some kind of a weird unit of energy from IR that doesn't affect things from visible light. Later, I realized that, he was far too unsophisticated. I understand it has to do with wavelengths to be absorbed. If you don't absorb it ….*
ROSALYN:	*It's irrelevant?*
NATHAN:	*Then it's irrelevant. We will make many new materials, and discover many new techniques. Many new things can happen. Ultimately, it's not inconceivable that we will make a new form of living system. Not based exactly on the traditional system, but a protein-like system that does not evolve. My guess is it won't compete very well with the traditional system, at least for a while.*
ROSALYN:	*There were some scientists and engineers I have talked to, two or three who feel these new life forms will compete with humans.*
NATHAN:	*Well, maybe some day. Now we're getting to the genre of nanofiction.*
ROSALYN:	*Great stuff.*
NATHAN:	*Yes.*
ROSALYN:	*And the newest one, what's his name … the guy who wrote* Jurassic Park *has one out now. He is a scientist.*

NATHAN: *No he isn't. He is an MD.*

ROSALYN: *I thought he was an MIT engineer. He is an MD?*

NATHAN: *Crichton is a Harvard BS and MD from someplace.*

ROSALYN: *He certainly does have some concerns.*

NATHAN: *I haven't read his latest one, probably won't, but that's not a bad example of where all of the hype could come from. You know, you are not going to go from the few things that we know about DNA and molecular biology to* Jurassic Park.

ROSALYN: *Oh, he made it seem so plausible.*

NATHAN: *He made it seem so easy. Nothing is easy. The things that seem easy to use turn out to be hard. That's the nature of doing science. You can never get at* Jurassic Park. *Never is a long time. It's not easy. I tell my students here, "Go do this, it will be easy." Well, we spend years getting these easy things to work. So, getting a hard thing to work is really hard.*

ROSALYN: *OK, so when you do your work, you are interested in basic science. Are you interested in applications?*

NATHAN: *Very much.*

ROSALYN: *Such as?*

NATHAN: *Well again, the organization of biological macromolecular systems in crystalline forms is going to have applications for generating drug leads. The organization of electronic components will be another step in Moore's law. As I envision it, nanorobotics is likely to facilitate the building of nanoelectronics, and the testing of nanoelectronics.*

ROSALYN: *To what end? I don't understand what nanoelectronics would be used for. Would it be to make computers faster and smaller?*

NATHAN: *Faster, smaller computers, and conceivably cheaper. But, at some point, some of that might be incorporable into biological systems.*

ROSALYN: *Detection systems for cancer cells?*

NATHAN: *Well, that sort of thing. I would like to be able to have the encyclopedia or maybe all of the books ever written in a certain cubic centimeter of my brain.*

ROSALYN: *Would you?*

NATHAN: *And tackle it when I need it, without actually having to look it up. I would like to have it or the computing capacity of a Cray.*

ROSALYN: *What's a Cray?*

NATHAN: *Well, I don't know what the super computers are today, that was a super computer a few years ago.*

ROSALYN: *Um hum.*

NATHAN: *Why not? Wouldn't I like to be able to solve partial differential equations in my head, to have them stored up there so I wouldn't have to learn all the math. I'd like to say: "Gee, I really want to know what that shape is going to be if I do this, Oh I see, it's X to the whatever."*

That type of thing, why not? We will get there eventually. I remember joking with one of my friends when I was a post doc. In the laboratory we had a calculator, and we kept it locked down. Whenever any of us needed to do a relatively simple calculation, but more complex, but too simple to waste your time coding a computer for it, you know X square, plus Y square plus Z square that type of thing to get the distance. We would go in and we would use it. I remember telling one of my friends, "Some day your kids are going to have one of those things on their wrists. They won't need to know how to calculate, they will be able to do it." He replied, "Oh, be serious." I have been wearing a watch like that for over 10 years now.

ROSALYN: *Does it calculate?*

NATHAN: *Sure, it's the only calculator that I own actually except for the one in here. I just carry this one with me and it has a little calculator on it. It's very primitive. This doesn't do science or science things.*

ROSALYN: *So you are describing technology as enhancement to human life.*

NATHAN: *Very much, yes.*

ROSALYN: *So does it screw up human life, too?*

NATHAN: *Of course, because we are human, we are going to have problems. I was in China last summer, and I wanted to check my e-mail there and it was a pretty primitive hotel and they didn't have a convenient connection, but across the street was an Internet café and I would go over to the Internet café and do my e-mail. There were maybe 15 other people, mainly adolescent males, in the Internet café. While I was checking my e-mail, every one of these teenagers was playing one of these video games where they were trying to assassinate somebody else who was trying to assassinate them.*

There are other types of investigators like yourself who worry about what people will do with the stuff that we discover, the stuff that we make. Is life significantly better now than it was 500 hundred years ago?

ROSALYN: *I don't know.*

NATHAN: *I don't know either.*

ROSALYN: *There are people who claim that we have a longer life span.*

NATHAN: *Yeah, we do have a longer life span.*

ROSALYN: *All of which is socially determined.*

NATHAN: *It probably helps to be upper middle class in America.*

ROSALYN: *Yep.*

NATHAN: *We wouldn't be having this conversation 40 years ago, because I would have been dead for 10 years then.*

ROSALYN: *Right and I would be of old age. One of my students just wrote a paper on heart disease and talked about some remote community—in Tibet actually—where they don't die of heart disease.*

NATHAN: *They probably don't even eat meat.*

ROSALYN: *It made me wonder whether people actually did have long lives that weren't measured. But anyway, that's neither here nor there. Are we better off? We are living longer. Does that make us better off?*

NATHAN: *I would rather live longer than shorter. And we're also living better.*

ROSALYN: *Because we have fewer diseases?*

NATHAN: *Fewer diseases, more medical advances. I don't know if we have fewer diseases, we just cure the diseases that we have.*

ROSALYN: *Right.*

NATHAN: *I lost three teeth once upon a time. And I have three implants now. I had a bunch of fat clogging my coronary arteries and now I have replaced that with stints. That's what I mean when I say that I wouldn't be here now.*

ROSALYN: *Oh, you really wouldn't be here.*

NATHAN: *I wouldn't be here at this age, if I had been born when my father had been born.*

ROSALYN: *There was nothing we could have done about those arteries then.*

NATHAN: *No, no, I mean my father died at 65.*

ROSALYN: *Heart disease?*

NATHAN: *He had his first heart attack at 46.*

ROSALYN: *Gee.*

NATHAN: *He was the kind of guy who was ... I am not so different from him.*

ROSALYN: *Sum it up in the genes. We are going to say that longevity and health is a social good.*

NATHAN: *I think so. Do you disagree?*

ROSALYN: *No, I want to live long and well. I just don't know what the limits are.*

NATHAN: *Well none of us know what the limits are. But we are working on it.*

ROSALYN: *I guess we will know them when we get there.*

NATHAN: *I guess there is always some way around it. There may be a Moore's law of that too, who knows? It's clear to me that we are programmed to die.*

ROSALYN: *And to age.*

NATHAN: *From an evolutionary standpoint, it's probably good for it to be that way for us.*

ROSALYN: *I think we are using our science as a vehicle to combat those two features of humans.*

NATHAN: *Oh, absolutely.*

ROSALYN: *That's why I am just not comfortable with it.*

NATHAN: *Well, let me look at it. Yes, we are changing nature in sort of minimal ways. Now, given a couple of dental implants as a minor change right.*

ROSALYN: *For now.*

NATHAN: *At some point we may be getting rid of the genes that would wipe me out.*

ROSALYN: *Right.*

NATHAN: *Probably not in my lifetime, but we may do that. If you think about it, how different are we from say dogs or cats, in that they live a fraction of our lifetime, a small fraction of our lifetime. It can't be that different, right? So, it's obvious that there is some programming in there.*

ROSALYN: *If we can figure out how to reprogram it, we are going to.*

NATHAN: *Oh absolutely, no question.*

ROSALYN: *Those are the kinds of things I am curious about.*

NATHAN: *Will that stop us from evolving? Probably. Is that a good idea? I don't know.*

ROSALYN: *Do we have any ability to direct the way we use our scientific knowledge? That's why I am asking the question.*

NATHAN: *Well again, I don't know.*

ROSALYN: *I don't know either.*

NATHAN: *Part of that has to do with the kind of society that we are living in at the time. It has to do with the fact that we actually have the knowledge.*

ROSALYN: *Should we seek to direct the outcomes of science?*

NATHAN: *I don't think you can direct the outcomes of science. You can direct the outcome of technology.*

ROSALYN: *Should we seek to?*

NATHAN: *Much of what we discover is kind of accidental. We don't know what's going to happen. You pursue one field of inquiry and all of a sudden you spread something open and there is a hole.*

ROSALYN: *You don't even know you were looking for it. And there it is.*

NATHAN: *Yes, the whole universe obviously just keeps going.*

ROSALYN: *Amazing, beautiful element of that.*

NATHAN: *Right.*

ROSALYN: *But, should we seek to direct the outcomes of nanotechnology?*

NATHAN: *Well, technology is there to be directed. The question is whose purpose is served. That's a function of the society that we are going to be living in and we should be very careful about the kind of society that we establish.*

ROSALYN: *OK, so I understand that nanoscience is chemistry.*

NATHAN: *Yes. Conceivably we, as you know chemistry and physics, have traditionally stuck their noses into biology.*

ROSALYN: *Right and when they do, it becomes nanotechnology, that's where I sort of want to ask where are we going?*

NATHAN: *So now we are talking about ethical questions right?*

ROSALYN: *Sure, when with me, ultimately all roads lead to my questions about ethics.*

NATHAN: *Right. I wouldn't be surprised. So there we are getting a little bit out of my valley, needless to say. The kind of society that we are living in is perhaps not the optimal one for having tremendous power over the fate of both our species and our environment. Nothing that any scientist does, nothing that any of us ever has done, even Einstein, is something that wouldn't ultimately have been done by somebody else.*

ROSALYN: *Now everybody I am in discussion with says that. Everybody says: "If I don't do it, somebody else is going to do it."*

NATHAN: *I try to do things a little differently.*

ROSALYN: *Oh, sure. To create an expression.*

NATHAN: *Exactly.*

ROSALYN: *Because ultimately you want to do it first. I am also learning that.*

NATHAN: *That's true.*

ROSALYN: *I have heard that 35 times.*

NATHAN: *I don't compete actually. I try to do things that are so crazy that nobody else would think of doing them. You probably haven't heard that from the 35.*

ROSALYN: *I really haven't. I have heard that about you, though.*

NATHAN: *I try to do things that are really crazy because I am only going around once. I really don't want to do something that the guy down the hall could do or the guy down the road.*

ROSALYN: *I truly understand that. Still, what you leave me with is a pointless inquiry because if you don't do it somebody else will.*

NATHAN: *Well, that's probably true. There is a line in this book of quotes from scientists, and one of them says: "The most incredible invention of science is the scientific process itself." The scientific process is something that kind of has a life of itself. It turns out that I had an idea, pursued it, and then saw many ramifications from it. Well, somebody else at the same time had a similar idea; he just didn't pursue it. He had other fish to fry. Whenever I have gotten into an area, there has always been somebody else who wasn't too far away from where I was, or who was doing something close to what I was doing. So, maybe it's a cop out, but it's also an observation that science will keep doing things. I mean, I was around during the recombinant DNA hearings. Are you familiar with those? You would have to have been there that day.*

NATHAN: *OK, so I was actually at the hearing. It was as my boss said: "It's the only game in town tonight." The city council was wondering whether it should allow this to happen at the local university. I said to myself, this is so incredibly pointless, even if they ban it, you know not just in this city but in America, somebody with a few bucks would just go somewhere else.*

ROSALYN: *Then, humanity marches on.*

NATHAN: *Humanity … the enterprise of science marches on.*

ROSALYN: *Marches on inside of the human quest for mastery of the environment.*

NATHAN: *Mastery and control for sure, but also improvement.*

ROSALYN: *Oh, improvement.*

NATHAN: *Yeah, yeah. Again, longer life, better health.*

ROSALYN: *Peace?*

NATHAN: *Scientists aren't going to generate peace.*

ROSALYN: *It's really wonderful to be able to talk to you about these things. It just gives me the biggest thrill and although it seems I am getting nowhere, I am really having a lot of fun. There is something very, very particular I wanted to ask you. I ask scientists and engineers, what are the social and ethical implications of nanoscience?*

ROSALYN: *The federal government is talking about $10 million for a center on the social implications of nanoscience.*

NATHAN: *Are they going to have it?*

ROSALYN: *There is a mandate there. In the political arena they are prioritizing that question and it looks like they are going to fund it.*

NATHAN: *Well, it's good to know, if there is $10 million for a nanoethics center, they're likely to have a couple extra billion in there for nanoscience.*

ROSALYN: *Are there implications, social implications, anything that comes to mind, other than sort of these long-term questions about ...?*

NATHAN: *What I see is only long-term stuff. The only sort of semi-mission-oriented thing that I have from DARPA is "make your 3-D crystals so that we can figure out the structure of some potential protein that our so-called war fighters are going to be exposed to." That's as far as it goes.*

ROSALYN: *That's caring for people who are soldiers.*

NATHAN: *Exactly.*

ROSALYN: *If they are sick, you can try to make them better.*

NATHAN: *There may be other nano things I don't know about that they are interested in doing. I have seen much more of the home security as coming under issues of virology, that kind of nano-yeah viruses are also nanoscale objects.*

NATHAN: *A little while back there was some guy (I forget what he did), who was a big fan of so-called nano. I never met him, but he went to the early Foresight meetings. He would call me up from time to time and he would say, "Well, don't you think nano is going to solve the environmental crisis?" I would say, "I don't know, maybe. How would you do that?" He would answer, "I don't know. I am not a technical person."*

ROSALYN: *See, that's the rhetoric that is in some of the public discourse.*

NATHAN: *I don't know where it is coming from except from Foresight.*

ROSALYN: *There's talk about eliminating pollution.*

NATHAN: *See, I don't ...*

ROSALYN: *Major changes in health care.*

NATHAN: *I don't see where any of that is coming from except insofar as to control the structure of matter may allow you to do this. But, again most of that sounds to me like chemistry or biochemistry.*

ROSALYN: *But then, there is no difference. We always hoped that chemistry and biochemistry would do those things.*

NATHAN: *That's right. But when I was a kid there was a show on TV where the sponsor used to say, "Better things for better living through chemistry." After about the middle 1960s or the late 1960s they said, "Better things for better living." Chemistry became a dirty word. I think nanotech is now becoming the sanitized word for chemistry.*

ROSALYN: *Ah interesting.*

NATHAN; *I could be wrong about that.*

ROSALYN: *OK.*

NATHAN: *I think I do have to go to my seminar.*

ROSALYN: *Thank you. Goodbye.*

PART III

BELIEF

BEN: *I think that the people right now can't imagine living to the age of 150, they can't imagine having their organs replaced, and they can't imagine having no pain. They can't imagine being able to communicate with anybody on the planet, instantly. They can't imagine these things that perhaps in 30 years will be taken for granted. I tell my daughter that there were no personal computers when I grew up. We didn't know what a computer was. She can't imagine being without one.*

ROSALYN: *So, alright. Then, is there any reason to even think about the future if it's so far off, if we can't even imagine it?*

CHAPTER FIVE

New Knowledge and Nature

> *In the old Egyptian days, a well known inscription was carved over the portal of the Temple of Isis: 'I am whatever has been, is or ever will be; and my veil no man hath yet lifted.' Not thus do modern seekers after the truth confront Nature—the word that stands for the baffling mysteries of the Universe. Steadily, unflinchingly, we strive to pierce the inmost heart of Nature, from what she is, to reconstruct what she has been, and to prophesy what she shall be. Veil after veil we have lifted, and her face grows more beautiful, August and wonderful, with every barrier that is withdrawn.*
>
> —Sir William Crooks, speaking before the Royal Society at Bristol, England, 1898

Science is distinguished as "the process, or the group of inter-related processes, through which we have acquired our modern, ever-changing knowledge of the natural world which encompasses inanimate nature, life, human nature, and human society" (Richter, 1972, p. 1). Closely related to science is technology wherein the emphasis moves from discovery to application. Richter distinguished the two by their relation to nature, where science accepts and even seeks a clear sense of nature's control over humans, and technology wrestles to release humans from the limitations placed on them by nature and seeks ultimately to control it. These distinctions fade in nanoscale science and engineering, which together are pursuing the study, control, manipulation, and assembly of multifarious nanoscale components into materials, systems, and devices to serve human interests and needs. At the meeting of the American Physical Society held at Caltech on December 29, 1959 Feynman (1959) spoke about an "expansive yet undiscovered world which exists beyond the reach of our hands and eyes, but that would soon come within reach of science." He said:

> What I want to talk about is the problem of manipulating and controlling things on a small scale .... A biological system can be exceedingly small. Many of the cells are very tiny, but they are very active; they manufacture various substances; they walk around; they wiggle; and they do all kinds of marvelous things—all on a very small scale. Also, they store information. Consider the possibility that we, too, can make a thing very small which does what we want—that we can manufacture an object that maneuvers at that level!

What about this notion of manipulating and controlling the material universe? The idea that nature is to be mastered, managed, and used for the benefit of human life was expounded by Francis Bacon, the recognized father of the modern research institute and founder of the inductive method of scientific inquiry. Merchant (1980) explained that "Bacon fashioned a new ethic sanctioning the exploitation of nature" (p. 170). What were once constraints against searching too deeply into God's secrets about the world, Bacon turned into sanctions to "stretch to their promised bounds" the "narrow limits of man's dominion over the universe" (p. 180). Bacon wrote in his New Atlantis about scientists as fathers and high priests, who had the "power of absolving all human misery through science." Merchant elucidated Bacon's treatise explaining:

> Not only was the manipulation of the environment part of Bacon's program for the improvement of mankind, but the manipulation of organic life to create artificial species of plants and animals was specifically outlined. Bacon transformed the natural magician as "servant of nature" into a manipulation of nature and changed art from the aping of nature into techniques for forcing nature into new forms and controlling reproduction for the sake of production. (p. 182)

Merchant explained further how Bacon's mechanistic utopia meshed completely with mechanical philosophy of the 17th century, a reconstruction of the prevailing cultural awareness that reduced nature to passive, inert atomic particles. His new ideology shifted the dominant paradigm away from sanction against tampering with nature to an ideology that sanctioned and even encouraged the control and dissection of nature through experiment. Centuries later, nanotechnology researchers such as Carroll embrace the ideology of mastery over all of matter, living and inert:

CARROLL: *It's useful to imagine being able to engineer things on the molecular and cellular scale. That presents all sorts of exciting possibilities in terms of being able to do research on kinds of disease mechanisms and disease progression. But I think the idea that we are ever going to have individualized little nanobots or something like that running around and doing things semi-autonomously in living humans, is a long way off. I wouldn't want to be on the FDA panel that passed judgment on products coming onto the marketplace and having inanimate objects making decisions inside of your body, other than at the level of "your insulin level is low, please add insulin." For things that are fairly automatically easy to check (simple kinds of replication of mechanical functions), I think will be a huge market for diabetes, cancer, and various chronic disease management. There are some real opportunities there to take diseases that formerly were fatal and put them into the category of manageable. I think that understanding things on the nanoscale will help that process. So, those are the things that I look toward—gaining insight and understanding and increasing the kind of toolkit for being able to do relatively simple tasks in an easier and more cost-effective basis.*

ROSALYN: *That's practical.*

CARROLL: *Well, I tend to be a pragmatic person. I tend to try to focus in on what works.*

ROSALYN: *So do you think our ability to nourish good cells and demolish bad cancerous cells in a practical way is within our reach?*

CARROLL: *It's being done today. Things that are being done today in a haphazard and improving fashion will be able to be done in the next 10 to 20 years in a much more targeted and efficient way. I am sure that we will do some new things along the way. So, for every thousand ideas that are being talked about, probably 35 or 40 of them will be realized—breakthrough ideas. It is just hard to know which 35 or 40 will make it, unless you work on the full thousand.*

Ethan is similarly inspired.

ETHAN: *We want to understand the dynamic structure. And this is just the beginning, because we are not really going to get the whole story from individual molecules, and how each one of them operates separately. Most of the central functions of the cell, of the leading cell, are performed by highly organized self-assembled molecule machines. Those machines are actually aggregates of complexes of several such proteins that come together to perform a particular task. There is self assembly and there are dynamic interactions between the different components so that they can build those machines. This is the Xerox machine of the cell .... We don't know what the machine is like. We know the compo-*

nents, and we have some models. We make this very artistic picture but not only do we not have a dynamic structure of individual molecules, we also don't know how different the molecules who come together to build the machine are.

We are actually very much interested in transcription and in general the whole process from signals transduction, to the signaling, to the commands, to the synthesis of the message, to the translation of the message to the protein and so on.

NEW KNOWLEDGE, MYTHS AND MANIPULATIONS

Myth relates to narration about events that happened at the beginning of time and establishes forms of action and thought by which humans can understand themselves in the world. The biblical book of Genesis has been referred to for centuries in multifarious considerations of humans and their relationships to nature, perhaps because it is such a powerfully effective, widely perpetuated myth that appeals to some very deep human longings. Whether or not Genesis might reasonably be interpreted literally as an accurate account of human origins is not under consideration here. Rather, its mythical properties are what are relevant to this inquiry of how certain aspects of nanotechnology pursuits might be understood. Humans and other mammals reach for and cling to the nurturing breast from the moment of birth, comforted in the reassurance that warmth, protection, and sustenance is available with relative ease. One way to interpret the Genesis Garden of Eden is as representing such a maternal giving of care. As the story goes, humanity's roots are there. It was there, in Eden, that the first humans were formed of the actual Earth. In the garden, all that humans needed for material well-being was readily available: no competition for resources; no laboring over the land; no uncertainty as to the whims of nature and its vagaries; no hunting, gathering, or wandering over the rugged land threatened by predators and other dangers to bodily integrity; no dire condition of potentially compromised health; no dropping dead. The story of Eden tells that human material needs were minimal; no clothes were required. What few needs there were—for food and drink, physical beauty, and peace—were provided for through nature. Nature was trusted. Humans could depend on nature for support and nurture.

The Genesis myth speaks about much of what defines Western civilizations' struggles over control of the material world. Once humans were ejected from the garden of material satisfaction and plenty, our condition became horrifyingly vulnerable. Nature became the object of fear, curiosity, trepidation, and manipulation. "We" responded in part with technology, as a means to gain control over that which once provided so completely for us. Humans had been weaned from mother's breast. In response, we gave up the trust of infancy, and became demanding, independence-seeking, needy toddlers. Perhaps humanity has matured into adolescence. As adolescents, much of technology-consuming humanity appears to harbor conflicting feelings of resentment toward the nature that was once caring Earth mother, now seeking independence and freedom from her. For some peoples of the world, nature's power and fickleness are dealt with through negotiations and appeasements. For others, the responses have been toward the control, even mastery over nature while also seeking to understand it and uncover its secret powers. Bacon gave scientists permission, even the moral imperative, to uncover nature's secrets. As they continue to do so, probing deeper and becoming more skillful and knowledgeable, where are scientists in relation to the mythical garden origins of contemporary, Western civilization? In other words, how might nanotechnology development be understood in light of that persistent myth of Western civilization?

As the story further unfolds, Eve and Adam have everything they need in the garden. But Eve's curiosity becomes their doom. The variation of the same story is told centuries later in Faust's bargain. Again, an evil force of temptation leads man to fall to his own curiosity, seeking knowledge to his own detriment.

Speaking in terms of history about the "program of science" from the 17th century on, Toulmin (1962) wrote:

> The political analogy implied in the term "laws" was not entirely idle: the Laws of Nature were regarded as expression of the Almighty's sovereign will and design. The objects of the brute creation had not received the gift of free will, and had no option but to conform to these laws: their 'obedience' was automatic. But for the scientist this was a lucky dispensation, since it meant that he could infer the Divine Laws directly from the behavior of natural things, without having first to ask whether they were obeying or rebelling against their Creator. (p. 168)

Centuries of thought and reflection have pondered the role, responsibility, and purpose of scientists in their pursuit of understanding and controlling the material universe. As such, science has long been associated with the distinctive view of nature as operating within laws, which although largely hidden, can be uncovered. In terms of nanotechnology pursuits, the word "matter" can be understood as that which is permitted by and under the laws of "nature." Some nanoscience / nanotechnology researchers are hopeful in their determination to mimic, manipulate, and control matter with precision. Others, such as Allen, have their doubts about just how far we can go:

ALLEN: *Human nature is about curiosity to explore the limits. We want to understand whatever we can. Theology asks the same questions. I was recently at a meeting where there were a bunch of really smart people from all the different branches of NASA, including microbiologists and astronomers, all kinds of people who worry about how life started, is there any other form of life elsewhere, setting a mission to Mars to explore life, and beyond Mars, how to observe and find new planets of other suns, other stars, and tests for visibility of life on those planets. Those kinds of questions have been asked from the very beginning.*

ROSALYN: *Of course.*

ALLEN: *What is life, how was life started? I believe that we will never be able to answer those questions. Science helps us redefine and rephrase questions. That's the purpose. It's really interesting because NASA has just finished a big project that they call Road Map, or whatever, "New Roadmap for space exploration," and this is what science and human knowledge is all about. Those very basic questions are not answerable. They are redefined. As we acquire more knowledge, we can ask them better, but we are asking the very same questions that Greek philosophers were asking, how life started, what is life, is there life anywhere else? We're not any closer. It's fascinating. Life is really a very fascinating form of creation because it doesn't leave too many tracks behind.*

ROSALYN: *What do you mean by "it doesn't leave too many tracks behind"?*

ALLEN: *We can find some fossils dating back to billions of years ago but you don't really know. We have some models, we have some speculation and we'll never know, and it really hurts. If anybody goes into the scientific disciplines, they are people just working on how life started and when somebody goes into this field there is a vague notion that you will be able to explain this thesis and some of them attempt to try to create life. OK so they managed to make some amino acids in a test tube, which is frightening, but they haven't created the cell, they haven't created cell replication, they haven't created anything.*

ROSALYN: *We're trying.*

ALLEN: *We'll never get there.*

ROSALYN: *Because?*

ALLEN: *There's maybe something more than ...*

ROSALYN: *Than us?*
ALLEN: *Than us.*

SCIENTIFIC UNDERSTANDINGS OF NATURE

Beliefs that were once held as immutable in the early sciences receded in the wake of the emergence of physiology. Toulmin (1962) explained:

> Physiology could now undo the damage to our world—picture earlier wrought by the dynamics and astronomy. The discovery that all living individuals were composed of cells—units so contrary in all their properties to the lifeless chaotic atoms of Democritus—could restore our deep faith that Nature was not essentially inert and mechanical, but had about her something creative, something fruitful. (pp. 335, 336)

Nevertheless, society continues to be embroiled in a cultural, epistemological quagmire over what is nature, what we believe about her and what she means to our sensing, perceiving selves. It may be a matter of real importance that awareness of human interdependence with and connection to nature is incorporated into the nanotechnology quest. In his foreword to *Understanding Nanotechnology* (Editors, 2004) Roukes affirmed that "Stepping back from the perspective of things molecular to those global—it is obvious that within our biosphere, at each and every moment, this kind of "mass production with atomic specificity" is ongoing with astronomical multiplicity. This is nature's awe inspiring nanotechnology; her machinery is already and always in motion" (p. vii).

Richter (1972) identified nature (encompassing inanimate nature, life, human nature and human society) as the primary concern of scientific inquiry. He wrote, "Scientists in their research are, in effect, asking questions of nature, and they commit themselves in advance to accept whatever answers nature may give, no matter what these answers may be" (p. 6). That relation between nature and science becomes very interesting in the nanotechnology quest. Richter viewed scientists as "seeking to surrender" their freedom of choice as it pertains to their interpretations of the natural world, as derived precisely and completely in their observations of that world under controlled conditions. The following interview excerpt gives credence to Richter's view:

ROSALYN: *Would you say that if we can build it, it's part of an order? And everything that is materially possible is so precisely because of that order?*

ELAINE: *If it's permitted by nature, then it's part of nature and you have to just choose which things you think will make life better and which won't. In a lot of cases you don't start with that kind of a value judgment because it's just a big "don't know."*

ROSALYN: *Right.*

ELAINE: *So the same technology that could be used to enable people to move limbs that have been paralyzed could be used to control someone else's actions remotely. The former we presume is good; the latter we presume is probably bad, although you could imagine cases where it might be good.*

ROSALYN: *Right. So the guiding principle is that if we don't violate the laws of nature, then we are working within the natural order*

ELAINE: *That's right ...*

It seems to be that in nanoscience there may again be a surrendering to the laws of nature, as they are discovered and learned. There is also a contriving to use those laws to control that nature. It appears that the "surrender" Richter referred to is lessening as scientists and technologists refine the search, and as the tools they use become more sophisticated. It would appear that nanoscale science is a continuation of the ongoing ancient human quest to uncover secrets of the natural world— to raise the metaphorical layers that obscure the workings of nature from human observation and keep it out of the reach of human hands. Nanotechnology researchers speak in terms of seeking to discover and understand nature's processes and laws:

TATUM: *We had a visitor here a few weeks ago who is director of an institute in Germany. He came here to talk about a paper that he had accepted to the proceedings in the National Academy of Sciences. He is a guy who really doesn't know very much about biology, and was just intrigued by the mechanical studies that have been done on a whole bunch of different hierarchically ordered, stiff and strong materials in nature, from seashell materials to teak, to bone, to certain types of other structures. It turns out that all of these have a brick and mortar structure that looks very much like this, except imagine each one of these is shrunk down tremendously so that the aspect they show is a very high ratio between the extent and the width. So if you look at abalone shell, there are protein molecules linking these individual platelets. If you look at dentin in the abalone shell it's calcium carbonate, in bone it's hydroxyapatite, and the aspect ratios might be very similar across these different structures, although the width might vary say in bone the hydroxyap-*

atite is only 4–10 nanometers thick and in the abalone shell it's actually 100 nanometers thick. So, his mechanics modeling suggests that there has been convergent evolution here across different minerals and across different forms of life, the rigid structures that are important for that organism are made the way they are because it turns out that even if the platelets have defects in them, at that aspect ratio and that length scale, they are going to exhibit their maximum strength, so it's very interesting, even if they were defect-free, they wouldn't do any better and the reality is probably that there are some defects in them due to the way nature operates. As this crystallization of the calcium carbonate was being directed by proteins, maybe there was an error and one of the proteins got caught, and so that's definitely an error in that calcium carbonate crystal. So, I was just astonished because we have ideas about how to layer up the graphite sheets once we get them, into just sort of brick and mortar structure and now we know from this model, what aspect ratio we should be aiming for with the graphite sheets, and that tells us what type of graphite to break up, and what type of graphite we should start with. Do we really need it to be a third of a nanometer thick, or can it be just 20 nanometers? It turns out it can't.

His treatment of all these other structures led us to a realization that in terms of the brick and mortar lay-up of materials where the rigid component is the brick and the mortar is something like a polymer, biopolymer or synthetic polymer, we now have a kind of an underlying mechanics that tells us what we can get away with and make it extremely strong.

The excitement comes when what is learned of nature's processes and laws can be used to manipulate and control matter (nature) itself.

NANOTECHNOLOGY AS ANOTHER RESPONSE TO NATURE

My ultimate concern is not about the quest to reveal the laws of nature per se, but about the belief systems that encourage that pursuit, and from those beliefs, what is done with the new knowledge and abilities gained. I am interested as well in what the nanotechnology quest might mean in light of evolving human consciousness. Knowledge does not assure wisdom. The historical tendency of some elements of humanity has been to apply knowledge in potentially destructive or benignly ignorant ways. There are many complex reasons why various individuals and organizational systems apply knowledge of nature to technological development.

Some of those reasons are connected to an awe of nature, a respect and admiration for it. The creative, instinctual drive to improve on and craft one's own environment comes into play, as do existential and aesthetic pleasures. Personal or culturally construed feelings of fear and powerlessness are also at times entangled in quests to use knowledge about matter for creation of technologies.

Through engineering, knowledge about nature's laws makes possible the creation of new devices and processes. Engineering serves to better the human condition and to solve problems that confront human beings and their societies. In both principle and practice, the human ability to engineer has been of untold significance for the well-being of humanity. For people living in technological worlds, life would no longer be possible without engineering. Everything on which we have grown to depend for our very daily lives—from roads to homes, foods to forms of transportation, modes of communication to educational systems, methods of worship to medicine to the entire economic system on which our societies have been based—have been engineered. Does that dependency on the engineered world represent a lessened dependency on nature? Who do researchers perceive themselves to be relative to the nature they are exploring, manipulating, and seeking to control? How do they understand their roles and responsibilities in relation with it? Do they see themselves as stewards of it, distinctive from it, subject to it, at odds with it, or as nature itself? Most importantly, how do their varied beliefs about nature and matter, especially those that are mythological or metaphorically conceptualized, influence their understanding and conceptualizations of nanotechnology?

These are the kinds of questions I pondered during a visit to the majestic and active volcano Arenal, located in a northern Costa Rican cloud forest. Ever since it first erupted in the 1960s, curious volcanologists and tourists from all over the world have traveled to see this volcano in its active state. There are perpetual cloud formations around the volcano, and extended periods of weeks, even months, when the volcano is obscured from view. Periods are relatively brief and unpredictable when it can be seen in its entirety. This uncontrollable characteristic of the volcano can be a source of great frustration and disappointment to its visitors. There is seismological technology available and in use to enable some detection and prediction of Arenal's activity. But those instruments provide only limited access to the goings on of the volcano itself. I wonder, if there were tools available to peer deep inside of Arenal's active crater, would people use them to do so? I think the answer is yes. Why not? It would be of scientific utility, and a great tourist attraction. In fact, there are two helicopters sitting crashed on the side of Arenal at the time of this writing. They crashed during an expedition to look into the crater. Local community members say that the magnetic forces around the volcano interfere with the reading of plane instruments. Local pilots know this, and generally avoid flying near the volcano. Apparently, at least two were convinced by tourists to do otherwise. The decision was fatal.

If the technology existed to harness the energy produced by Arenal's geologically astounding activity, then might this be engineered? Undoubtedly, some engineers would be motivated to try it. And, ultimately, if through technology humans could control its eruptions, at will, many people would probably consider this an ultimate achievement of science and engineering, because in large measure the control and manipulation of nature is the hallmark of applied science and technology.

What would it mean for that much power to be in human hands? Which hands would have access to it, and to what ends? Why would it be good, and what harm might come of it? Where might the desire to see, manipulate, and control something like an active volcano come from: explicit fears, implicit terror, needs for domination, drives to competition, curiosity, excitement, and novelty? Perhaps, all of these are factors.

The emphasis of technology is on man's control of nature, whereas the emphasis of science is on understanding nature's control of man (Richter 1972). The advent of nanoscale science and engineering has made such distinctions much more difficult. The nanotechnology initiative evolves from the close association of scientific inquiry into how nature works at the nanoscale for manipulation and control of nature at that scale. I use the volcano as an illustration because it represents nature's great majesty, unpredictability, and raw power. When Arenal erupted, it came as a total shock to nearby residents, many of whom were killed as a result of its spewing lava and toxic gases. Towns once thriving are today under cover of miles of dense, hard black lava with young green growth coming up in its crevices. Those people who are alive to tell the story of Arenal's eruption still stand in awe of its smoking active cone and its deep rumbling sounds. Humans can use technology, to some extent, to predict volcanic eruptions, but we have not yet learned how to control them. And available knowledge and technology does not permit us to use the incredible energy of volcanic activity to our advantage. But many nanotechnology proponents and researchers seem to hope and believe that this is exactly what can be done at the atomic level of matter. Getting down to the atoms, mastering and learning to use them with specification, promises ultimate human control over nature, and seems to be one primary thrust of the nanotechnology initiative.

CONTROL AND FEAR IN THE MANIPULATION OF NATURE

I suspect that many researchers in my study would be unhappy with my interpretation of what it may mean that nanoscale research and development are fundamentally about the control and manipulation of nature. Although they use that language themselves, and may agree in principle, I think most believe that this is a good thing. Most researchers in my study have described their work as being about curiosity, the good work of bringing benefit to human life, and improvement to human well-being through the mastery I have spoken of. They generally see that matter (nature) abides by very specific laws. Most of these researchers further defend that any science or technology done within those laws is not only moral, but that there is a moral imperative to learn those laws and to use the knowledge acquired for the improvement of humankind. I do agree with this supposition. If there is any tone of criticism in my reflections, then it is about the extent to which we pursue those understandings and controls for other kinds of unrecognized reasons.

Let's go for a moment back to Costa Rica to illustrate this point in a different way. While visiting there, I was in a car riding down a dirt road in the same cloud forest region of the Arenal volcano on nationally protected conservation lands. A very poisonous snake, the Fer-de-lance, was slowly sliding into the road. My husband stopped the car and our family watched as this very beautiful, potentially deadly creature stared at us with seemingly equal curiosity. We were all thrilled for the opportunity to see such a wonder of nature, while also coming to terms with the absolute fatality of its poisonous bite. We and the snake watched each other for what seemed like about 5 minutes. Then another car came up behind us. We pointed excitedly to the snake so that the person in the car behind could also enjoy seeing it. Unfortunately, a man who had no such eco-tourist wonderment drove that car. He promptly and intentionally drove over the snake and then sped away. We then watched the snake in shock and dismay, as it struggled and then died in the road. A few moments later, we came upon a naturalist tour guide watching Howler monkeys in the nearby trees. We stopped our car and I requested an explanation for what we had just witnessed. Besides being against the law to kill an animal on protected lands, it seemed arbitrary and immoral to me. The response was, "People here kill snakes." I asked, "Why?" The guide tried to console me by explaining that the mother snake of this species produces thousands of offspring each year. Beyond that, there was no reason given.

In fact, these snakes are not generally known to chase people without provocation, or aggressively attempt to poison people. This one was in the wilds of national forest preserve, not in or near someone's home or school or business. Killing one or even thousands will not make them go away or be less poisonous. Without them, the eco-balance would be disrupted in ways that would not make life better—but on the contrary—much less pleasant for humans. Very few cases of snake bites occur each year in the area. The only fathomable explanation for killing the snakes was the one not given: Many people are terror stricken. Some have an emotional reaction to snakes that precludes all reason. See a snake. Kill a snake. This is true all over the world. It has been true for centuries. Remember Genesis. The story presents the snake as all that is evil to human life; it is the snake, which is held responsible for human's rejection from Eden. The nurturing Earth that was Mother Nature herself turned on humanity, ejected humanity from the breast, tricked humanity with one of her own creatures. That ancient myth penetrates the collective unconscious of the human psyche and its lingering influence can be witnessed today in so many human encounters with the wild natural world. Had the man who killed the snake been willing to reflect honestly about his own fears, motivations, and beliefs before his action, then the snake might be alive today.

What would happen if nanotechnology researchers all over the world explored such tacit levels of belief with one another and in the domain of public discourse, using dialogue to bring forth beliefs and motivations, which may be engaging the researchers in their work? Engaging dialogue such as this (at the level of Third Dimension Nanoethics), offers an opening toward deeper understandings of why nanoscience is being pursued, and to what possible ends. I do not mean to suggest that the intention of nanoscience research is toward the destruction of nature. I mean to say that the snake is an example of how mythical imagery can represent deeply and widely held beliefs and fears about nature, which if identified, discussed, and understood may become less threatening. Without the perceived threat, perhaps creative, life-affirming possibilities can emerge more readily from nanotechnology development.

During one of our conversations a researcher commented:

> I mean, knowing what we know now, knowing the fact that some microbes would become resistant to antibiotics, would we have chosen not to develop antibiotics? I don't think so. However, it worries us that if we breed a supergerm that is resistant to all known antibiotics, we better have a different approach to be able to combat that. That's nature: struggle and progress.

Nature is seen, most typically, as a source of continual struggle, against which we must use the combat of scientific knowledge. Nanotechnology enables the development of more powerful weapons against the opponent. This, of course, is not the only motivation for nanoscale research understanding. As Clifford explains, human nature has a great deal to do with it:

ROSALYN:	*What are you doing? What's your work?*
CLIFFORD:	*What is my work?*
ROSALYN:	*What meaning does it have for you? You've devoted a large measure of your life to science and engineering.*
CLIFFORD:	*Yes.*
ROSALYN:	*Why?*
CLIFFORD:	*It's curiosity, it's human nature.*
ROSALYN:	*OK.*
CLIFFORD:	*It's the nature of the way we are. That's a nutty question.*
ROSALYN:	*No, it's not. Something about it fits within your own belief system, and you know it.*
CLIFFORD:	*It's like asking, "Why did our ancestors back in Africa insist upon going up to the top of that next hill?"*
ROSALYN:	*OK, alright then, are your pursuits of scientific understanding part of an evolutionary process? This curiosity of human nature, is it leading us in some particular direction?*
CLIFFORD:	*I don't know, I hadn't thought about that. I don't know, maybe it's a survival skill, I haven't thought about it.*
ROSALYN:	*OK.*
CLIFFORD:	*I really haven't, that's a good question. Why are we curious? Could you imagine a species that had no curiosity?*

Again, Feynman's (1959) famous talk included the following statement: "I am not afraid to consider the final question as to whether, ultimately—in the great future—we can arrange the atoms the way we want; the very atoms, all the way down! What would happen if we could arrange the atoms one by one the way we want them ...?"

It interests me to note that Feynman connected the emotion of fear and the notion of finality to the aspiration of arranging the atoms. What is there not to be afraid of, when considering the question of whether or not science is capable of such control? And, what is there that represents finality? Well, death is something final, to fear (Becker, 1973). At least, it is something that humans have always feared and have traditionally conceived of as final. (Interestingly, over 30 years later, Feynman faced his own impending death from cancer with remarkable acceptance and peace of mind.) Could it be that through the conceptualization of our omnipotence, that perhaps death's clutches can be loosened just a bit?

There are multiple elements of Richter's thinking that are relevant to researchers' considerations of nanotechnology development. Richter saw that the control of nature by humans, which scientific activity seeks, is drastically limited by two factors. One, it involves control over man's beliefs about nature itself, and two, it involves control of nature as observed only under certain special conditions. As scientists are asking questions of nature in their research, they commit to accepting whatever answers nature may give, no matter what the answers may be. This presents a dependence of science on the natural environment where the content of scientific knowledge is determined, in principle, by forces or conditions, which are beyond human control. Richter (1972) considered science to be a cultural, cognitive, and developmental process of rapid transition, wherein one knowledge system replaces another:

> The direction of scientific development is similar to that of individual cognitive development. The starting point of scientific development is traditional cultural knowledge. The structure of scientific development is similar to that of the evolutionary process in general and the process of cultural evolution in particular. Science is an extension of cognitive development from the individual to the cultural level, and a developmental outgrowth of traditional cultural knowledge, and a specialized cognitive variant and extension of cultural evolution. (p. 58)

What might killing snakes have to do with nanotechnology? If there are elements of motivation in nanotechnology pursuits, which project and express subconscious responses to the banal fear and trepidation of nature, then they need to be understood. Any hope of directing nanotechnology development in conscientious and life-affirming ways necessarily means coming to terms with those exact elements of tacit belief.

NATURE AS INSPIRATION AND MASTER

> So, from the seventeenth century on, the program of science was dominated by the search for "Laws of Nature." The pattern of Divine Craftsmanship that Newton had revealed in the solar system presumably extended to the design of the whole universe; and the Creator's specification for the cosmos, as perpetuated in the Laws of Nature, must reveal itself to the devout and methodical enquirer. (Toulmin 1962, pp. 167, 168)

According to Marburger, "Nature has blessed us with magnificent examples of nanostructures to stimulate our imaginations."[1] Researchers speak of the great creative imagination in terms of being inspired by nature. Marburger continued, "We have produced some things that do not exist in nature. That nature had no time yet to evolve." That perspective has emerged with consistency in narratives about nanotechnology. For me, one who has little scientific orientation, the notion of creating things that do not exist in nature has been a curious, albeit uncomfortable, concept. For researchers, such as Cecelia, Anne and Patal, creating new, otherwise nonexistent structures is part of the thrill, and goes directly to the creativity of their work at the nanoscale:

CECELIA: *I was sitting around with my mom ... so this idea really came out of nowhere.*

ROSALYN *Yes?*

CECELIA: *It's the epitome of where I get really crazy about using my ideas. So my mom is watching a video of sperm on TV and she wanted to know how they knew where to go.*

ROSALYN: *Hum.*

CECELIA: *That's how I came up with the whole idea of making nanoscale analogs, where they first self-propel and then deliver information.*

ROSALYN: *Going to a precise target?*

CECELIA: *The other idea I came up with is totally wild. I was just looking at DNA, so it's again a biological system. It's really kind of neat that we draw inferences from that.*

* * *

---

[1]As said by the Honorable Dr. John H. Marburger III, White House Director of the Office of Science Technology and Policy, at a nanotechnology meeting in Washington, D.C., March 2004.

ROSALYN: *You used the word* manipulation *but you didn't use the word* control*, although I hear that a lot too, the control of matter. Are those two words different for you?*

ANNE: *Yes, I think they do have a slightly different definition in nanoscience. Manipulation tends to mean—you have heard the phrase "Top down and bottom up?—I think manipulation tends to go along with top down. I think control tends to go more with bottom up. My background is more in dealing and living through Moore's law and seeing things getting smaller and smaller from the top down.*

ROSALYN: *Are they both neutral terms?*

ANNE: *Top down, bottom up?*

ROSALYN: *No, control and manipulation.*

ANNE: *Control and manipulation.*

ROSALYN: *Of matter and energy.*

ANNE: *Um, yes. I think they are neutral in the sense of good/bad, sure. I think control has a little more sex appeal to it. In control, there are new concepts yet to be discovered that will allow unusual control. Top down manipulation is pretty much an extension of where we have been. So I think our abilities keep getting better and better, but manipulation in many senses is a continuation of Moore's law. Whereas control implies to me that there are new concepts yet to be developed and discovered that will allow this control, allow the rearrangement of matter from the bottom up.*

ROSALYN: *Is that exciting to you?*

ANNE: *To me? Yes, very much so.*

ROSALYN: *Why?*

ANNE: *Because I can see the course of the potential for new properties.*

ROSALYN: *New properties?*

ANNE: *New properties of matter.*

⋆ ⋆ ⋆

ROSALYN: *I am asking you what you dream of. You seem very restrained in your imagination, so far.*

PATAL: *Oh, I don't think so.*

ROSALYN: *Tell me please, and then help me understand.*

PATAL: *Alright. So what it is it I want to do?*

ROSALYN: *What is it you want to do? What is your dream?*

PATAL: *Well, I dream of putting these devices to use. You know, like having motors stretch across a wound, and heal it.*

ROSALYN: *To act as some kind of repair mechanism?*

PATAL: *Or, maybe they just provide the tracks for other pieces to come in. You could have this cross link network and direct where it goes. You figure out what to feed it to make it go in that direction. There are plenty of examples in nature where that happens.*

STRIVING FOR KNOWLEDGE FROM AND ABOUT NATURE

The ability to put together structures, such as molecules that humans (not nature) have formed, is a true source of pride and joy for researchers. Interestingly, nature is credited by many of them (Maurice and others, for example), as the original model for the creative inspiration of going beyond it:

ROSALYN: *You use, I think, strong language. You say, "If we learn to do better photosynthesis we can change the world?"*
MAURICE: *Absolutely. I feel that really strongly.*
ROSALYN: *I would really like for you to talk about that.*
MAURICE: *Well, there are four things that it will do. The first thing that it will do is solve, once and for all, the energy problems of the human race. That's a pretty serious issue, right?*
ROSALYN: *Because the energy is always there through the sun?*
MAURICE: *The amount of energy that falls from the sun, on this country, just the United States, in one day, I forget the exact numbers, but it is more than a thousand times the amount of energy that we use.*
ROSALYN: *More than a thousand times.*
MAURICE: *Than the energy that we use.*
ROSALYN: *In what form?*
MAURICE: *Sunlight.*
ROSALYN: *When you say falls to the Earth?*
MAURICE: *Total energy from the sun. What I mean is, photons are coming; they are coming at different wavelengths.*
ROSALYN: *Right.*
MAURICE: *Einstein told us that the energy is H Nu. Nu is the frequency, if you know the frequency, you can compute the energy. If you know the solar intensity all around the country, you can figure out exactly how much solar energy falls on the United States, or on any other place, in one day.*
ROSALYN: *And what you are saying is that we are not using it.*
MAURICE: *We are using a tiny, tiny, tiny bit.*
ROSALYN: *Lying on the beach, sucking it in.*
MAURICE: *Yes, well that gives you a bit of energy. We are using it in several indirect ways.*
ROSALYN: *Passive solar receptors, is that what you are talking about?*
MAURICE: *Well, there is passive solar, and there is some active solar.*

ROSALYN: *Right, a little bit.*

MAURICE: *And there is some, not a lot that is solar powered or light powered. I've got this little solar ...*

ROSALYN: *That's interesting.*

MAURICE: *Almost everybody has a little solar powered calculator that you can walk around with.*

ROSALYN: *Sure, sure.*

MAURICE: *There are things like that.*

ROSALYN: *I see.*

MAURICE: *More importantly, there is a lot of hydro in this country. That's all direct solar. Basically that generates wind, which carries rain, and then the rain falls down. The rain has a high energy, a high gravitational energy, and it comes down to a low gravitational energy. It is all powered by the sun!*

ROSALYN: *That part I don't get. I've been to hydroelectric power plants.*

MAURICE: *How does the water get there? It comes from rainwater. Where does the rainwater come from?*

ROSALYN: *Sure from the evaporation and then the sun sends it back. So you are really attributing most of our energy to sun?*

MAURICE: *All except nuclear and geothermal.*

ROSALYN: *Which is us messing with the nucleus, it doesn't occur with nature, or does it occur?*

MAURICE: *It is not so much us; there is natural nuclear stuff too.*

ROSALYN: *Up there?*

MAURICE: *No, down there.*

ROSALYN: *Subterranean?*

MAURICE: *How did people build bombs, back in the forties?*

ROSALYN; *Well, that's right, good point.*

MAURICE: *Or radium. There are a couple of elements that are naturally radioactive. There are a lot that are artificially radioactive. We know a lot about nuclear energy.*

ROSALYN: *So you are saying that except for nuclear, everything else comes from the sun?*

MAURICE: *And geothermal.*

ROSALYN: *And geothermal. Right.*

MAURICE: *Basically, the reason the Earth is hot inside is nuclear decay. And the reason that geothermal sources exist is nuclear decay. So except for those two, everything else is sun powered. So, the sun is probably providing 95% of the energy in this country now. It is just that the way it is mostly providing energy is through plants and animals that decayed thousands of years ago, and became fossil fuels. We burn those. There is no reason why we would have to do that if we could do efficient, photovoltaic energy genera-*

tion. I know I sound like a preacher, but it's because I really believe in this.

ROSALYN: *I'm stunned actually, well for a few reasons. First, your perspective of the sun as the primary source of life is very interesting.*

MAURICE: *Oh yeah?*

ROSALYN: *It is interesting metaphorically too. But that's a different discussion.*

MAURICE: *I think the first word out of God's mouth in the Old Testament is "Let there be light."*

ROSALYN: *My point exactly. Alright, back to our energy discussion. You are committed to capturing this falling energy and using it in ways that are more sustainable?*

MAURICE: *I would love to see that happen.*

ROSALYN: *Do you think it is possible?*

MAURICE: *Oh yes. It works now. It just doesn't work efficiently enough.*

ROSALYN: *So, it's not affordable, industry doesn't want to invest in what you are saying?*

MAURICE: *Right now there are three problems: The single crystal silicon stuff, which is the standard stuff, is just too expensive. It is very expensive.*

ROSALYN: *The processing?*

MAURICE: *Just making the stuff. I met a guy at one of these Foresight meetings a couple of years ago who is an independent investor. He is a real, believing, ecologist, who made a bunch of money on an investment, and built a summer home in the Sierras in California. His home has got active solar all over and hence, has full power independence. He sells his power back to the California grid. It cost him $200,000.*

ROSALYN: *It's very expensive. We just built a house, we looked into it, but we decided not to go solar. The best thing we could do was geothermal, we did that but it is a small contribution, the solar was outrageously expensive.*

MAURICE: *So how do you get around that? Well I'm a chemist. I think you get around it by using molecules because they are cheap.*

ROSALYN: *OK, under certain conditions I suppose you can get them to do what you want to?*

MAURICE: *That is how it all works. What you want to do is a process called artificial photosynthesis. You want to build molecules that do what nature's molecules do, but more efficiently.*

ROSALYN: *You mean, the ones in here? (Pointing to a plant)*

MAURICE: *Yes, the problem with the ones in there is they have to do so many other things. They have to make it through the winter.*

ROSALYN: *I was going to say, they have to be nourished and hydrated and …*

MAURICE: *… and their primary obligation is not really to provide energy; it's really to provide …*

ROSALYN: *It's to feed.*

MAURICE: *To make the plant work. Right. The notion of energy transduction from the sun using molecules is what we want to do. That is what artificial photosynthesis is about.*

ROSALYN: *Does artificial photosynthesis create molecules for this purpose?*

MAURICE: *Yes, that's one of the things people who call themselves "scientists working on artificial photosynthesis" do. Did you talk to my department chair?*

ROSALYN: *No.*

MAURICE: *Well, he is one of the real heroes in this business. He has been doing artificial photosynthesis for 10 years and he has some new ideas on robust molecules. It's going to happen. The nice part about these things is for some of them you can say this is going to happen. It's going to happen.*

ROSALYN: *Wow.*

MAURICE: *From my point of view, there are two real issues. One is longevity. Say you put this stuff on your roof, and suppose instead of $200,000 it only cost $12,000. You don't want to do it every year.*

ROSALYN: *Right.*

MAURICE: *So it has to be stable long term, that's one problem. The other problem is just efficiency. If you can make a molecule system that is as efficient as the current, expensive solar cells ... Go home, that's it, you're done! So, it's doable, we know it's doable; it's just a matter of doing it.*

ROSALYN: *What do you use to build these molecules?*

MAURICE: *Well the original things that people started to use were quite similar to what is in current use actually. They used norphyrins and phthalocynnines which are very similar to what's in a leaf. Now there are better molecules. Better in the sense that they are easier to make, they are easier to process, and they are probably a little more robust.*

ROSALYN: *Are there molecules that exist elsewhere in nature or are they completely fabricated by humans?*

MAURICE: *They are totally synthetic.*

ROSALYN: *Is somebody looking at what these molecules might do besides generate energy? In other words, once they are created, you've created something new, so then what?*

MAURICE: *The fundamental scientific issue is how does this process, photovoltaic charge transfer work? Why is one molecule different from another?*

ROSALYN: *That's one.*

MAURICE: *Which molecule is more efficient? Which molecule responds to different frequencies? Which molecule is faster? Which molecular is slower? Which molecules undergo what the chemists don't want to see, which is structure reorganization?*

ROSALYN: *Structure of the molecule itself?*

MAURICE: *You don't want it to reorganize. As soon as it reorganizes, it's in trouble.*

ROSALYN: *How do you keep it stable, predictable?*

MAURICE: *The fundamental science is, what is the process?*

ROSALYN: *Alright.*

MAURICE: *And, can I design it? What I'd love to do is design materials to have a function. For instance, the dye in your dress, this was known at the end of the 19th century by Holtmann and other people. Make a molecule that is a dye, make small modifications of it and it'll change its color. The first job I ever had was making molecules that would be different colors. We were trying to design a really nice orangeish color. And you make molecule modifications. You test it and see what color it is. So, that's a science that we know and understand rather well. We would like to be able to do the same thing for photo capture molecules. We'd get to the point where the principles of organization are so well understood that you can say, "OK you want to get blue sunlight. Well use this dye."*

ROSALYN: *So, that's the basic science.*

MAURICE: *We'll understand the whole process. How do you form the electron and the hole? How do you separate them so that they don't recombine in a nonradiative fashion? In this case you would have just heated it up, which doesn't help you.*

ROSALYN: *I still don't understand what happens to the molecule when you are finished with it or when it's finished with its task.*

MAURICE: *Oh, the molecule acts as an agent. It goes back to its ground state and it starts again. That's the whole point. If you can't use the molecule at least a million times, don't bother.*

ROSALYN: *You use it a million times and then it goes back to its …*

MAURICE: *Each time. Here is what happens. Take that one.*

ROSALYN: *OK.*

MAURICE: *This is energy going up. So here is my molecule. Let me just call it DA, Donor Acceptor. It's a molecule, one side we'll call D, the other side we'll call A.*

ROSALYN: *OK.*

MAURICE: *Now for the photo exciting part. We get some energy from the sun and it goes up here to a different state, and we'll call that D star A, star meaning …*

ROSALYN: *It's got energy.*

MAURICE: *Lots of energy.*

ROSALYN: *Right.*

MAURICE: *Now charge transfers, and it goes to a molecule we'll call D+ A–. So the charge has separated, the electron has gone from the D side to the A side. Now this thing is really interesting. It's a dipole. It has a big posi-*

tive charge here, and a big negative charge there. Now it comes back down to the A side. That's all that's involved.

ROSALYN: *That's it?*

MAURICE: *That's it.*

ROSALYN: *Because you have taken the energy from this state? Is that what happens?*

MAURICE: *That's what you want to do. There are two ways that you can do this. It can go by radiative decomposition, which is to say that light comes back out. But, you don't want that.*

ROSALYN: *I would think not, we'd heat up wouldn't we?*

MAURICE: *Well, it would just be a lot of heat waste.*

ROSALYN: *Right.*

MAURICE: *Or, it can do something even worse which is nonradiated. Then it is just heat. That's what happens.*

ROSALYN: *But it is just heat?*

MAURICE: *When it's in the sun for a long time it gets hot.*

ROSALYN: *It absorbs it, and then it is just there.*

MAURICE: *That is what's happening. That doesn't do any good. What you really want is, and this is where it is artificial photosynthesis now rather than just charge transfer, you want to install it in some structure that looks like this. There is the A there and the D there, and the A is negative, the D is positive, this is an electrode and this is an electrode. And what happens is the electron comes off there, the hole goes in there. That comes around to an external circuit which you can use to drive your electric train or make your air conditioner work on.*

ROSALYN: *Right, I understand now.*

MAURICE: *So now the idea is I shine light on it, it goes up there, it goes to that, I bring the electron and hole together to make it work, and what do I wind up with? I wind up with DA back again. And I do it again.*

ROSALYN: *Source of light?*

MAURICE: *Sun.*

ROSALYN: *OK.*

MAURICE: *You could do it with just a bulb though.*

ROSALYN: *But what's the point, because you are trying to generate electricity not use it.*

MAURICE: *Now for the science point of view. That's what you do. You use the bulb. The reason you use the bulb instead of the sun is you're doing it at night.*

ROSALYN: *Right, because you are doing research.*

MAURICE: *But this is the application. The photovoltaic application is to use the sun to make that, and use that in this environment to make current flow. Then you can sell the current, use it, store it, or do whatever you want with it.*

ROSALYN: *Store it, really?*

MAURICE: *That's where the battery comes in.*

ROSALYN: *Sure.*

MAURICE: *Now, suppose I've got this thing and it's a bright sunny day in Tucson, and I'm making a huge amount of electricity, I don't need it. The power grid doesn't want it, my house is cool and there is nothing else for me to do. I'd love to be able to store it.*

ROSALYN: *Store it or send it to somebody else.*

MAURICE: *That would be even better … If you could balance it dynamically and send it off to another state.*

ROSALYN: *Yes, and then you've got distribution of resources*

MAURICE: *But storage is really a good thing anyway. I mean, suppose it is night time. You need to be able to store energy. And storing electricity is difficult. I don't know if you have been to Niagara Falls, the way they store it there is, they store it in water. At night they pump water uphill, using this power. Then, during the day, the water comes down and they generate hydro. It's a way to store it. It's environmentally unpleasant, because you wind up with hideous looking places where you store water.*

ROSALYN: *I know that when I charge this battery if I don't use it within about 12–20 hours it's gone. I don't know how …*

MAURICE: *It couldn't be.*

ROSALYN: *Really?*

MAURICE: *No, I mean they should have a longer shelf life than that.*

ROSALYN: *Maybe I'm exaggerating.*

MAURICE: *20 hours?*

ROSALYN: *Maybe it's a week.*

MAURICE: *It ought to be at least a year.*

ROSALYN: *Alright, so is most of your theoretical work around these questions?*

MAURICE: *This is one of several questions that involve charge transfer.*

ROSALYN: *So it is all about charge transfer?*

MAURICE: *Charge transfer here, charge transfer here, and charge transfer here. Three different kinds of charge transfer. For the batteries, it is not electrons anymore. It's ions moving. That's charge transfer.*

ROSALYN: *You are looking at that, too?*

MAURICE: *DNA uses charge transfer for repair purposes, for broken DNA and things like that. But as I said, the underlying theme of what I do is charge transfer.*

ROSALYN: *OK. If this didn't have so much promise for altering the way we consume and collect energy, would you still be interested?*

MAURICE: *Sure, I am a scientist.*

ROSALYN: *Because of the basic science?*

MAURICE: *Yeah sure. I mean, these are puzzles. The scientific drive here is to understand the mechanisms, the quantum mechanics, how the struc-*

ture relates to the properties. What the chemists call structure—function relationships. That's my puzzle. My puzzle, that's what I enjoy. And I do other stuff. I'm doing some stuff on proteins that has nothing to do with this. And a little bit of stuff on water that has nothing to do with this.

ROSALYN: *On water?*

MAURICE: *Oh, water is basic.*

ROSALYN: *I'm really interested in water. What are you doing on water?*

MAURICE: *Well, the fundamental issue with water is, how does it affect everything else that happens? If you take the water out of a plant, it's gone. Take the water out of us, we're gone. Water has enormous effects on everything it touches. Life is water based. This is the planet that has water.*

ROSALYN: *This is the blue planet. What is it you are looking at?*

MAURICE: *I want to understand how water changes the processes in chemistry.*

* * *

ABDUL: *So we modify the protein and attach it at a particular site, such as an organic DNA molecule, OK? So imagine that these are the two organic DNA molecules, tethered, with some kind of a soft tether, so that they fluctuate. The distance between them is fluctuating and we can collect those green and red photons, and get this kind of signal. From that signal we can get the instantaneous distance, which we need to know. But if instead of looking at one pair we have eight pairs, those fluctuations in the signals are much reduced, and if you go to a really true ensemble then we don't get fluctuations anymore.*

ROSALYN: *So, what's going on?*

ABDUL: *We are summing 128 green signals onto one detector and 128 signals of the red molecules onto the other detector. OK, let me go through this again.*

ROSALYN: *I don't understand why the fluctuation changes so dramatically when the number goes up.*

ABDUL: *This is a critical point. The fluctuations here are presented as changes, the instantaneous changes in distance. But now if instead of one pair of molecules, I have 8, the intensity of each of the green and red is fluctuating. When I take 8 of them, I'm averaging those fluctuations and when I take 128 of them, I don't have fluctuations anymore at all. What I'm measuring in this case is the average distance between red and green over 128 pairs. But when I'm measuring this I'm measuring the instantaneous fluctuations. This is a big fundamental point. By being able to have the sensitivity to isolate and look at one, you can look at the changes, but if you have the whole ensemble you cannot see changes, you just see averages.*

ROSALYN: *Completely different data.*

ABDUL: *You got this one?*

ROSALYN: *Yes, this is exciting.*

ABDUL: *So what this means, for instance, is that we can take polymeric change, (this is basically spaghetti which makes it protein) and let it fold. And you can follow that process. Or, we can take a molecule, and you can think about this as the energy point of the cell. Then you provide it with a protein gradient. It's very much like the motor that drives the hard disc in your computer. It can work as a motor or as a generator. So you see that there are changes, fundamental changes of the structure. The structure is dynamic during the work of that machine. So this you can think about, like "Pacman," or a locomotive riding over its molecular track. In this case it will represent the molecule that you call a nucleus. A nucleus is a molecule that digests DNA. If you have DNA, it basically breaks in part to individual pieces. If you put donor and receptor on two particular sides, on the jaws of that Pacman, you can follow the changes in this structure during work. So, for example, for starting protein folding we take the gene sequence, the amino acid sequence. (There are basic questions, questions that were postulated four decades ago, that we still don't know how to solve.) We begin with the structure of the protein. The protein will fold into a very particular structure that can perform and jump. So instead of doing crystallography, nowadays we have the gene sequence of every gene. Imagine that we had a computer program that could tell what is the structure because we have the sequence. Well, we cannot do this. For 40 years we have tried to do this. We are getting better, but this is such a complicated job. Another question associated with the protein folding field is how the protein folds.*

ROSALYN: *Backup for a minute. What is the significance of structure?*

ABDUL: *Everything is structured. There is hierarchy of structure. We have the body, we have tissue, we have the organs, the cells, and in cells you have all kinds of molecules that come together into all kinds of structures until you get to the individual unit of proteins or DNA and those have structure.*

ROSALYN: *What I'm trying to understand is if that structure is predictable given the work that has to be done or the environment it is in?*

ABDUL: *No it's not.*

ROSALYN: *Is it random?*

ABDUL: *We have models and we ask how it is changing. We have end points of structure. We don't know what happens in between.*

ROSALYN: *So you are interested in that, what happens in between?*

ABDUL: *Correct.*

ROSALYN: *And what is your hypothesis? What's your working assumption?*

ABDUL: *I'll give you an example. For every particular machine, if you know enough, then there is some model. We test to see if that model is correct or not.*

ROSALYN: *Alright, so would that vary with each and every protein?*

ABDUL: *Of course.*

ROSALYN: *And each and every job it is sent out to do would change even that one protein structure?*

ABDUL: *Correct.*

ROSALYN: *So, perhaps there are infinite structures?*

ABDUL: *Well, hopefully we can learn some ground rules. Structural biology was an amazing contribution because we could design drugs to inhibit enzymes or whatever, but this is all based on fixed structures at end points. And sometimes you might need to target the drug to an intermediate structure.*

ROSALYN: *Or a changing structure?*

ABDUL: *In between, or stop the machine on the way for example.*

ROSALYN: *I see.*

ABDUL: *OK?*

ROSALYN: *OK. Go ahead please.*

ABDUL: *So the other question is how proteins fold and this is a case of a very famous paradox named after Leventhal. Forty years ago, he said if you take 100 amino acid proteins and every amino acid is bonded to its neighbor, that bond has two degrees of freedom. So if you start with a "spaghetti," an unstructured spaghetti, and we collapse it into a particular structure and allow every bond to explore all possible angles for a very small protein, just 100 amino acids, it's going to take longer than the age of the universe. So proteins do not fold this way. Obviously, they must have some kind of an energy landscape to encourage them to tumble down, and lie down right at the lowest energy very quickly. So, in other words, we can put this part in a graphical form. Think about the spaghetti lying here on a very flat surface, which we'll never find its potential with all the singularity. It will sit here and nothing happens, but if you have this funnel shape thing it will fold down.*

ROSALYN: *OK, structure is everything.*

* * *

STEWART: *You are holding a piece of material in your hand and you can attach electrodes to it or you can just look at it and watch what happens, and that's what happened a lot in the early stages of hydrogel research or polymer research; you would make a material and you would look at the hydrogell and see what it did.*

ROSALYN: *Are you creating structures that don't exist normally?*

STEWART: *Right, they are synthetic structures.*

* * *

LAWRENCE: *Well, hopefully you would know enough about control.*

ROSALYN: *About control?*

LAWRENCE: *To keep what you do in hand. You know what happens when the body goes out of control. The result of a lot of cells growing out of control is called cancer. Similarly, if we develop the means to say, turn this process on or off, then we have the ability to control it.*

ROSALYN: *That's exciting.*

LAWRENCE: *I think so.*

* * *

GERARDO: *Rust on iron occurs naturally due to oxidation and at some stage it has nanometer dimensions. Not interesting. Dull. Not nanoscience or nanotechnology. If, on the other hand, because of your understanding of the nanoconcepts, single molecule concepts, you could take that piece of iron and in some way modify it to prevent oxidization, to prevent rust—that would change the world.*

ROSALYN: *I've heard nanotechnology described as the precise manipulation and control of matter.*

GERARDO: *Right. Those are just the right words. Precise manipulation and control at the nanoscale. That's science.*

ROSALYN: *So rust appearing first at that monolayer is nothing. That's just the way it works?*

GERARDO: *That's just the way it works. Iron plus oxygen wants to go to iron oxide.*

ROSALYN: *And it goes one layer at a time, one molecule at a time?*

GERARDO: *That's exactly right.*

ROSALYN: *OK. And through knowledge of that process at the nanoscale we might affect that reaction?*

GERARDO: *Changing the corrosion property of the metal is not my field at all. But, what a big impact that would have in the world.*

* * *

ASHLEY: *You can use that as a little bar code. For example, you could put a strand of DNA on the outside. Then you will know what the strand of*

DNA is that you put there, because you know what type of particles you put it on. You can look and say "Well, that's the silver or gold particles, so I know this sequence on that type of particle. Whereas if I had sitting next to it another particle that had gold, gold, silver, silver, I would know that that was another sequence of DNA in that one. Because when I made them separately, they had separate strands of DNA. I could mix them, and still tell who's who."

* * *

JOHN: *We are getting to the point in biology (and I think this is fascinating) that we can think more globally, and can start to think of things as whole organisms. We could start to think even in terms of taking a look at the microbes in their own environment and look at the interplay between the environment and the microbes or look at the interplay between a host and the pathogen. Or, at the interplay between how the immune system is affected by the neurosystem.*

* * *

JARED: *So they finally get down to the point where they have made the smallest thing they could possibly imagine. They have finally made things that are the same size of the things that are the biggest things the chemist can possibly make. And so it has been this sort of union of efforts that has finally connected within the last 10 years or so ... That has made this into an interesting field for a whole bunch of different reasons. It was there all along waiting to be discovered, but now it's being pushed out.*

ROSALYN: *Is that true in general? That things which come to be studied, in science, were there all along waiting to be discovered?*

JARED: *Well sometimes, but you have to have the tools to discover it. Sometimes you don't even have the tools like all of this business with the scanning probe microscope. I was thinking about what life was like before scanning probe microscopes came along. There was a lot of theory. There were these other kinds of microscopes that had nanoscale resolution associated with it. In the 1800s people understood that there were atoms—which everything was built out of atoms and so on. Even the structures and molecules were worked out in a way. It's amazing that they figured out all of this without ever having seen any of these things. Finally, with electron microscopy and later on with scanning probe microscopy, it became possible to see all of these things in detail. In spite of the fact that it was sort of there all along, OK, sometimes it takes the right tool to be able to realize what you are looking at and what's there. But other times it's something where somehow or another all the smart people never figured out that something was a certain way, and so on.*

ROSALYN: *So in a way it's almost as if science is an inquiry about things that already are.*

JARED: *Yes, right. But then you could go further. A lot of our science is figuring out what's there, but another answer to science is making things that you have never made before.*

ROSALYN: *That didn't exist before?*

JARED: *Right. That's another thing that is a significant activity of what nanoscience research goes on here—making new things. We are always, at least the scientists that I have associated with, are kind of doing a little of both: figuring out what's already out there and, also asking what can you make that's new?*

* * *

MAGGIE: *You are getting into shades of grey again. OK, so we are imagining that what might be possible with nanotechnology is to restore emotional or mental functions to someone who has lost them—or something like that?*

ROSALYN: *Yes.*

MAGGIE: *OK, this would be a really neat thing to be able to do. There will be some folks who won't like the idea of neural implants, or who may say that we should embrace our limitations and use those to grow or some such. But I am not a big fan of that mode of thinking. When you start talking about being able to ask, "would I like to have all of the resources of the Library of Congress wirelessly linked to my head?" Maybe. "Do I want someone to be able to hack into my brain enhancement system and put thoughts into my head that I didn't choose to have?" I don't think so. You have to start thinking really hard about whether we can have one kind of enhancement of functionality without having greater risks of abuse? There would have to be a lot of firewalls in there.*

ROSALYN: *So, is it possible that this type of scientific evolution of our abilities is inevitable?*

MAGGIE: *It's life.*

ROSALYN: *Do you think we have any control over where we are going to end up, where we are going in the process of discovery?*

MAGGIE: *As a species?*

ROSALYN: *No, as a species endeavoring in nanotechnology.*

MAGGIE: *I think that in the end we will do what is possible.*

* * *

VINCENT: *Imagine you have this kind of model. You see this is one nanotube and this is another, which fits one shell inside of the other like that. It is usually the case that if those two nanotubes are perfectly chiral*[2] *then sliding one against the other is very smooth and it's a very easy process. The reason might be surprising because, when I slide one of these things against the other, the atoms will bump against each other.*

ROSALYN: *Friction.*

VINCENT: *It turns out that the bumping is actually averaged out very well. The nanotubes have very high symmetry and any place you have two atoms like this, you have two atoms like that somewhere else, they cancel out.*

ROSALYN: *It's a commonality.*

VINCENT: *The commonality comes into it also.*

ROSALYN: *Um hum.*

VINCENT: *This, you can make it like a spiral like this, like this is a right- or left-handed screw. So if you had two layers of a nanotube that both had a screw attached to them, you might think that if you tried to slide one inside of the other, it would be like trying to slide a nut against a screw that would actually left spin when you do that, which would be interesting to have a transducer between when you are in the rotary motion. It makes something move and it would spin. It would be an interesting thing to have. It turned out it doesn't work because although there is a tendency to twist as you slide it, it's incredibly weak because it's incredibly smooth.*

ROSALYN: *Um hum.*

VINCENT: *The gear, the threads strip very easily because everything cancels out.*

ROSALYN: *Because of the part that is symmetry …*

VINCENT: *The symmetry, yes.*

ROSALYN: *Yes.*

VINCENT: *But if you break the symmetry somehow, if you take one of those, and instead of having it be perfectly concentric, you make one of them a tittle bit to one side, not perfectly concentric, neither one falls to the side of the other one, then you break that symmetry and they don't cancel out any more and probably—we are still working on it—the threads get stronger at that point. So, you can do cute things. Not only do you have a screw on that scale, but you have one that you can turn on and off of those threads, so that if you put on some electric field, if the tubes are*

[2]A molecule is chiral (and said to have chirality) if its overall structure and overall three-dimensional configuration is always chiral in accordance with the preceding geometric definition regardless of how the molecule is conformed.

like this, your electric field and the symmetry breaks because they stay rolled into another and then the threads are on and then the actuator has threads and then you turn off the electric field and it pops back to the middle and the thread's off. Very neat! Right now we are working on the theory for that. The dream would be that we would actually get to use it for something.

BACK TO EDEN

A conference entitled National Nanotechnology Initiative: from Vision to Commercialization was held in Washington, D.C., in spring 2004. Noble Laureate Dr. Richard E. Smalley spoke there about an immediate and severe energy crisis as being a primary impetus for the rapid development of nanotechnology. His main message seemed to be that civilization has a serious and urgent problem: worldwide energy consumption needs in the face of rapidly depleting oil sources. Basing his presentation on the research of another scientist, Smalley asserted the claim that the Earth will run out of accessible, crude oil, really, really soon. We are also running out of time, he emphasized, to address the problem. He delineated, with great detail, the urgency quantitatively, using charts and graphs to further reinforce the truth of his claim. He did not speak at all about the problem in terms of meeting needs for human life, per se, but focused his imperative on energy for human prosperity. Smalley reminded the audience that "a lot of people got crazy rich" on oil and he asked, "What will do that now?" He claimed that virtually all of the prosperity experienced by civilization in the past century was due to one factor: oil. Only nanotechnology, he asserted, will provide us with the tools needed to get energy from new sources—mostly from the wind and the sun. But only nanotechnology can bring access to that energy, he said. And to get there, to a place of prosperity and abundance, much more federal money needs to be allocated to chemical and energy research.

Smalley showed a slide about CO^2 emissions causing a rise in atmospheric CO^2 associated with a rise in global temperatures to make the point that we have to be careful with how we go after energy. He remarked, "We'd all prefer it to be just a little warmer for the weekend here in Washington D.C., but the world's climate is complex. If you perturb the climate past a certain point, it may decide to do something different, which could be devastating to civilization." What is the "it" that has the volition to decide to change to the detriment of humanity? So often, research scientists and engineers, and others who are proponents of nanotechnology, speak about it—nature—as if humans are somehow separated from it. The language they chose to use describes it as something other than self; something willful and potentially harmful. I heard the following subtext in Dr. Smalley's talk: Energy brings prosperity. At the current rate of consumption, oil (like mother's milk) is going to soon dry up. Humans will be forced to find new ways to get mother to provide. But we have to take great care—for mother can be fickle, easily angered, and create havoc. We also need very refined, sophisticated tools, no more rough drilling at the ground for the crude flowing sustenance. Now, we've got to probe the very atomic structures of life. As another speaker at the conference said, "At the end of the day, you have to go to the nano world. The engineers have to finesse insertion into that world." Similarly for Smalley, through nanotechnology, nature (the Earth) will give us what we want, which—regarding the argument he presented—is to find a way to "make energy to be as cheap as dirt." Before nanotechnology, the threat of oil depletion was met with rhetoric about conservation efforts—it won't last forever so we must use it wisely, now. Conservation? No more. The view being advocated here is that science and technology will find a way to assure abundance and prosperity; all it takes is money and ingenuity. Nature will finally be revolutionized, her secrets will be revealed, her "gifts" will be for the taking. Perhaps we are not that far from getting back to Eden after all.

At the same meeting in Washington, Dr. Marburger ended his speech with the comment that there are two strains of concern regarding nanotechnology: the sacredness of objects and potential harm. He felt confident that through scientific study, the later could be addressed. My question is, what about the former? What is sacred, and to whom? After listening to the researchers in my study, and to many other scientists and engineers who have spoken in conferences such as the one cited earlier, I have delineated what appear to be consistently held, implicit beliefs and assumptions about nature:

1. If nature is not controlled, it will destroy.
2. Nanotechnology provides a means by which to control nature.
3. Nature is governed by laws, which must be learned and adhered to.
4. As long as what is done is within the laws of nature, there are no moral or physical limits.
5. Nature is like the Holy Grail. It holds the keys to understanding the secrets of life.

6. The power of nature is awesome.
7. That power can be in human hands, through nanotechnology.
8. Nature is slow to evolve.
9. Humans can improve on nature, even surpass it using nanotechnology.
10. Humans can be free from dependency on nature. Eventually, aging, sickness and material limitations can be bypassed with nanotechnology.
11. Humans can and should create structures that do not exist in nature.
12. Nature doesn't provide for human needs, and doesn't always take good care of human beings.
13. New things must be made so that humans can take matters into their own hands.
14. Nature can be exploited to human benefit, but only if it is preserved. For now, humans still need the Earth as the primary resource of nature. (One day, outer space may prove to be resourceful.) Therefore, it must be protected and preserved.
15. Nature is intelligent and masterful as exemplified in such nanoscale phenomena as self-assembly.

Commonly, researchers refer to their experiments or findings as either "not very sexy" or "sexy," suggesting that there is some type of attraction and stimulation involved in the work. The problems they work on become "sweet." They describe certain science as "beautiful." One researcher commented, "The representation of quantum mechanics is exquisitely known. It is the most beautiful of all science." Nature becomes the wild and beautiful lover to be dominated and controlled for her resources.

Looking a bit deeper, there are a number of paradoxes inherent in the way researchers use language to reveal their beliefs about nature, for example:

16. Nature is mysterious but can be penetrated.
17. Nature is controlling but can be controlled.
18. Nature is the creator of life and the destroyer as well.
19. Nature is unpredictable yet law abiding.
20. Nature is flawed and yet perfect.
21. Nature is inspiring and threatening.

22. The individual human self is of nature but nature is other than self.

Following those precepts, the objectives of nanoscale research, as pertaining to beliefs and perceptions about nature, might be denoted in this way:

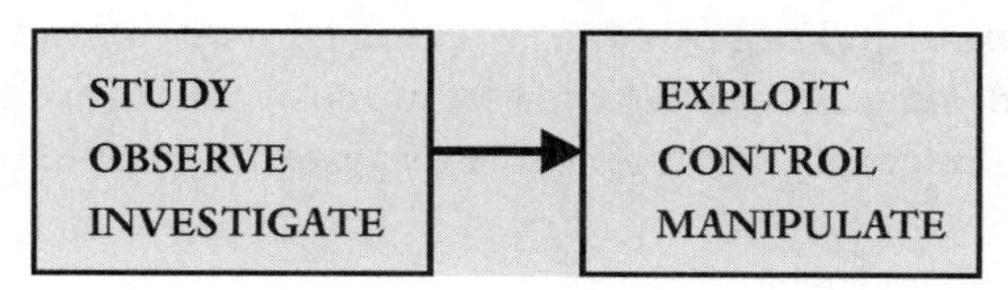

I suggest an alternative:

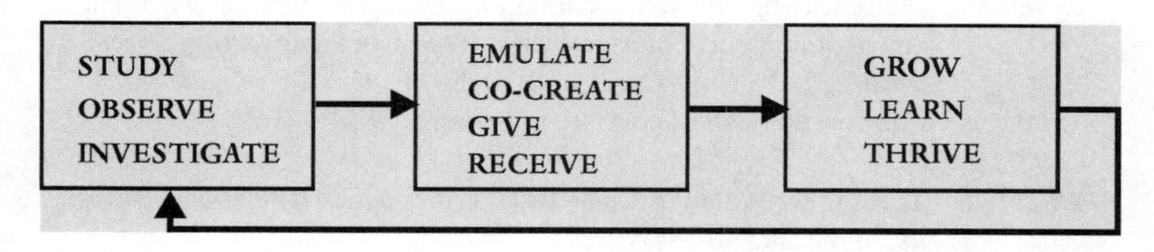

As nature's laws are more fully learned and understood, humans will better understand why and how things happen as they do. If nature is perceived and experienced as a living, life providing, sacred organism on which humans are dependent, then it will more likely be approached with respect and humility. From deep and abiding respect, nanotechnology research and development can approach nature as the source of all material sustenance and nourishment, with gratitude and appreciation for all that it provides. Nanotechnology might then be alternatively seen as a highly sophisticated, enormously powerful, potentially dangerous tool to be used with great care, in concert with the rhythms and evolution of nature itself. Finally, if somehow in the process of probing nature, in researching and developing nanotechnology, humans can return something back to nature in the way of its sustenance, then from that sense of relationship, mutuality, and responsibility, material prosperity will come.

An audience member at the Nanotechnology Initiative Conference asked four panelists a question. He wondered if scientists had considered the giant redwoods or sequoias as an example because, "they are masters of organization at the nanoscale. They respond to seasons and to their environment. They are even self-healing. Are they being exploited as explained?" Having just been to visit Muir Woods, I appreciated his comment. The guide there pointed to one tree and said, "That was probably here during the time that Jesus was walking on the Earth." The gentleman's question was answered in this way: "If you want an even better example, take the brain. It is a more extraordinary example. It reacts in real time, randomly, like evolution. Its circuits are made in real time, exquisitely performing and organizing, working its way by trial and error .... Atom by atom nanotechnology is too much work. Most of us have resigned to self-assembly. Biology gives us a great example of that."

CECELIA

CECELIA: *I have scientific dreams; I am very different from some of the people you are talking to. I have scientific dreams and long lasting career dreams.*

ROSALYN: *Can we start with the scientific dreams?*

CECELIA: *Sure.*

ROSALYN: *Then, if we run out of time, when I come back I will remember to pick up on the long lasting career.*

CECELIA: *Absolutely. I think the long lasting career thing would be similar for most people.*

ROSALYN: *Right.*

CECELIA: *I have been here a year and a half, so my personal career dreams and goals are mostly dreams of really making my name and establishing myself. Part of that is getting tenure on the way.*

ROSALYN: *Sure.*

CECELIA: *So a lot of the focus of my energy, whether I am dreaming or not, is that characteristic of myself, as a young scientist, to make an impact in the chemistry community. When you are young, it's a big deal. It's "how do you do that?" "How do you make yourself known and not be obnoxious about it?"*

ROSALYN: *I hear people say: "blah, blah, blah, do you know her work?" and "she is really young."*

CECELIA: *It's a big deal. It's not a fame thing for me so much. In fact, since the last time I talked to you a lot has happened around here and I think my whole approach as to why we are doing what we are doing is much more centered on the successes of my students and using them as the measuring stick at this point. Just watching the way they are learning and the way they are going. I think 90% will want to pursue careers in academia, which is really rewarding. My undergraduates are phenomenal and a lot of my curriculum has changed in the last little bit.*

ROSALYN: *Were you less student oriented before?*

CECELIA: *It's not that I wasn't student oriented. It's just that in the scientific business, your personal life is very much interwoven with your career. You were here six months ago?*

ROSALYN: *It was somewhere toward the end of the summer. You actually had boxes all around the office, and had just started a couple of months before. So it might have been July, maybe.*

CECELIA: *Since then, I have got three more graduate students that are postdocs and three more undergraduates. The lab has doubled since you were there.*

ROSALYN: *Oh my gosh.*

CECELIA: *My biggest hopes right now are just to do right by them and get them all focusing on where they need to be.*

ROSALYN: *Fair enough, because life changes our perspective and experiences change our perspective.*

CECELIA: *Absolutely.*

ROSALYN: *Do you continue to have scientific dreams?*

CECELIA: *Yes, absolutely. Part of taking care of my students is seeing them do outstanding things. I want to see them have major papers in major journals and just be able to say: "Look at what I am doing." Some of their projects are very high risk, very high impact studies, and they have really wide ranging readership for those papers, whether it's for information storage or curing cancer. There are many things they are working on, and they could all be famous. So my scientific dreams are to make these ideas and products work.*

ROSALYN: *But you are still talking about your students?*

CECELIA: *Absolutely, they go first.*

ROSALYN: *Do you have research that you distinguish and separate from what your students are doing?*

CECELIA: *No, they are the ones doing the work. Some of the stuff they are working on is stuff they came up with. I have an undergraduate who was doing a project and I had no idea she was doing it. She came to me in January and she said: "I have all of this data, what do you think?"*

ROSALYN: *Really?*

CECELIA: *It's awesome. A lot of them are like that.*

ROSALYN: *So does that mean your own science has "reached its height" at the graduate school and now your work is about something else?*

CECELIA: *Well, no.*

ROSALYN: *So where are your own scientific aspirations? I am really respecting how much you care about your students right now, but I am also trying to nudge you a little bit.*

CECELIA: *All of the projects that they are working on—I am talking about their successes—are all under umbrella ideas that are mine.*

ROSALYN: *Of course.*

CECELIA: *OK, and obviously I am super interested in them. I have to be because I have to write the paper, write the proposals. There are things that I would really like to go after that we are not going after because there have to be limits.*

ROSALYN: *I want to talk outside of the range of limits; what would really get you going, anything; any specific science projects?*

CECELIA: *Do you want me to come up with a specific one?*

ROSALYN: *I want you to come up with projects that lead to particular outcomes. Of course you don't always know where you are going when you do the search.*

CECELIA: *No.*

ROSALYN: *But there must be something in your imagination about where you could be going?*

CECELIA: *I think two things that we are working on are two things that were totally crazy ideas of mine, literally. One was sitting around with my mom and it was just a …*

ROSALYN: *You told me about that one.*

CECELIA: *The nanosperm thing?*

ROSALYN: *Yes.*

CECELIA: *And so it came out of nowhere. And now they don't even know that that's where it came from and they are doing totally different things, but that's the epitome of where I get really crazy about.*

ROSALYN: *OK, now this nanosperm thing. Refresh my memory. I remember the conversation with your mom.*

CECELIA: *So this is the one where she is watching video sperm on TV and she wanted to know how they knew where to go.*

ROSALYN: *Right.*

CECELIA: *And so the whole idea of making nanoscale analogs, where they first self-propelled and delivered information to a precise target at the same time. So, that's where that whole thing came out. The other idea is totally wild. It was just looking at DNA. So again it's a biological system, which is really kind of neat.*

ROSALYN: *Is this with the tracts and the motors?*

CECELIA: *Well, that's along the lines of the nanosperm. The DNA one is brand new for us.*

ROSALYN: *OK.*

CECELIA: *This is conceivably 6, 7 weeks now. OK. You know that the DNA has a double helix?*

ROSALYN: *Yes.*

CECELIA: *And it contains all of the information for life.*

ROSALYN: *Right.*

CECELIA: *And on top of being the ultimate in information storage, it is self-replicated.*

ROSALYN: *OK.*

CECELIA: *So the only problem with DNA is that it's really, really sensitive. It breaks down into different solvents, enzymes attack it, and you can't build computers out of it all that well. So we were thinking that we would like to build a structural and functional analog to DNA. Instead of using hydrogen bonding along the DNA backbone, we would use metal complex. Alright, so we have got short chains that bind metals and the geometry of that is all very metal specific. You can design them to do this and when you do that you get a double helix. If you design them correctly, then they become information storage, because you can throw electrons on the metals if you opt to. When they conduct, it's very determined what the metal is or how far apart it is from an electrode. If I go big on a chain coming up from an electrode, an atom that's a nanometer away is going to act different from an atom that's 10 nanometers away.*

ROSALYN: *So would you use DNA as the scaffold? Isn't that basically what you have done?*

CECELIA: *We are not putting it on a DNA scaffold at all.*

ROSALYN: *You are not?*

CECELIA: *It's a different polymer.*

ROSALYN: *It's a different polymer, but it does what the DNA does including the helix?*

CECELIA: *It does.*

ROSALYN: *Why would you use DNA as the model?*

CECELIA: *Because DNA is the ultimate in information and self-replication. There is no other system that does that.*

ROSALYN: *Oh, so you are trying to get the same functions in something that doesn't have such sensitivities?*

CECELIA: *Exactly.*

ROSALYN: *Like a metal.*

CECELIA: *Exactly.*

ROSALYN: *And that can also conduct electricity, so that it can be used with a computer.*

CECELIA: *Right, so that's one of the questions; if DNA's information is storage for us, in using the information you have to have a way to get it out.*

ROSALYN: *Right.*

CECELIA: *In DNA analysis they do those electrophoresis gels and they have the DNA sequences. That's not very good for our kind of information, so the best way we know how to do it is to conduct electrons, or to use magnetic stuff.*

ROSALYN: *Right.*

CECELIA: *That's how everything works. We don't have to use these electrophoresis gels to do that. We know how to read that information out easily. So now we can choose metals and we can choose sequences and we can place them in certain directions and they all have different electric signatures as a result. As an example for how stuff like this affects my head, I have literally started doing molecular modeling on Friday nights. So, I am in front of my computer. I have three computers at home, doing different models all the time, for about a day and a half. I kept e-mailing my student saying, "Check this." One of the files was actually named "Holy Shit Brian." I get so hyper about it that I will call them at 3:00 a.m.–4:00 a.m. in the morning and say I am really sorry to interrupt your night, but you have got to see this.*

ROSALYN: *You are wild.*

CECELIA: *Stuff like that is just great.*

ROSALYN: *That's OK, because that's your imagination. What could it mean if you could actually do this, if you could make it work the way you want to? I know you are a chemist, not a chemical engineer.*

CECELIA: *Correct.*

ROSALYN: *So maybe I need to ask that of a chemical, or a mechanical, or an electrical engineer, but what do you imagine the engineers would say? "Thanks so much because now I can build X, Y, Z for this purpose?"*

CECELIA: *I think we have talked about nanofabrication and nanoelectronics before.*

ROSALYN: *Yes.*

CECELIA: *These are really nanoscale devices where you have single metals lined up, essentially atom wires, and they have precisely defined geometry. Because of this, we can stick them where we want to, and that's the new way to go toward molecular electronics.*

ROSALYN: *That's what I was thinking. It might be more efficient …*

CECELIA: *It could be.*

ROSALYN: *It could be faster.*

CECELIA: *It could be, because in soft assembly it's about having selectivity. Of course you need to get the information lined up in the right way and attach the electrodes in the right way and then you can build those into the system. So there is a long way to go. It's only been 6 weeks so far. It could potentially be really interesting and the other really cool thing about this is we know how to do the soft replications. So we can mimic both functions.*

ROSALYN: *Both functions?*

CECELIA: *Both of them yes. We can do the information storage and we can replicate them. For information storage it's really important, because when you make this stuff, you have only the tiny, tiny little metal right. Now, let's say you want to make a boatload of it. You will need to replicate it and you need to replicate it without any mismatches. We think we know how to do that, so we can mimic that function of DNA as well, which is what I think really sets us apart. Other people know how to define single module devices and module devices but the fact that we can replicate it and make oodles and oodles of materials is really important. So, I am very excited about that.*

ROSALYN: *And because it's metal, it can be put in all kinds of environments right?*

CECELIA: *Yeah. These polymers that we have got them on can hold up just about anything. The only thing is that it's not really happy in acid because it removes the metals. But actually, we want to remove the metals. We use a weak acid and the metals go away as part of our replication process. If you want to unwind the double helix, you just throw on a little bit of acid and the metal comes out and you can break these two strands apart. So it works out well.*

ROSALYN: *So you are having fun?*

CECELIA: *Oh absolutely, it's a blast; it's a total blast. It's interesting that you asked me about my dreams, because the last 2 weeks have been nothing but that inorganic DNA. I just can't get it out of my head; the things that we can do; the many, many, many steps in between here and way down the road.*

ROSALYN: *Inorganic DNA?*

CECELIA: *That's what we are calling it, because we had to come up with a catchy title.*

ROSALYN: *OK, what about the applications?*

CECELIA: *I think it's important that you do a very careful step-by-step study of all of these different things. You have to make a binding conference. You know all of this different stuff, but you have to tell people in the end why ...*

ROSALYN: *Why it's important?*

CECELIA: *Molecular electronics is the main thing.*

ROSALYN: *So when your colleague says "I want to put electronics everywhere," does it make sense? This notion of ...*

CECELIA: *Chips in your shoes?*

ROSALYN: *Yes. What kind of DNA did you call it?*

CECELIA: *Inorganic.*

ROSALYN: *Oh inorganic, does that make it more possible?*

CECELIA: *It might.*

ROSALYN: *Possibly?*

CECELIA: *Well, we always go back to my mom. But my nonscientific mom said: "Are you going to do genetic engineering?" My God I don't think so, but these would sure be compatible with DNA.*

ROSALYN: *They would be?*

CECELIA: *Absolutely.*

ROSALYN: *Even being metal?*

CECELIA: *Because the nature of the proteins is very compatible, in fact it's made of peptides. So they are much more stable than DNA and the nucleotides, and they would be completely biocompatible.*

ROSALYN: *Really?*

CECELIA: *Absolutely.*

ROSALYN: *Oh, now I am worried.*

CECELIA: *We can't ...*

ROSALYN: *No I know, not the gene part. I am talking about a hybrid maybe.*

CECELIA: *I don't even know why it would be useful. I don't know enough biology or biochemistry or why anyone would want to do that.*

ROSALYN: *So what do you imagine dealing with, with the inorganic DNA?*

CECELIA: *Ah, everything that we have been talking about has been electronics.*

ROSALYN: *Oh, back to electronics!*

CECELIA: *I haven't thought at all about biology, there are so many complex functions.*

ROSALYN: *Well, we don't need to go there. Even just working with electronics ... it's phenomenal what would be possible.*

CECELIA: *There is a lot. I keep telling my students to get back to the lab. It's Saturday night, it's 1:00 am, why am I in the lab? Actually, I came into the lab to work because I just don't have time to do lab work.*

ROSALYN: *Why not, because you are doing teaching and writing papers and running around, or recruiting?*

CECELIA: *All of the above. I just don't have time.*

ROSALYN: *What happens if you don't get into the lab, would you be sad?*

CECELIA: *Normally, I am too busy to even worry about it. It's now 13 people in the lab, it's enough for me just to have them come to me and tell me what they are doing, to fill up my day.*

ROSALYN: *But when you come up with a notion as wild as DNA, are you comfortable with just turning the lab over to your students, and saying "let me know how it goes"?*

CECELIA: *Oh no, I am in their face all of the time.*

ROSALYN: *There you go.*

CECELIA: *That's right, I am not actually doing it, although I came really close on Saturday to coming in and actually doing an experiment. But yes, I walk through the lab all the time and it drives them crazy and then they are all online. Thank God for instant messenger, because I can reach them any time of the day. And it's like: "OK, this is great, how is it going? Did that actually work?" It drives them nuts. But they are OK. They have to be.*

ROSALYN: *It's your creation. You have to be there.*
CECELIA: *Exactly.*
ROSALYN: *That's a really exciting new development.*
CECELIA: *That was actually something that my colleague had thought about.*
ROSALYN: *One of the things you stayed up late at night talking about?*
CECELIA: *Absolutely.*

CHAPTER SIX

Imagination, Metaphor, and Science Fiction

Human ingenuity, especially in unfamiliar domains of inquiry such as nanoscale science and engineering, is a complex, symbolic, and highly metaphoric engagement of the human, thinking mind. Humans respond to novelty and other forms of ambiguity through the imagination. Some objects of human engagement are reflected on only and most effectively in indirect ways, such as through metaphor. Metaphor functions as a primary and critical means of expression for otherwise ineffable human feelings, visions, concepts, and beliefs. And, it is not necessarily an impediment to clear thinking. Sometimes, it makes clear thinking possible by dealing tacitly with the sublime and the unknown.

The exponential increase in the human ability to control matter, which nanotechnology represents, points toward the increasingly powerful ability of humans to experiment with and even alter the fundamental constitution of living organisms, of nonliving matter, and of human bodily experience. Where does the desire and ambition for such awesome power come from? What kinds of beliefs are held about what it means to be a human being living in a nanotechnological world? How might those beliefs effect perceptions about what changes may come to human experience with an emerging nanoworld?

Why do so many people in technological society hold paramount the precise human control over matter (the aim of nanotechnology)? Is it a matter of evolution, divine promise, or human will? To even approach such a question requires unpacking the metaphoric, imaginative engagements of the human mind with notions of self, other, and life, in the quest for control over matter: Third Dimension Nanoethics. One means of access to the conceptualization of nanotechnology, and its quest for control of matter, is through the "hidden hand" of science fiction.

SCIENCE FICTION AS PREDICTOR

There are only two ways, according to Dyson (1997), to predict the progress of technology: economic forecasting and science fiction. The problem with economic forecasting is that it is good only 10 years forward. Science fiction is a more useful guide, he espoused, because it does not pretend to predict, it only tells us what might happen. Both miss the most important developments of the future: economic forecasting because it has too short a range and fiction because it has too little imagination! However, Dyson explained that science fiction can be used to discern patterns in the rise and fall of engineered technologies because it is the same rhythm that repeats itself in the evolution of a species, only a thousand times faster in the evolution of human technology. But, Dyson warned, the cyclical surfacing and fading of new technologies does not guarantee that all engineering ideas come to fruition. For many different reasons, some of the dreams of technology will never be realized. Others will. Some discoveries will be a complete surprise, and some devices will be used in ways never intended.

Predictions of the past confirm that attempts to predict the future are exercises in folly. Science fiction is no more equipped to do so than any other attempts at such. The exercise of human free will keeps the future both dynamic and alluring. However, science fiction has an important function in conceptualizations of the technological future. Empowered by imagination and vision, human will is a powerful force of design and creativity. In fact, science fiction can be used as an important source of understanding about how the imagination is being used to express beliefs, feelings, and perceptions.

CULTURAL CRITICISM OR ANALYTICAL PHILOSOPHY?

Science fiction is often misunderstood. For example, Florman (1997) pointed to postwar European cinema, "with its images of factories, office towers, and other buildings standing ominously in the background as a lamentation and reproach" as representative of the demise of the glory of engineering. He used the fiction novels of the 1950s and 1960s which give unsettling visions of the future, to illustrate how the protagonist is "struggling for liberation from an oppressive technological environment." "Even in the field of science fiction," he explained, "apprehension is replacing the optimism of earlier times" (pp. 15, 16). Categorizing science fiction in that way overlooks the philosophical function of science fiction and its important role in conceptualizing both what is and is not desired in the future. And there are other explanations for the grave tenor of science fictional images. For example, could those powerful images help us to come to terms with human material conditions, and the various possible futures we might create? Could they also help individuals and the collective to face the primordial human encounter with the sublime? Storytelling is fundamental to human communication about the sublime. It may be that science fiction addresses deeply held beliefs, fears, and ambiguities over society's relations with technology and its effect on individual lives.

Literature, such as Stephenson's *The Diamond Age* (1995), suggests that despite the stated good intentions of nanotechnology researchers, there is nothing that nanotechnology can do to improve the quality of human life. While telling the story of a little girl named Nell, *The Diamond Age* weaves the reader through a world that is at once familiar and strange. Much as with today, in that world human beings have divided themselves into rather isolated societies, living in self-ruling territories that are determined primarily by racial identity and common ideas and rituals. Wars are fought over material resources, land, and power. Criminals are judged according to the severity of their deeds, and entertainment is accorded significant value. However, unlike today's world, in Stephenson's (1995) fictional world, nanotechnology is able to provide for all of the basic human needs, and many other material desires. "Matter Compilers" use "feed" to build from molecules any object that is programmed into the system, from food to clothing to entire buildings (p. 57). Scarcity of resources is a political maneuver by powerful territories to both limit and control the use and possible abuse of technological power. This is one reason war persists, such as in the "Chinese Civil War" where people in the North have no access to nanotechnology (p. 253).

Stephenson's world combines highly developed technology with traditional, sometimes archaic, social conditions. In some territories, social justice is suppressed under the pursuit of information acquisition. There are world governances, but the independence of each territory leaves social conditions up to the individual governments, so that in the "Chinese Coastal Republic" the old ("Confucian") laws prevail and criminals are sometimes punished with physical torture or death. In one such occurrence, a judge from "New China" orders a magistrate to "revert to the time-honored methods of his venerable predecessors." The prisoner is strapped to a heavy metal rack that was normally used for canings, stripped from waist down and situated over a bucket for elimination. Earlier, the court physician "thrusts in a spinal tap," introducing a set of "nanotechnological parasites." These parasites had "migrated up and down the prisoner's spinal column" through the cerebrospinal fluid, and "situated themselves on whatever afferent nerves they happened to bump up against." These nerves, used by the body to transmit information, such as (to name only one example) excruciating pain to the brain, had a distinctive texture and appearance that the "sites were clever enough to recognize" and the ability to "transmit bogus information along those nerves" (p. 125). Despite their extraordinary technological abilities, the inhabitants of this future world are troubled, competing, and perpetually in struggle.

In his study of science fiction, James (1994) cited Edmund Burke, who argued that it is magnitude in nature that gives rise to feelings of the sublime: "Infinity has a tendency to fill the mind with that sort of delightful horror which is the most genuine effect and truest test of the sublime" (p. 102). He further asserted that scientists inexplicably share this passion for the sublime. That is certainly true of both the engineers and the scientists of my own study. James saw the essence of the sublime as "a feeling of helplessness and terror when humans realize their frailty and small size in the face of the might and magnitude of the universe" (p. 105) and explained how in the Romantic era, people believed there to be no divinity to protect mankind. The sublime was a consequence of the liberation of humanity, by the Enlightenment, from the protection of revealed truth. Human struggle over encounters with the magnitude of nature is evident in science fictional imagery and story.

Science fiction is a medium that projects beyond known or accepted facts or theories, as one that assumes and reminds us that the orderly universe can be exposed and exploited by rational endeavors, and that man can change reality. In science fiction, everything that is, all the givens, are open to modification. In its essence, it projects into contexts that are at variance with what is now taken to be basic, and depicts the consequences of countersuppositions. Smith (1982) explained that through science fiction, a person can remain essentially the same while undergoing change in a world that forms a coherent whole, and is accessible to rational inquiry.

At the same time, it depicts worlds that are at various removes from the world we know. Smith understood that science fiction shares philosophy's goal of discovering what is essential and valuable in reality. To that end, it has an important aporematic, or problem setting function. It can also have an important persuasive function. Often intended as thought teasers or puzzles, science fiction is uniquely suited to apply pressure to common sense. Science fiction can become an effective means of grappling with that which we do not rationally understand.

If philosophy seeks to discover what is essential and valuable in reality, then science is its handmaiden. Its hands and eyes are the nanoscale science and engineering researchers who are forming critical questions about novel phenomenon, and seeking answers to those questions in their laboratories. Individually and collectively, humans discover and appreciate what is essential and valuable when meaning is made of what is perceived. The psychosocial functions of science fiction are rich and wholly provocative in their capacity for enticing reflection about technological creations, and how human beings might live in relation to those creations. Whereas Florman took science fiction to be a sign of the times, and Dyson felt it was an inadequate but valuable source of prediction about the future, I offer the suggestion that science fiction is a means to make meaning of the feelings of the sublime that humans sometimes seek to reconcile though technology. And, in the case of nanotechnology, it is a reminder that the future is yet to be created.

SCIENCE FICTION AS CULTURAL NARRATIVE

A number of the researchers in this study have referred to science fiction in their responses to my questions about the nature of nanotechnology, and about the inspiration for their work in it. About one third of those references to science fiction have been unsolicited (totally without my prompting). Sometimes it seems they are reaching into the imaginative processes as an indirect way to express their particular beliefs or to convey their personal values. When they refer to science fiction it is usually with excitement or animation. Two of the most common references made are to *Star Trek* episodes and the book, *Prey*. *Star Trek* is usually cited as a source of moral or social provocation. *Prey* elicits three distinctively different reactions. First, there is the concern that the book could cause a problem for them personally, if the general public absorbs it as an authoritative source of understanding about nanotechnology. Second, some wonder if it is accurate in depicting how little is actually known about the potential hazards of the nanoscale materials being used in laboratories around the world. A third reaction is total dismissal, as conveyed in the following excerpts of discussions with Tyler and Helen:

TYLER: *For the first time, in the midst of our species we are really in a position to at least think about changing our nature to some extent, from the chemical level. We were talking about it over 25 years ago and nobody has really done much talking about that since. Mostly they were concerned about the dangers of pathogens or destroying actual strains. I've thought about that a bit. In terms of what I do myself, there still seems to be no real likelihood that we're going to do anything that is going to have any impact on life that I know of. The publisher sent me a copy of* Prey, *so I read it. Not only is it Michael Crichton and his usual crap, but it also has nothing to do with nanotech. It was just stupid.*

ROSALYN: *I didn't quite understand the science in it.*

TYLER: *There was no science in it.*

ROSALYN: *OK.*

TYLER: *At least, there was no physical science, I don't know about the systems stuff. It was just crap.*

ROSALYN: *In my opinion, the only really good science fiction I know of that's way out there for nano is more about assemblers and it's sort of ...*

TYLER: Diamond Age?

ROSALYN: *Right, Neil Stephenson's.*

TYLER: *It was a crappy book too, but the concept is sort of nano.*

ROSALYN: *Have you found a well-written book with interesting nanoconcepts in it?*

TYLER: Blood Music *was a good novel, but it was a long time ago. There was also something else, the* Nano Chronicles *or something, I forget.*

TYLER: *So, you read science fiction?*

ROSALYN: *A little.*

Some of the individual researchers are embarrassed by their references to science fiction, classifying it as unrealistic fantasy and therefore unworthy of their serious reflection and consideration. I try to convince them otherwise, because in fact, it does seem to serve a very important function in our discussions. Humans must have some way of making meaning of encounters with novel phenomena and with unfamiliar, perhaps uncomfortable, curious, and intriguing experiences. Through science fiction, the metaphoric image takes the mind beyond recognized material boundaries into domains of unexplainable sensory experience and psychological knowing. It can be used to reach beyond what is known and understood in order to make meaning of those elements of human awareness and perception that otherwise elude ones' grasp. It also can help to break down conceptual limits to awareness, and in turn, to inspire the imaginative projections of otherwise unachievable dreams.

HELEN: *I've probably done this before, but if I could go back to* Star Trek, The Next Generation ...

HELEN: *In* The Next Generation *there is this one—oh I sound like an infant here or a high school student.*

ROSALYN: *Not really, you're not the first scientist who has referred to* Star Trek *in these discussions.*

HELEN: *My hero, Jean Luc Picard, is recently divorced but unfortunately recently re-engaged.*

ROSALYN: *He's quite something, isn't he?*

HELEN: *He is the cat's meow. Anyway, they go to a world where they encounter a society that's sort of a primitive earth, well, maybe not primitive earth, but maybe 1600 AD. The point is that this one woman realizes what Picard can do and she takes him up to the ship, she starts to worship him, and they create a culture around him. All of a sudden there's a new religious symbol system and he's god because the technology is so far advanced from where they're at. Right now, we're developing technologies that, especially when you think about reproductive endocrinology, and what you can do with reproductive medicine, we're doing things that 200 years ago or even 100 years ago were godlike, but a hundred years from now we'll look back and say, that was pretty simple.*

When I have asked directly, a number of the researchers say they have watched or are familiar with the films *Gattica, The Matrix, Minority Report,* and *A.I.* One person, in turn, asked me about *The Lord of the Rings* and expressed surprise and dismay at my lack of familiarity with it, saying that he owns all the film episodes and texts, and sees the material as the "best literature ever written." Science fictional references have been so common in these discussions that I wonder if there is something particularly alluring about science fiction for nanoscience and engineering researchers, but I can only surmise as to why. One possible answer to the question of "why?" might be found in Smith's thesis about its philosophical function. Another possibility may be in the search for meaning. Researchers of nanoscale phenomenon are moving into previously inaccessible domains of human curiosity and understanding. Such endeavors into realms of the novel simultaneously entail the making of meaning about what they discover, and also about what they fail to understand or accomplish in their laboratories.

As individuals, a number of researchers have been quick to defend that the search for scientific understanding at the nanoscale is driven primarily and simply by human curiosity. Perhaps this is true. Nevertheless, the nanoworld being explored must somehow be conceptualized by its explorers. It may be that science fiction serves this purpose.

Benjamin is a participating researcher. He was trained as an engineer. I offer the following four excerpts from a conversation with Benjamin as a case in point of how enticing and provocative science fiction can be as a tool in the conceptualization process. The context of the excerpt is a discussion about his work and life as a research scientist. We seem to have diverted a bit from the subject of nanotechnology, and moved on more generally to the human condition:

Conversation With Benjamin, Part I

BENJAMIN: *I know that if I were in charge I would really be wanting to, OK ...*
ROSALYN: *OK, what?*
BENJAMIN: *I would want to get off this planet.*[1]
ROSALYN: *We talked about that before.*
BENJAMIN: *We did?*
ROSALYN: *Yes, you did.*
BENJAMIN: *We have been stuck here. All of our eggs are in one basket. If we were an investment, the best answer would be to sell.*
ROSALYN: *You talked about the asteroid ...*
BENJAMIN: *We are highly vulnerable here on this planet. We're not only vulnerable to external influences. We're vulnerable to internal things.*
ROSALYN: *Because we have made a mess of things?*
BENJAMIN: *We are vulnerable to malice, we're vulnerable to ignorance. We're vulnerable to want. We're vulnerable to disease, to ...*
ROSALYN: *How is getting off the planet going to change that one iota? We're still going to be us.*
BENJAMIN: *Oh, because it just means that there are multiple locations and it's unlikely that any single cause will destroy them all at once.*
ROSALYN: *Oh, I see. But we're still taking all that stuff with us.*
BENJAMIN: *Of course, unless you leave certain people behind. I can imagine, a few folks that I wouldn't want on my ship. I'd be happy to have you on my ship.*
ROSALYN: *Thank you.*

[1]Not an uncommon assertion among some scientists. For example, see Kaku, M. (1997). *Visions: How science will revolutionize the 21st century.* New York: Anchor for discussion on the dependence of human survival on finding a planetary home beyond Earth.

BENJAMIN: *And some others.*
ROSALYN: *I don't think there would be too many people on my ship, either. Alright, that's one. If you were in charge, you'd get the hell out of here, what else?*
BENJAMIN: *No, I wouldn't necessarily leave. I'd develop the capability to leave.*
ROSALYN: *OK, so you want the freedom ...*
BENJAMIN: *What I actually believe is that you could easily find people to colonize.*
ROSALYN: *Absolutely. So what else would you do if you were in charge?*
BENJAMIN: *Well, I don't know. I haven't thought it through. Except that particular thing, that solar flare going on out there, one of the biggest ever ...*
ROSALYN: *That we've been aware of.*
BENJAMIN: *Yes.*
ROSALYN: *Maybe there were many big ones before we had the instruments and we didn't worry about those things.*
BENJAMIN: *But, OK, what's the point? The point is that we live in a place that is inherently dangerous and we need to be ready. We're not ready. We're not even getting ready. We've lost vision as a nation, as a people. Next question.*

Back to Nicholas Smith for a moment, Smith (1982) demarcated three types of science fiction, each of which can be further distinguished by two subcategories:

1. *Natural science fiction*, which emphasizes physical sciences and biology (not psychology). The novel *Prey* is an example. In it, life is familiar and the plot was centered on an unanticipated biological problem.
   a. Extrapolative asks, "What will happen if this goes on?" It presupposes current scientific knowledge projected onto environments unlike the current.
   b. Speculative natural science fiction projects fundamental advances in science and technology, which supersede theories currently in force. It asks, "What would happen if this were to be the case?"
2. *Cultural science fiction*, which focuses on knowledge of human culture and social life. The film *The Matrix* is an example in that it spoke to questions of consciousness, as well as to mystical notions of existence.
   a. The extrapolative-type projects humans or intelligent life forms in unusual situations, but their thoughts and behavior are explicable in terms of existing theories.
   b. Speculative cultural science fiction projects situations in which social and human sciences undergo significant advances that cannot be anticipated at present.

3. *Metaphysical science fiction*, which projects the deepest level modifications. It employs philosophical assumptions, which are at variance with enlightened common sense. In *Star Wars*, entire value systems are foreign, life forms are strange to our common notions, and the notion of the "Force" speaks to metaphysical questions of creation, awareness, and existence.

Each of these three forms and their variances project characters into novel situations to isolate essential facts about human beings, society, space, time, history. When Benjamin spoke of leaving the planet Earth, he used the genre of natural science fiction. The point he seemed to be making is based on current knowledge about the vulnerability of humans and our genetic isolation on Earth.

Conversation With Benjamin, Part II

One way to interpret his use of science fictional ideas is that his awareness of inevitable personal annihilation is projected onto the whole of humanity, as a life-affirming value for the perpetuation of the human species. There are precious few provisions in our technological societies for directly and consciously facing ones' own mortality. Without ritual, metaphoric imagery functions to help individuals and communities to do so. It may be that Benjamin's use of science fiction allows for him to explore the inevitable and perplexing fact of human mortality through imaginative domains of other lives in other places. The practical imperative of establishing human colonies on far away planets and stations gives credence to fantasies of immortality. In the face of what is scientifically known and affirmed—that humans are indeed mortal—Benjamin was able to construct meaning around the meaninglessness of that knowledge through technology, and its capacity to project life into other material realms:

BENJAMIN: *There was a wonderful* Star Trek *episode. It was the old* Star Trek.
ROSALYN: *That was the best* Star Trek.
BENJAMIN: *There was a planet that was so full of people that they literally couldn't even lie down, they were just standing around and they didn't believe in any kind of killing, but in order to alleviate their population they had to capture Kirk because he had had some kind of fever and therefore he could make them sick and a bunch of them would die off. It was a statement of our population growth, explosion, whatever. It will*

level off. The technology will mature, I just don't know when and where, but it will change, it will be different.

ROSALYN: *OK. So why is it hype, to suggest that going down to the nanoscale will ...*

BENJAMIN: *Because people want to get funding.*

ROSALYN: *Well, right, that's a given. But given that, people get funding and then they bring these circuits down to the nanoscale and then they can do very new novel things with them like extend life or whatever we're up to.*

BENJAMIN: *One of the things we need to do is to cure diseases.*

ROSALYN: *We need to cure diseases?*

BENJAMIN: *Get off this planet.*

ROSALYN: *Get off the planet. Do we need to extend life?*

BENJAMIN: *I don't think it's a question of whether we need to or don't need to, that's a value judgment. I don't think anybody will be sorry to the First Dimension.*

ROSALYN: *OK.*

BENJAMIN: *I don't want to be old. Well, that's not the point, the point is not that we extend life in such a way that you live your last 400 years as if you were 100 years old now, the point is that you extend it in such a way that everything will change.*

ROSALYN: *Everything will change, that's right.*

BENJAMIN: *Everything would, if we have a longer time. I don't know how I got back on this, but if we have a long lifetime, it isn't going to be like it is now. It's going to be fundamentally different, some things will change.*

ROSALYN: *A lot of things will change ...*

BENJAMIN: *... I think the time will come when people get to be 60 and instead of retiring they will take 5 years off, go back to graduate school, pick up a degree in anthropology or archeology, go dig in the middle of the Sahara, or maybe take off 20 years and just meditate or go on some sort of humanitarian sabbatical or religious pilgrimage or something, and then they will come back and maybe decide to be a medical doctor because they've been helping people. We will structure things very differently ...*

ROSALYN: *Family will change, belief systems will change, educational systems will change, birth will change, the brain will probably change. Yeah?*

BENJAMIN: *I don't know.*

ROSALYN: *You don't know. OK. It's just going to happen. It doesn't matter what we think about it, is that what you're telling me?*

BENJAMIN: *You're making a value judgment already.*

ROSALYN: *Yes. You know what I'm doing. I'm trying to figure out whether we should ...*

BENJAMIN: *You can't stand the thought that maybe there's uncertainty.*

ROSALYN: *No actually, uncertainty is comfortable, more than it ever was. It's freeing, it's amazingly freeing. What's not comfortable is the notion that we're not consciously setting a course for ourselves.*

BENJAMIN: *Well, I'm setting a course for me. I can't influence all those idiots downstairs that smoke cigarettes.*

ROSALYN: *Alright, they're the ones I'm worried about.*

BENJAMIN: *Well, you know.*

ROSALYN: *I want there to be a conscious decision-making process about where we're going, what we are doing. What is uncomfortable is this sense that well, it's all just unfolding and it's going to happen whatever it is, but we don't know what it is. Well, if we're intelligent maybe we should put our minds to what it is and have some say over it. That's what I'm uncomfortable with. It's because I don't buy that it's simply a matter of uncertainty. I think we have got a whole lot of will behind what we're doing. I try to keep my opinions out of these conversations but, you've got me right up against the wall here. I do ... I wish we would wake up.*

BENJAMIN: *Make a suggestion, say something concrete.*

ROSALYN: *I'm sorry.*

BENJAMIN: *That I can respond to.*

Here Benjamin moved to the cultural science fiction of *Star Trek*. In this part of our conversation, he commented on the human propensity to self-defeating behavior. His passing remark about people who smoke cigarettes is linked to his *Star Trek* episode reference where overpopulation problems were taken to the absurd. A social problem that is scientifically understood and should have relatively simple technological solutions is rendered impossible to correct because of belief systems and behaviors that preclude them from responding as they know how. We know cigarettes kill, but we smoke. We know what causes high birth rates and yet we overcrowd to the point of having no room to lie down. Benjamin put his trust and hopes in the maturity of technology, and its stimulus to social and cultural change. Using extrapolative, cultural science fiction, he took us to a future where everything is essentially the same (and therefore familiar), except that we live for a very long time. And the human species is safe from the threat of external natural disasters, because our genetic seed is adequately dispersed. The moral imagination at work through this use of science fiction is in direct reconciliation with the fundamentally human denial of death (a subject elucidated by Becker's (1973) research. It affirms over and over, in various ways, that what is essential and real is the bodily life we now know.

Conversation With Benjamin, Part III

In the third excerpt of the conversation, we are back to talking specifically about the nanotechnology future. At the start, I asked Benjamin to talk about the future and potential of nanotechnology to answer the question of where it might be leading. Here, well into our 90-minute conversation, he strove to conceptualize the possible annihilation of the human species, and most importantly, the human will to engineer an alternative:

BENJAMIN: *You've got to distinguish the hype. There are going to be some things to cure people with. In general, people's motives have been positive, that hasn't always been the case. If you had announced back in 1930 that you were going to develop chemicals that had the ability to selectively kill organisms ...*
ROSALYN: *Right.*
BENJAMIN: *Penicillin.*
ROSALYN: *Right.*
BENJAMIN: *Where is that going? Medicines they are counting on because you've got medicines back at the dawn of history. You depend on people to try to do the right thing individually and then you depend upon people collectively to try to do the right thing as best they can. If it works, we'll all be better in 150 years from now, and if it doesn't work, if somebody screws up, we may have a major population collapse. I think that's something that we all should be prepared for.*
ROSALYN: *OK, OK.*
BENJAMIN: *It's something that doesn't seem to come up in most people's conversation, but it's certainly a significant possibility. There could be a major population collapse on earth, to the tune of 90+%. It could go to zero, but then of course, then it makes no difference to the future ...*

In each of these three excerpts, Benjamin made some allusion to the prospective collapse of humanity. Whether induced by human accident, or by the random events of a cold and arbitrary universe, he recognized that humans are vulnerable to extinction in some form. This story is told in different ways, presented as both a moral and a technological challenge. But the solution is always the same: Find a home beyond the planet Earth. I do not know whether Benjamin truly believes that humans are susceptible to catastrophic onslaught, or whether what he was saying was actually that his own death is inevitable. Either way, science fiction makes available a replacement for the lost home, and a new possibility of survival.

Benjamin's narrative is echoed by popular scientist-writer Freeman Dyson, who wrote about "the dream of expanding the domain of life from earth into the universe" as one that makes sense in the long term. One reason, he explained, is like the desire to climb Mount Everest because it is there. But he offered another practical reason. That is, as Benjamin pointed out, human beings are vulnerable here on Earth. As Dyson (1999) put it,

> Recently the inhabitants of Earth have become aware that our planet is exposed to occasional impacts of asteroids and comets that may cause worldwide devastation. The most famous impact occurred sixty-five million years ago in Mexico and may have been responsible for the demise of the dinosaurs. During the next hundred years, as the technologies of astronomical surveillance and space propulsion move forward, it is likely that active intervention to protect the Earth from future impacts will become feasible. (p. 103)

Dyson specified the imperative to protect Earth. (Of course, Earth did just fine after the impact of 65 million years ago. I think the Earth doesn't need us to protect it from anything, but ourselves.) As far as both Everest and space go, as magnets of human conquest, perhaps those motivations and the values they represent might be understood a little differently. Climbing Everest and forming space colonies are not simply matters of humanity's needs. They are matters of personal meaning and ambition. They are metaphoric in character, representing the need to exceed established limitations. Conquest, fueled by the desire to overcome physical and psychological frailty and weakness, is well disguised behind the protective thrust of magnificent technological power. A sense of cosmological irrelevance and worthlessness haunts the psyche. Science fiction, be it in literature or film, at least provides an avenue of escape—a way to seek meaning, place, and purpose as projected out to the imaginative possibilities of the beyond.

Conversation With Benjamin, Part IV

Ideas of personal transformation emerge and are addressed through the promises and possibilities of nanoscale manipulations of matter. Benjamin's imagery and fictional science language are by nature transformative, from earthbound humanity to a more expansive, less vulnerable population in the universe:

BENJAMIN: *We are just about to gain control (at a limited level) of the molecular structure that life is built on. The ability to control that eclipse isn't anything that I can imagine anybody is going to do with inorganic nanotechnology.*

ROSALYN: *What about when that inorganic nanotechnology is fused with carbon-based science?*

BENJAMIN: *Well, I am interested in that, too. But I do not have too much knowledge in that area. I am collaborating with someone else's group in the limited way that I am able to do, on something that is related to that biological/organic, inorganic biotechnology interface. I think that's very important. Like, for example, it would be very nice to avoid eating. I eat too much. Instead, we could run off whatever chemical that the biologist will tell you to and just go and plug your fingers into the wall and charge up.*

ROSALYN: *Won't you miss the taste of cheesecake and bacon?*

BENJAMIN: *No, because you can just synthesize that.*

ROSALYN: *Tricking the brain into believing that the tongue tastes flavors and senses textures that it's not?*

BENJAMIN: *Well, if food tastes good, then you could get just as much enjoyment out of being charged.*

ROSALYN: *Is that possible?*

BENJAMIN: *Almost anything that you can imagine is possible. I think that doesn't violate known principles, known physical laws. I really believe that the capacity of life for diversity of structure and function is just about unlimited, especially as our imaginations go. All you have to do is just look at weird stuff that goes on. I am an engineer. All you have to do is just look around. The birds manage flight in their brain cells where electric currents are generated. The electric signals that are generated when they fly are navigational devices. Why can't we build GPS devices into our heads? It's in the combination of computer technology to manage information. The computer is not a calculating device, fundamentally it's a complexity managing device is what it really is; it's not just a number thing. It's a machine that is a tool for managing complexity and the complexity that needs to be managed of course is this enormous DNA thing that controls the structure of life. And progressively, also the way the proteins get made and all of that stuff. The first computer of any serious kind is about my age. What's going to happen in 150 years? This Macintosh thing down here; this thing has more computing horsepower than existed in the world when I was a child. What's going to happen when this technology has been around 500 or a thousand years? Disk space is galloping forward at an astonishing rate. So is the amount of information that can be put on magnetic disks. There was a time when people thought magnetic disks were going to run out and there were some technological discoveries that made it possible to continue the scaling of disks. The scaling of information storage has been even faster than the Moore's law scaling of the speed of computers, more and more stuff. What's going to happen after 150 years, 300 years, 500 years, if we don't destroy ourselves one way or the other? We used to worry about nuclear weapons, now we have not just nuclear weapons, but we have biological weapons. Chemical weapons, I am not too worried about, but the biological weapons are potentially really serious. People are going to figure out how these big molecules work. Eventually maybe, although I am not sure about this, but maybe they will, given the knowledge of a specific biological molecular structure. Maybe they will be able to figure out what the global implications of that are for the organism. In other words, predict the characteristics of a creature from knowledge of its DNA. Be able to take a sample of your cells from scraping your cheek, get the DNA, decode it. Take it up to the lab.*

ROSALYN: *I have spoken with scientists who are working on that now.*

BENJAMIN: *Of course they are.*

ROSALYN: *Yeah.*

BENJAMIN: *I am not sure it will ever happen.*

ROSALYN: *As I said, there is at least one serious scientist who says that is exactly what he is doing.*

BENJAMIN: *I believe he thinks he is trying to do the science. There is no doubt people are trying to do the science.*

ROSALYN: *Um hum.*

BENJAMIN: *The question is whether he will ever be able to take a computer and put in the decoded genome and then say "OK, What's this going to be?" with no knowledge and punch the button and it comes out and it shows Rosalyn, or me, you know? If I had to bet I would say that the only thing I am worried about is that I believe at some level computing may have limits, in that ultimately the computer may be the system itself. I don't think you will be able to compute the world faster than that world computes the world.*

ROSALYN: *No.*

BENJAMIN: *You can simplify systems and compute them faster, but I am not sure when you reach great complexity whether machines like this will … I am not sure where that ratio is. It's true that sometimes if you have a very complicated event, you can't really simulate it practically on a computer. The question is when does that inability mature in the sense of you are not ever going to do better. I am not sure about that. But to close this point, I think the real power will be the molecular level engineering of biological structures. To me, the power associated with that is almost unimaginable, and it conveys with it the natural*

> *ability to interface with other kinds of systems, directly. There are already fish in the sea that communicate or whatever they do visually by chronophers in their skin. Way down deep there are fish that glow and they have Ripley light and all kinds of stuff, so that self evidently is possible for living creatures. Once you grab that, what's to prevent almost anything?*

As for a decreasing appreciation for the profession of engineering that Florman lamented, and especially for an engineer such as Benjamin whose research takes place at the nanoscale, its existential pleasures still seem to be enjoyed, especially as enjoined with the meanderings of the imagination. Through science fiction, researchers can have powerful access to the tacit elements of cognitive processes toward conceptualizing and making meaning of technological creations in response to perceptions of the human condition. In explicit, conscious terms, nanoscale researchers speak about their work as a response to curiosity, a drive to knowledge, an imperative to bring novel solutions to fundamental, material problems. Although the "hidden hand" of science fiction provides them a way to work out the alternatives, beyond known and accepted laws of nature, in the dialectic exchange of science fictional imagery, researchers and laypeople can share in the process of searching for answers to the value and meaning of nanotechnology development. In general, scientific practice and engineering are inaccessible to the layperson. The general public has few resources for processing and communicating about what science and technology mean to collective and individual human life. Separated by language barriers, the intellectual walls of professional training, and physical walls of laboratories and universities, researchers, and the public appear to be separated. Science fiction is one mechanism that provides dialectic access, one to the other. It allows individuals and communities to cross the barriers with shared, understandable language and provocative images. Through exchanging and sharing those images, individuals can reach past intellectual domains into the collective unconscious of metaphor toward otherwise inaccessible realities.

I agree with Dyson regarding our inadequacy in predicting the future. However, I believe that human beings are fully capable of both negotiating and conceptualizing the future, and of expressing our fears of and hopes for it, through the imaginative process. The nanoscale research engineer/scientist can be a deeply imaginative person. It is their creativity of thought, grounded in knowledge of physical laws and the experimental method, which lends itself so beautifully to investigate the novel world of nanoscale phenomena, and thereby make possible various new technological futures. Their personal responses to science fiction, and the rich imaginative visions nurtured in their minds, are fertilizing those futures.

TECHNOLOGICAL PROJECTIONS OF SELF

Human beings construct technology variously as an expression of human nature and toward the distinctively human search for meaning. As with other technological endeavors, nanotechnology developments will express human beliefs, ambitions, and ideals, both symbolically and materially. For example, our televisions project images of ourselves, mirroring human activity and offering a sense of companionship and security. Humans construct buildings reflective of our own bodies, with outer layers (walls) that function like skin and hidden, interior systems that mimic neural and vascular mechanisms, interior plumbing, heating ventilation air conditioning (HVAC) and electrical systems are held just below the exterior, skinlike walls. Waste pipes, which leave the building from down below, and vents, which carry airborne wastes up and out, represent elimination systems in the human body. Similarly, humans tend to design robots with humanoid attributes such as upright positioning, moving arms, and expressive faces with "seeing" eyes, and automobiles with headlights placed in a position that mimic illuminated eyes on a face. Likewise, human self-awareness will contribute to framing the blueprint of nanotechnology's construction. For example, Geoffrey's concept of human intelligence fuels his imagination about humans creating a superior species:

GEOFFREY: *We don't know how it is going to come out. I have no clue what will happen if we create a competitor or superior species. For one thing, I think it would take a while (I am trying to imagine this), if it were a silicon-based intelligence. How long would it take me to fathom that out there, whatever there was, where my world consisted of streams and bits emerging from something, how would I infer from those streams and bits the existence of something which took other things that I couldn't imagine and put those things into this mouth whose function or existence I couldn't imagine and chewed, and mixed it with fluids and dissolve, and reincorporate it to make more of those things by a process of binary division? This would seem to be inferring with existence, from the point that sentience would seem to be such a difficult job that it would keep it occupied for several hundred years, or forever. For all we know, there is some intelligence whose nature I absolutely can't infer, that is pulling the strings right now. If we were to create a peer competitor of that sort, could it really compete? I don't know. It's an interesting thing to speculate. I think we are attracted to the idea of intelligence and we don't find it in humans, so that in a platonic sense, we look elsewhere. Therefore, we are not an intelligence species. We are an intuitive species. What we can't find in us, we will construct.*

ROSALYN: *All we have to do is look inside of the nature that exists separate from the human to the extent that it does, at the immense complexity of interconnections of every system that there is on the planet. Amazing interdependence and codependence, and complexity is right in front of us, and yet we try to improve on it.*

GEOFFREY: *The interconnection yes, but codependence we don't know about.*

ROSALYN: *Alright, I am making an assumption here.*

GEOFFREY: *For all I know you can eliminate 90% of the species on the planet and things will get along just fine.*

ROSALYN: *Alright, that's a point well taken. We couldn't do without 90% of the trees, but perhaps we could do without say, mosquitoes.*

GEOFFREY: *You could surely eliminate us. I think the lower down in the system you go, the more careful you want to be. My colleague makes a strong case for ants. But bacteria, if you would eliminate streptomycin ...*

ROSALYN: *You think we would?*

GEOFFREY: *I think other things would.*

ROSALYN: *If you eliminate us, not a problem?*

GEOFFREY: *It might be better [without humans] but eliminating ants would be a problem.*

ROSALYN: *As with spiders, I would say.*

GEOFFREY: *Elimination of spiders I am sure would be a problem.*

ROSALYN: *Major problem at least to us if they were gone.*

GEOFFREY: *Major problem to everything. My colleague makes the point that the bio mass due to ants and the bio mass due to humans is about equal and if you eliminated us, the world wouldn't even notice, and if you eliminated spiders and certain ants, everything would collapse. According to him, they are absolutely crucial in the interconnecting sense of tying all of the birds of the ecosystem together.*

ROSALYN: *Interesting.*

GEOFFREY: *And we are not. We are just the top food chain and so we are basically parasitic on the system.*

In the allure of science fiction, imagination becomes fodder for the dialectic process of sensing and encountering the new worlds that might be created through nanotechnology (or, the worlds we fear we may create). Such is the sentiment expressed by Po-Kei:

PO-KEI: *Certainly some of the nanofantasy came out of science fiction.* Star Trek, The Next Generation, *was one of the most fantastic, well-educated science fiction series ever written. It extrapolated the idea of having little machines, and then extended that to little machines that could fix themselves, and then little machines that could replicate, and little machines that could communicate and form a collective intelligence. I think that this fiction is a really good fantasy about where nanoscience research could be headed. I think that you ultimately could have simplistic nanomachines that actually have abilities to send signals to one another. But I don't think that you'll see anything representative of intelligence between little fabricated nanomachines. I do think that having little nanomachines that can do little functions is a definite possibility.*

ROSALYN: *Levers and pulleys and motors, and such?*

PO-KEI: *Simple, little motors, anything that you think of, any sort of machine that you think of on the macroscopic scale, and I am not even sure how you do all of this, but the nanofabricated guitar is a good example.*

ROSALYN: *What is it?*

PO-KEI: *Cornell has these fantastic nanofabrication facilities to make nanoscaled devices and they made a nanoguitar and it had little strings on it and they were actually able to pluck the strings and register the sound.*

As an element of human cognition, science fiction about nanotechnology's future is an expression of diverse imaginative possibilities offers to the public discourse a way of engaging and participating in the creation of our nanotechnology future. As both laypersons and trained scientists and engineers engage the genre of science fiction, they share in the making of meaning about our possible futures.

MYTHS OF METAMORPHOSIS

In her study of the theme of metamorphosis in literature, Warner (2002) offered:

> Metamorphosis is a defining dynamic of certain kinds of stories-myths and wonder tales, fairy stories and magic realist novels. In this kind of literature, it is often brought about by magical operations; but as I discovered in the course of my reading, magic may be natural, or supernatural, and the languages of science consequently profoundly affect visions of metamorphic change. (p. 18)

It may be that nanotechnology and science fiction may at times be intertwined in the human pursuit of meaning, through the myth of metamorphosis. My discussions with researchers about nanotechnology are often enriched by narratives of positive transformation in which the body, or sometimes the whole of nature, is subject to creative transformations made possible by material manipulation at the nanoscale; but why not? Tales of metamorphosis are ancient and persistent in the human search for meaning, science being one source of such tales. Myths of metamorphosis are at the heart of the various visions that arise in science fictional accounts of possible nanotechnology futures, like Julian's:

JULIAN: *There's this nano sci-fi book called,* Prey. *It's fun.*
ROSALYN: *Is it fun? OK. What if you could write your own version of a futuristic nanoscience novel?*
JULIAN: *Yes.*
ROSALYN: *More fiction science than science fiction, meaning there is a lot of basic, scientific knowledge in it but also incorporating your fantasies about nanoscience and what it will allow us to do. What would your book be?*
JULIAN: *What a great question! I think I would hit to the bio side. What was that movie? There was a movie about exploring the innards of a person, some science fiction movie. They shrunk something that looked like a spaceship down to the size of a sail. I would build nanocrystals that would go into the body that could transmit information to the outside world via wireless connections. Maybe even to the person or to the doctor or whatever and that person would be able to inspect, even fix the body parts. These would be wonderful sensors to say what's going wrong.*
ROSALYN: *Hum.*
JULIAN: *Now I still need a good-looking woman and a hero.*
ROSALYN: *Go for it.*
JULIAN: *And a conflict.*
ROSALYN: *It's the concept that's the beginning point that's important, right?*
JULIAN: *I would have an array of these electronic devices in the body to inspect the body, giving the signal wire out. They would have the right biological connections to these nanocrystals which would be semiconductors, and to other crystals that are relative to the right organ of the body.*
ROSALYN: *That would be incredible, so these nanocrystals would be intelligent?*
JULIAN: *They would be.*
ROSALYN: *And, they would know what kind of tissue they were looking for?*
JULIAN: *Right, because before I would insert them, I would attach something that knows how to find liver cells, for example.*

ROSALYN: *Liver cells?*

JULIAN: *Or heart cells, any cells ... One of the causes of aging is cosmic rays.*

ROSALYN: *Sure, that makes sense.*

JULIAN: *Cosmic rays are also going to damage these nanocrystals in the body.*

ROSALYN: *Well, if they are in the body, would the cosmic rays be able to penetrate to the crystals?*

JULIAN: *Absolutely, the same way they do the cells. And so there are all sorts of things you have to do. You can easily imagine one atom out of place in this nanocrystal, causing a signal—hey, there is my conflict. I was looking for a conflict for my novel.*

ROSALYN: *That's it, you have got one.*

JULIAN: *So, so.*

ROSALYN: *I just thought of another one. What if you could inject into the body something that repelled the cosmic rays, kept them from actually penetrating the cells of the body?*

JULIAN: *Very hard to do from a physics point of view.*

ROSALYN: *Because these rays are so powerful?*

JULIAN: *More interestingly maybe is to ask, "How can I heal the damage of these nanocrystals? I know that they are defective but can I heal them?*

ROSALYN: *OK. Could you have them rebuild themselves if they are damaged?*

JULIAN: *That's right.*

ROSALYN: *The way you imagine it, does this nanocrystal become a permanent part of the body? Of course, we are still just playing in the wheel of the fantasy.*

JULIAN: *I haven't thought about that. Probably not. I think you would like them to dissolve.*

ROSALYN: *Well, the notion of the cyborg comes to mind when you have these intelligent nanocrystals in the body. There is something to me personally, that is horrifying about these crystals being permanent, and remotely controlled—something horrifying about it.*

JULIAN: *It's interesting about the brain. I think in the next 50 years, maybe next 100 years, the society will really begin to learn and understand the brain. And then I am much better prepared to sort of think about how I would use nanotechnology, OK, but the plumbing of the system, the plumbing of our bodies, blood flow, digestion, I think we understand pretty well. So I am quite prepared to imagine how nanocrystals can be used.*

ROSALYN: *You mean from the neck down, right?*

JULIAN: *From the neck down, that is exactly right.*

ROSALYN: *Because we don't know enough to mess around with what to do in the brain?*

JULIAN: *I don't know how to mess around with the brain, that's right, that's right. So I can't even put together the scenarios.*

ROSALYN: *Yet there are those researchers who actually do mess around with the brain.*

JULIAN: *Yes, such as with nerve transport. That's current, moving from this neuron to that neuron.*

ROSALYN: *So you feel secure enough with our knowledge of the plumbing of the body, to imagine using nanocrystals inside of it?*

JULIAN: *Absolutely.*

ROSALYN: *To repair it, to restore it, heal it?*

JULIAN: *Yes.*

ROSALYN *OK, what an amazing dream.*

JULIAN: *I think it is an amazing dream, but I really do believe that it is possible to make some combination of electronics transmission, and so on and have it be biologically sensitive to what's going on in the body and compatible to the body.*

CHAPTER SEVEN

Noble Revisited

Ricoeur (1995) said that to be human means being estranged from oneself. Although destined for fulfillment, all humans are inevitably captive to an "adversary" greater than themselves. He saw that although free and determined, human beings are both responsible and captive. What might be the "adversary" that captivates the human soul? What might be the sources of alienation from self? Ricoeur located the estranged human being of today in a world that is empty of meaning and hope, in contrast to early humans who held fast to belief through powerful symbolism. I want to suggest that the adversary causing the estrangement can sometimes be relationships between humans and the very technologies they have created for purposes of fulfillment and control. He suggested that through the power of myth, the nature of the human being is elucidated, and contemporary technological humans can recover the sense of the sacred that has been irremediably lost.[1] What it means to be human is a matter of constant and evolving social negotiation, as well as a matter of private, personal investigation. The meaning of being human, as a dynamic perception and ongoing search, is determined in part through belief, including the mythical and imaginative elements of it. Ricoeur maintained that wholeness of the soul is achieved through using metaphor as an ally for understanding and articulating faith. He saw the imagination as generating new metaphors for "synthesizing disparate aspects of reality that burst conventional assumptions about the nature of things" (p. 8). In its aim to more deeply understand and control material reality through language, Nanotechnology research has begun to burst conventional assumptions about the nature of things.

[1]Ricoeur, P. (1995). *Figuring the sacred: Religion, narrative and imagination*. Minneapolis: Augsburg Fortress.

Ricoeur (1976) stated, "To mean is what the speaker does. But it is also what the sentence does" (p. 19). Ricoeur saw poetry and scientific language having in common that they "reach reality through a detour that serves to deny our ordinary vision and the language we normally use to describe it" (p. 67). The literal sense is left behind, collapses, so that the metaphorical sense can emerge and work its redescription of reality. He saw this redescription guided by "the interplay between differences and resemblances" that gives rise to tension at the level of utterance (p. 68). It is from there that a new vision of reality arises. Metaphoric language, used to describe and explain nanotechnology, reshapes perceptions of material reality and thus, of the self.

Nanotechnology quests, like all newly developing technological ambitions, are also quests for personal fulfillment. As such, they entail the search for selfhood. Desire for fulfillment leads humans to create new technologies, but new technologies and devices cannot provide answers to the fundamental and perpetual question of humanity, "Who am I?" Technology-driven worlds can thus become adversarial, captive but not fulfilling, when serving as a primary source of selfhood. The way out of this paradoxical problem is to possess the self, as Ricoeur (1995) suggested. How might the nanotechnology quest either address or hinder the possession of self? In other words, when technology is used to obtain precise control of matter through human will, is desire fulfilled, is freedom acquired, or is the self lost even deeper in the captivation of technological prowess?

In Ricoeur's assessment, the individual person longs for the values and forces once felt by primordial people but lost to technological life. In agreement with Ricoeur, I suggest that the condition he spoke of is not inevitable. It is possible to recover what has been lost to technological life (i.e., keen awareness of and sensitivity to the body the individual's delicate yet vital connection to nature / Earth / human other; awe of the sacred) through engagement with the symbolic elements of belief that function in the process of meaning making about being human in a technology driven world. Myth can serve that function.

Myth, for Ricoeur, is traditional narration, relating to events that happened at the beginning of time. Its purpose is to provide grounds for ritual actions, and generally to establish forms of action and thought by which humans might understands themselves in the world. His treatise suggests that by elucidating the nature of the human being through the power of myth, humans can recover the sense of the sacred, that has been irremediably lost.

THE MYTHIC RELIGION OF TECHNOLOGY

Some of the more public proponents of nanotechnology (i.e., its visionaries, venture capitalists, and government sponsors) tout the fields of nanotechnology as offering potentially transformative new ways of confronting some of the otherwise insurmountable problems before humanity. Rhetorical claims suggest that, through nanotechnology, we may have finally found the answer to the likes of cancer, pollution, energy needs, keeping up with Moore's law, military vulnerabilities, aging, and a slowing economy. Most behind-the-scenes laboratory researchers are more cautious and less dramatic in their proclamations about what nanotechnology might offer than the more "public" expert figures. But nearly all point to significant improvements in an array of applications, whether they are in physical or mental health, the economy, military operations, or creating novel materials and new consumer markets. Perhaps the nanotechnology future will, in fact, turn out to be as rich as it is variously and fantastically projected. Is that what this nanotechnology revolution is about, meeting human needs and providing for improved qualities of living? Or, is it about something else altogether?

In *The Religion of Technology*, (1995) Noble defended the conviction that some technologies have not met basic human needs because "at bottom, they have never really been about meeting them." Instead, Nobel indicated, they have been aimed at the loftier goal of "transcending such mortal concerns altogether" (p. 206). As examples of how technological pursuits develop inside of mystification, Noble pointed to the development of the atomic weapons program, the nuclear arms race, the space program and NASA, artificial intelligence, and genetic engineering. According to Noble, religious aims such as eternal life become associated with technology; making the technological enterprise an essentially religious endeavor caught up in the quest to recover humankind's lost divinity. Noble characterized modern technology as "the religion of technology," a faith that "fires Western technological imagination," a rational pursuit that is driven by "spiritual yearning for supernatural redemption," and a pursuit that is "wise about the world, but inspired by other-worldly quests for transcendence and salvation." The problem, as Noble defined it, is that technology is purported to be designed to advance humanity's material position within the world, but in fact the merging of religious belief with technological promises diverts resources away from the use of technology for the material betterment of humanity. Noble further surmised that a secularist polemic and ideology obscure the connection of religion to technology: "Thus, unrestrained technological development is allowed to proceed apace, without serious scrutiny or oversight without reason" (p. 207). For Noble, as long as technology is tacitly connected to religious myth, there is the potential for injustice and disregard for authentically humanitarian concerns. And so he sought to identify the religious mythology enmeshed in technology so that we might "disabuse ourselves of other-worldly dreams" and redirect our technological capabilities toward more humanitarian ends.

I put forth my own hope that nanotechnology development will evolve ethically, with earth-respecting, humanitarian intentions. That entreaty may be confounded if Noble's thesis is correct; that we have been abiding by a system of blind belief; and that our technology portrays a "disdainful disregard for; indeed an impatience with life itself" (p. 208). What if the rapid emergence and development of nanotechnology is in fact steeped in mythological fantasy, caught up in the ancient human search for meaning and something in which to believe? Would that mean a tremendous waste of resources? Would that spell increased prosperity only for a relative few? When nanoscale research manifests as appropriated technology will it, too, represent a disregard of humanity, a disdain for life itself? Will its applications and uses be unjust? Challenged by Noble's assertion, the ethical development of nanotechnology requires that any pseudoreligious aspirations that may be embedded in the development of nanotechnology be acknowledged, identified, and dispensed. In particular, a great deal could be learned from making explicit any religious mythology that may be embedded in the rhetoric of policymakers who are leaders and advocates of nanotechnology development. If those people with the power to appropriate public funds to nanotechnology development have entangled personal faith aspirations with technology policy, then this should be known and understood.

For example, California Congressman Mike Honda (2004) said in a speech about nanotechnology that "words don't bridge the valley of death. That takes money." One could consider that statement as a reference to the 23rd psalm or other mythical use of a "valley of death," to be meaningful and symbolic. Senator George Allen referred to the Nanotechnology Development Act as "the single largest, federally funded, multiagency project since the space program of the 1960s." With a potential worldwide impact of $1 trillion, he touted nanotechnology as "the key to the future," and said that he was "thrilled with the endless potential, the endless possibilities" it represents. What is an "endless possibility" other than a mythical language alluding to notions of the infinite? Might "endless potential" reference mythical promised lands? Might the rich rhetorical features of his speech point to an entangling of myth with technology policy? Senator Allen himself is concerned about the misuse of myth. He spoke of the myths portrayed about nanotechnology in books like *Prey*, and their ability to undermine research much like GMO[2] research was halted by the loss of public confidence. To keep myth at bay, there will be, he said, a National Nanotechnology Preparedness Center to educate the public, advocate nanotechnology, and assure that the public will not be frightened. It is fascinating that, in this rhetorical situation, myth is on the one hand being portrayed as a threat, whereas on the other it is being used as a tool of persuasion.

The power of nanotechnology and its unbridled development in the direction of anywhere is contained in its tacit, mythic elements. "Unmasked, it begins to lose its power." Stahl (1999) extended the conviction:

> If technology practice is seen as "objective reality," as determined by the needs of the machine, then we have no options. All we can do is to accept our fate and "adapt" or "get ready for" what the machine has predestined for us. But if we see technology as an implicit religion, as wrapped in myth and mystification, then it becomes discourse. It becomes one way of talking about the "real world" alongside of others. We become free to weigh, evaluate, and ask questions. What are our ultimate values? What kind of future do we want for ourselves and our children? (p. 34)

And this further reinforces the importance of the narratives of individual researchers, most of who work behind the scenes, overseeing laboratories and graduate students and contributing to the global knowledge of science and technology. Is their work in the laboratory entangled with a religion of technology? What is the significance of their own personal faith, religion, or mythological fantasies in light of the work that they are doing both as curious scientists, and as the primary instruments of the nanotech-

---

[2]The acronym GMO stands for "genetically modified organism," and was first used years ago to designate microorganisms that had genes from other species transferred into their genetic material by the then-new techniques of "gene-splicing."

nology initiative? I continue to address these questions in my study. But for present purposes, it seems there may be a crucial role for mythology in the ethical development of nanotechnology, particularly if that enfoldment is to be conscientious in its direction.

The coming pages contain the complete transcription of a discussion with Kent, which is quite rich when understood in terms of mythology. Kent's narrative underscores the assertion that technology practice is discourse and, as such, it leaves open the freedom of choice about this venture into uncertainty—the enfoldment of our possible nanotechnology futures. But, before turning to that discussion, I first explain a bit about the meaning of myth.

MYTH

Ricoeur (1967) defined myth,

> ... not as a false explanation by means of images and fables, but a traditional narration which relates to events that happened at the beginning of time and which has the purpose of providing grounds for the ritual actions of men of today and, in a general manner, establishing all the forms of action and thought by which man understands himself in his world. (p. 5)

Myth provides a sense of security and a feeling that gives a sense of meaning. Jung (1993) suggested that mythology can be an expression of religious inclinations. As such, it is constructed to provide a sense of meaning and control to shared experience. It serves to guide, inspire, and enable us to live in an otherwise uncontrollable and mysterious universe. Jaspers (1978) saw science as the modern myth, but incomplete. He wrote that like so many earlier myths, science appears to explain the natural world around us. But it can only answer how things happen, not why. But—unlike Jung, Jaspers, and Ricoeur—for Noble, myth is conceived of as a threat to realism, a falsehood juxtaposed to that which rationally, truly is. For Noble, when portrayed in the guise of technology, myth can "blind us to our real and urgent needs." My own understanding of myth, for the purposes of this treatise, is as a cognitive processing tool, with the very important functions of ascribing meaning and purpose to human life, and to reconnection with the sacred, especially for those who are technologically dependent.

Myth is the natural and indispensable intermediate stage between unconscious and conscious cognition. Jung (1933) understood every myth to represent an "important psychological truth." (p. 217). Human seek to adequately frame such an ontology that explains human identity, purpose, and meaning in the universe. And, despite claims to the contrary, science seeks the same. Unfortunately, in its various searches for meaning, humans of modernity have attempted to supplant myth with science, believing that in doing so we can secure for ourselves a material explanation and meaningful connection to an existence that is otherwise unexplainable. Maybe the mythological elements of science are critical to that quest. However, that being said, identifying any myths that may be at work inside of nanoscale science endeavors offers the increased likelihood of forming ethics to accompany progress in nanoscale technology. Given its aim for precise control of matter, to include recognition of the mythological elements of nanoscience in formulating an ethics for its unfolding, is the only way I can see possible for conscientiously directing the development of nanotechnology toward truly humanitarian aims.

KENT

Kent is a physicist by formal training. He spent many years working inside of a U.S. National Laboratory. Now he is on his own, working independently with private funding for exploration of his own ideas. He is an advocate and spokesperson for the development of nanotechnology. Our discussion follows:

KENT: *I will give you a glimpse of some other things I am working on. I tend to have this notion that we are compelled to be emotionally driven by what I call the invocation of rapture index.*

ROSALYN: *Slow down, the invocation?*

KENT: *Invocation of rapture index.*

ROSALYN: *OK.*

KENT: *Of which nature is the highest quotient. Why is that? There is a reason for this. Very much in the same way that a mother responds to her child as superseding all other criteria.*

ROSALYN: *So that's a dependency relationship. Are you suggesting the same between us and nature?*

KENT: *Yes, that's what I am getting to. So, I will toss this out.*

ROSALYN: *OK.*

KENT: *There's a scene in* West Side Story *where Tony meets Maria and everything fades into the background. Same thing would happen in a case of a mother and child.*

ROSALYN: *Sure.*

KENT: *This is the absolute zenith of their attention. So as a collective organism, as a sort of a collective system, we as a species were designed on purpose. I find it extremely interesting that across all boundaries of social, racial, whatever—humans have these very specific common features. One which is important is that we all tend to respond to nature. It's a very compelling one. The shape of the land, the look of the clouds in the air, the ripples of water, just the way that flowers appear, just the whole composite of a nature scene. There is something incredibly engaging about this. Why would that be? I think it was designed as a protection system.*

ROSALYN: *A protection of us?*

KENT: *No, no.*

ROSALYN: *To protect what?*

KENT: *A mother wants to protect her child.*

ROSALYN: *Are you talking about a Gaia perspective?*

KENT: *Yes, exactly.*

ROSALYN: *OK, so we are talking about "Mother" then.*

KENT: *Yes, that's right. When they walk together, right, what's interesting to me—and I am not a theologian by trade, although I have studied a lot of religious philosophy, just because I wanted to find out more why this stuff is and how it works—OK, so if you look at virtually every individual's culture and if you have ever had any knowledge of going backwards in time, it's strictly the search in the area of language and ritual where you find almost basically the same common three elements. It's extremely consistent. A, that life extends beyond this physical expression we call a body, and B, all living things are somehow more or less connected in un-seeable yet tangible ways and, C, the planet itself is some kind of a living system.*

ROSALYN: *Of course.*

KENT: *Now in the land that you can see, or the environment that you can see as far as the eye can perceive, there is something alive out there and we are a part of that living thing.*

ROSALYN: *Absolutely, yes.*

KENT: *And so anecdotal evidence would kind of suggest rather strongly that despite the incredible variety of humankind, and all of these differences, we see the same thing and translate through their own mechanisms of appropriation. That's interesting. That's very compelling data. So why would that be relevant now? Here is the reason why; because we have had this 2.5 million years or so of fumbling around. Maybe the last handful of centuries was like a spike of development, in a bumpy way. You stepped across certain thresholds, and given the reasons of technological organization, the complexity, and the different kinds of sociological systems pro-*

> *posed, there is a vertical line in terms of nanotech, info tech, bio tech and what I might call conscious tech, cramming together into one spot. And yet the spiritual health index, by my words, is still kind of a flat line down here some place. In other words, up here is the acceleration of the technologies/social complexity group. Down here is this spiritual health index. For us to be able to graduate to this next evolutionary event plateau, or this next increment, which is rapidly pushing up problems, we have to see those two coming close together. It's a test ....*

All humans share some elements of awareness, elements expressed most directly and fully in the unconsciousness mind. Jung identified those elements as archetypal, which often expressed mythically, cross the domains of culture, race, and time. For Jung, the basis of myth is the collective unconscious province of archetypes. As a mythological externalization of unconscious subject matter, archetype designates only those psychic contents that have not yet been submitted to conscious elaboration. Jung placed a very high value on the function of archetypes, calling them "inalienable assets of the human psyche." He urged that we not attempt to destroy them, nor to deny then, but rather to dissolve the projections in order to restore their contents to the individual who has involuntarily lost them by projecting them outside of himself. I want to suggest the possibility that archetypes, which function to aid the human mind to make meaning of its perceived experiences in the body and world, are entangled in the quest for nanotechnology development. Kent is one scientist who provides a rich example of archetypal symbolism in meaning making about nanotechnology:

ROSALYN: *OK, is this your natural pace to talk this quickly or are you rushing?*
KENT: *Actually, I am slowing down. Sorry.*
ROSALYN: *I just wanted to check in with you.*
KENT: *I talk profusely.*
ROSALYN: *I just don't want you to rush.*
KENT: *No, no. I just blast off what's going on.*
ROSALYN: *OK.*
KENT: *I think that we are all connected.*
ROSALYN: *Of course.*
KENT: *And for those who choose to connect in this way, we have to go through the clumsy method of forming words.*
ROSALYN: *The body can be very inhibiting.*
KENT: *It's very slow; it's like a slow serial port.*
ROSALYN: *Yes.*

KENT: *As opposed to a very high throughput port that offers new paths and levels. So, within a minute that I saw you I said ...*

ROSALYN: *You knew.*

KENT: *OK, yeah there is a glow that is there.*

ROSALYN: *Hum.*

KENT: *There is a receptor site. So.*

ROSALYN: *OK.*

KENT: *So here we are, we can skip some. The next square—just to kind of see where there are hitches—goes as follows. It is imperative, not because of us humanoids being here, but because this is a frame of reference that transcends, too. OK, life is very common. I think life is as common as ...*

ROSALYN: *Why wouldn't it be in a material universe?*

KENT: *One could say: "Look at all the wasted space on your planet." I am actually a member of something called Contact Consortia, and we have a conference that we call Contact. It's a whole bunch of folks—biophysicists and other biology people and physics types, mixed up with theologians and philosophers and science fictionists. It's a very interesting ... you would like this group. It's very much into space. We have a 3-day long conference and in a very serious way we try to discuss what do we, how do we do this, and whether we are ready. You know, what is the test that we have to somehow pass before we can integrate this larger equality up there?*

ROSALYN: *Integrate with the larger ecology?*

KENT: *Integrate with the larger ecology, yes.*

ROSALYN: *OK.*

One archetype is "The Test," which is exceedingly difficult and determines ones' fate. In literature and other art forms The Test is ancient, persistent, and found in many forms. Here, in Kent's imagery of integration into the "larger ecology," it is both personal and universal. Rebirth is also a primordial affirmation based on archetypes. It includes two groups of experience, transcendence of life and ones' own transformation. Jung (1981) defined metempsychosis as a transmigration of souls, where life is prolonged in time by passing through different bodily existences. This is contrasted to archetypal *rebirth*, which occurs within the span of individual life where the idea of improvement is brought about by magical means. The subjective transformation is represented as fateful and an event that gives rise to the hope of immortality. Enlargement of personality is one feature of this process. Various visions of the human material condition being transformed through nanotechnology's ability to enhance bodily functions, to detect and eradicate diseases that otherwise kill the body, and so forth, could be understood as examples of the myth of archetypal rebirth. Kent continues in the vein of archetypal rebirth:

KENT: *So, here is the problem. Let's just pretend for a moment that life is common. That will be a defense argument. Second of all, let's just say that you had continents on this earth that were separated by great oceans and so that made it possible for us to be ready to travel at a certain point. Sometimes you would have the road lone traveler that would stumble across some other kind, but for instance, I think that North America was visited many more times than just when Christopher Columbus happened to wander over here. We have folks from every part in the world, miscellaneously cross, criss-crossing multiple times. In some cases, it was just a little blip and then it kind of faded away and in other cases the interaction was catastrophic. It caused horrendous disruption and complete chaos from various corners. So, I tend to view evolution as a time in this process that has provided periodicity and amplitude in the trauma cycles. Don't overexceed the system's capacity and it's fine.*

ROSALYN: *Is the trauma somehow necessary?*

KENT: *It is necessary. It is a disruptive process from which a larger, more robust one comes, whenever it comes out. Let's try this on a larger scale. Let's go to the cosmic ecology out there. Let's just say for a moment that every once in a while somebody goes completely under. They are still kind of wrestling around with that. So, let's just say that on our own accord, we have to cross a series of tests. Again, trauma cycles, before we are even in a sense allowed the privilege to kind of get out there, to present ourselves in some context. Because even right now, we can glimpse at some of the toys we could play with, with the string theory kind of coming to life, and all that kind of stuff. We are sort of playing with these things, like building blocks. We are not quite ready yet. But are we going back to the spiritual health index. Are we really ready for that? Would it be a traumatic experience from which we could not recover or which would cause us irreparable harm, or should we have to do our own internal pharmacycles to prepare us as a growth mechanism? I would offer furthermore that half of the majority doesn't make it. It's a filtering system. But what they want you to do when they come together, even though there might be some traumatic interaction more or less because of the ecology, does tend to become more or less. So we are kind of at the threshold now. It's kind of like driving up to Las Vegas where you see the glow of lights at every place in the city. We are sort of seeing the glow, in our own clumsy way, trying to figure out what to actually do.*

ROSALYN: *We are not ready, I suppose.*

KENT: *We are not ready, I know. Well, some of us can get ready.*

ROSALYN: *You are saying that collectively we can't go yet.*

KENT: *Collectively, yes. There is a very tiny fraction of us folks who are sort of nibbling at this in our own separate ways. But what I do feel is that there is a conversion syndrome occurring.*

ROSALYN: *Yes? Then what do you imagine or envision will be the destination?*

KENT: *Maybe not a destination, but a flow pattern.*

ROSALYN: *OK.*

KENT: *Inside of consciousness.*

ROSALYN: *With consciousness as an instrument that helps us get …*

KENT: *An instrument for influencing our future and our surroundings. So we are kind of stuck in these biophysical organs, called bodies. We are a great base level in the system. There is something else I call the base potential that allows us to have a certain power of abstraction, to conceive of them, and to perceive intangible physical means. You have to have faith in the improbable to understand that there is something beyond this, and that's another one that is designed and sort of featured. It gives us kind of a tool to get this far. So there are stages. We are still a 0, although we are being prepped to move up, like software being sourced. But, if you go to Stage 2 and beyond, then the boundary point between physical matter, like this body here, and this sort of quantum out there, becomes intertwined. And in fact quanta look like God to us. Because they materialize what they wish as they need to. The rules of time and space work very differently out there in that quantum as they do in this time universe here, and yet the ability to step into back and forth and translate from matter to energy and back and forth as a kind of whimsical thing to do, that's very dangerous if its misused. So the species in question has to be conditioned, like a child, to go through its probability before it gets to the point where it sort of deserves the capacity to do this. Now, to kind of put it in an ethnological perspective, you may recall back in 1946, 1947 there was an island off the Coast of Fiji; it was a businessman and an anthropologist who dropped in on a plane. The local island folks were there and they said: "Oh, my gosh!" The plane was kind of a God who just visited. They never saw a plane before. To them a silver burrow with people coming off, that was God, that was it. So to us, quantum beings from Stage 3 let's say, would be way beyond being sent by God. It's not that they are necessarily God in some mythical way, it's more like there is a hierarchal structure of different types of beings in different stages which in a sense is like parent to child in sort of a larger scale, you know that kind of hierarchy, and we are just at the very bottom as we speak. We may go through all kinds of trials and tribulations before we stumble on, but at least we have gotten this far. It's also interesting that we have been allowed the privilege at least to get a glimpse of what could be.*

ROSALYN: *That's the choice we have made?*

KENT: *It's a choice. And so you take this a step further and say OK, fine. Look at the religion systems. They are like a plug in the Windows desktop.*

ROSALYN: *Yeah?*

KENT: *No, I am serious. As children grow up, they no longer need to have this as a way to have structure in their life. If you become more mature, you are allowed the freedom to make a different range of choices.*

ROSALYN: *Yes.*

KENT: *So as a collective organism, we are migrating toward this flow pattern. Maybe we don't need a rigid religious doctrine as opposed to having a better understanding of how we fit into this type of mythology, of what our own personal responsibilities are, that kind of thing. I think some of us do, many still don't. It's a very uneven distribution.*

ROSALYN: *Yes.*

KENT: *So, so where might this be going? It is just my personal theory. Here I do nanotech business; out there I do something very different. Here on the West Coast is something called the Super Nuetic Science, they are a system.*

ROSALYN: *The Super Nuetic?*

KENT: *Yes, exactly. They are kind of like a parallel version of the Monroe Institute. I kind of dabble in these territories, because they are in various ways trying to look at the intersection between biophysics, state of mind, quantum physics, and ways to measure, translate, convey this into some kind of process that can be used not so much as a personal growth, but also to develop some larger framework. So I often use the bowl of Jell-O as a metaphor. Imagine a bunch of fruit in a bowl of Jell-O, you perturb the bowl, the Jell-O wiggles around, so does all the fruit in it. It may be that the fruit doesn't know that it is spinning. It doesn't matter. If there is a superficial continuity that happens to translate into a sort of quantum interaction, and if I happen to believe that all living things do respond to this quantum, it would tend to suggest that complex systems that behave biologically may not even be composed of biological things at all, but they have biological behaviors, so therefore they are also subject to this. That's the lesson that we may be just on the edge of learning.*

ROSALYN: *Which suggests that there is some kind of pulsation from the Earth that synchronizes all systems, including nonbiological systems?*

KENT: *It could be as a majority we are learning about how to take the pulse of consciousness, but also learn about how we can apply our consciousness directly as an instrument to sort of shape the perturba-*

tions that are going on this time. Because if we don't .... Well, we have two choices. We either proactively drive the history of our future or we can simply be the recipients of whatever is thrust upon us. At a certain point in time you are supposed to learn how to do this.

ROSALYN: *Choice one being my hope for nanotechnology.*

KENT: *Right. In ancient times it was done through ritual and through practice and we were prompted with the culture of the moment. But people lived a different life. I mean, life was kind of slower, it was attached to the ground and you were very sensitive to your surroundings. For most ordinary folks, it was pretty good. Now we live in a world where we are at an extremely fast pace, complexity keeps going up, we are time constrained downward and even productivity has increased overall. At the same time we are living in a world where we are exposed to every conceivable kind of energy field one can imagine. All the way up to 60 Hz. So it would be as if somebody looked through a 100 million candlepower aircraft. Somebody blasts it in your face, and says: "Look at this candle." You probably wouldn't see the candle very well. So we have to relearn how to see the candle and yet for most folks, we don't have the time. We are raising families. We have homes to pay for. We are traveling. Life is pretty complex. If you could wear a little gadget like a complexity meter, you know like a 1 to 10, you would be pegged at 10 all the time.*

ROSALYN: *Like an alarm system all the time.*

KENT: *Yes, exactly right. So to compensate for that is in fact part of the larger design structure. Now that we have these interesting clever tools to work with, we could use these tools to recapture what was organic knowledge of the previous era, but reemerging, amplified, sped up a little bit through tools we are just beginning to look at now.*

ROSALYN: *I have been afraid that what we are doing with our new devices will separate us even more.*

KENT: *It could go either way.*

ROSALYN: *Right.*

KENT: *And I understand this phobia, because believe me I ...*

ROSALYN: *It's not a phobia. It's a concern.*

KENT: *I was actually up to the nano workshop 2 years ago. It is interesting that we try to grapple with this concept in its own sort of interesting way. [The conference] is an enormous event with all these different little topics.*

ROSALYN: *Yes.*

KENT: *And yet there were people there from these different organizations, one called Psychological Systems Analysis, and people like Peter Russell, who works on the brain. All of them were trying to tie this altogether. The point is there is this rapidly growing awareness that has become an absolute necessity as we speak.*

ROSALYN: *For survival?*

KENT: *There is an urgency to say whatever it is however badly it is mishandled; we have got to try something, because this is it folks, the doors are open and the light is coming to us and here we are. So that's kind of the premise. Along with that I just kind of wanted to put this out there, because it might help answer "who the heck am I?" and "why am I here?"*

ROSALYN: *It does help a lot. You use the word "opening."*

KENT: *Right. There is a community here we call Damnhur. Damnhur is a community of about 5,000 people. It's a secret, mind you. They built this incredibly elaborate temple. They burrowed into a vein of one of the hardest minerals ever. But the point is, there are these supposed like noble sites on the Earth like this valley called Damnhur.*

ROSALYN: *Have you been there?*

KENT: *I haven't been there, but friends of mine have. Do you remember the Tesla society?*

ROSALYN: *No.*

KENT: *They have a chapter in our society, so I have been told. So about a year and a half ago, the co-founder of Damnhur made her first public appearance ever. She came to this Tesla meeting and I was there. She explained that 5,000 people are now at this thing, and the whole point was to create an alternative society. If you search their site you can download the whole portfolio. It's quite something. They acquired a valley, one that used to have a plant that then got closed down. The whole valley kind of just shut down. So they quietly came in and bought this valley, burrowing into this vein. (The local villagers just had no idea what was going on.) Their general vision of the Earth is kind of along the same lines that we are following here except that they have their own way of responding to it. Meanwhile, here in this neck of the woods we have an ever-growing number of groups coming together. What I find really fascinating is the cutting across all of these different areas, science and medicine and different areas of knowledge. Can you feel the draw like moths to a light, like there is something there? We don't know what it is, but we know there is some reason for us coming together. There is almost like a hunger, like a thirst that wants to be quenched in some way. And, it's amazing that in just 2 or 3 years, the caliber and the range of folks who have your level, who woke up all of a sudden. Whatever the hell was in it before, forget that. This is where we should be going.*

ROSALYN: *Yes. Can you connect it to nanoscience?*

KENT: *Yes, I can.*

ROSALYN: *How so?*

KENT: *So here is the answer. You know I give my talks on nanotech; the very first thing I say is go outside and have a look. It's like it's up a big tree, or maybe look at a whale or something, that's nanotech. Nanotech is not little gadgets or little machines or any of this annoying nonsense. Forget it. Just stop all of that. There is a lot of fluff and frills. Nature has been doing nanotech for billions of years.*

ROSALYN: *Well, it's the only master of nanotech that we know.*

KENT: *That's right, so I mean if you look at the cell, look inside, what do you find in there, you find ribosomes, mitochondria, protozoans, whose primary purpose is to exchange messenger proteins back and forth and then they rake up and reassemble more.*

ROSALYN: *Reassemble more. Yes.*

KENT: *So, guess what we are doing as we speak. We are sort of stumbling into the area, one of these areas reserved to God up until recently. In fact, let me see if I can use the old legalistic insurance clause. When you buy insurance for a house or whatever it is, they always talk about "acts of God," the weather, and some other unknown thing that we can't control. Well, as our technology tools get ever more prolific, like we are actually able to adjust the weather—both the U.S. and the Soviets have been doing this for years—we can actually create rain, we can make tornados go, all that kind of stuff. So, is that an act of God? So same thing here, we are just now unraveling the very biophysics of what makes this stuff work. So pretty soon we are going to be able to be God in that context.*

Jungian archetypes function to predispose humans to approach and experience life in certain ways, organizing percepts and experiences to conform to that pattern. This, Stevens (1990) explained, is what Jung meant when he said that there are as many archetypes as there are typical situations in life: "There are archetypal figures (e.g., Mother, father, child, God, wise man), archetypal events (e.g., birth, death, separation from parents, courting, marriage, etc.) and archetypal objects (e.g., water, sun, moon, fish, predatory animals, snakes). Each is part of the total endowment granted us by evolution in order to equip us for life. Each finds expression in the psyche, in behavior and in myths" (p. 39).

In chapter 5 I told the story of a man's reaction to seeing a Fur-de Lance snake in the road. In that case, it was the archetypal snake that was being encountered. Kent invoked numerous archetypes, including Mother, separation from parents, and outer space:

ROSALYN: *So have you read Genesis lately?*
KENT: *No.*
ROSALYN: *Have you ever looked at it?*
KENT: *Maybe I should.*
ROSALYN: *Take a look at Genesis. It's an incredible book. There are two creations in it. One of them is where God uses the plural pronoun to record itself and says: "They are trying again to be like us." Now, I don't think we were ready at the time.*
KENT: *Sure.*
ROSALYN: *So, were we put into a place to be controlled?*
KENT: *Well, we were small children.*
ROSALYN: *We were small children. We were not ready.*
KENT: *Small children.*
ROSALYN: *The book of Genesis suggests to me that there may be a time when we are ready.*
KENT: *Sure.*
ROSALYN: *To approach that reality for ourselves.*
KENT: *Right.*
ROSALYN: *It's a really beautiful writing, but very much misinterpreted.*
KENT: *I am sure. In fact I had a book given to me with the missing books of the Bible, some of the missing pieces. But, but the point is I think you are right in a sense, that we are still yet to figure out how this works, but for some reason we were given little blips of information.*
ROSALYN: *All the way along ...*
KENT: *All the way along. Little upgrades, just like software upgrades. We will adapt accordingly. And that's part of the test process.*
ROSALYN: *Yes.*
KENT: *OK, so here we are in this time. I call it the God boundary because ...*
ROSALYN: *The God boundary?*
KENT: *Yes, because we used to compare these multiple things to something out there versus what we can control here. Well, that boundary keeps moving further and further out. Now we can play with the weather, now we can make arts of life work, now we can break apart, reassemble matter in a way that doesn't necessarily abide with the goals and that kind of thing.*

Sometimes archetypal images of rebirth are borne out of the notion of regeneration and emergence, giving rise to personal commitments of scientific discovery and ingenuity. As such, metamorphosis becomes a promise of nanoscale science and technological development. Marina Warner (2002) cites metamorphosis as the principle of organic vitality. In this mythical image forms take on different forms and the whole of nature evolves through creative power. With each new transformation, the new shape more fully expresses and perfects the transformed subject. Through her study of fantastical literature, Warner finds that metamorphoses often arise in crossroads, or points of interchange. From these discussions with nanotechnology researchers I surmise a metamorphosis arising at the crossroad where the observation of nanoscale phenomenon meets the manipulation of matter toward technology. Warner's studies demonstrate that the languages of science profoundly affect visions of metamorphic change—an organic process of life that keeps shifting. Seeking to understand "what are behind these changes?" she finds that the new and the strange lure and delight us; that stories of change offer an intrinsic pleasure and a freedom to enter a new world.

ROSALYN: *If we aspire to God-like qualities and the ability to create or mimic creation, then I have concerns that if we do that prematurely we will really screw it up.*

KENT: *Oh yeah, you will screw things up really bad. That's part of what's supposed to happen. That's my point.*

ROSALYN: *I don't think it's been determined yet which way we are going.*

KENT: *That's right, OK; here is the analogy I have come up with. In the film* 2001 Space Odyssey *...*

ROSALYN: *Yes?*

KENT: *It starts out that the eight figures, they are all hanging around.*

ROSALYN: *Yes.*

KENT: *And then one of them, all of a sudden, picks up a bone, a tool. First thing he does is he bashes his neighbor's head with it, a metaphor of technology. So he has frozen the moment there, and all of a sudden there is a spacecraft. Well, we are at the point where the bones are spinning off.*

ROSALYN: *OK.*

KENT: *We don't know how long its going to be, but we know it's up there spinning about.*

ROSALYN: *OK.*

KENT: *A friend of mine has a house right up on the edge of the peak, and the only reason I mention that is when you go up, if you ever drive up there, you can get to this point you can see one direction, all open lands. It's really beautiful. Anyway, open land as far as you can possibly see. There are these big rolling hills and lakes. Look the other direction and you are facing cities and buildings and factories and oil refineries and smog. It's about as dramatic a contrast as you could possibly have. An artist with a canvas couldn't make a more dramatic thing. This is real. So I was up with a friend one night and we were looking at it and she goes: "Oh my God Kent, this is like the cancer civilization crawling across nature." We were created in this earth, we have gotten this far and now we are here. OK fine, so, we have been playing with our tinker toys, we have been prevailing upon the Earths' resource space, we have been in a very clumsy, sloppy way kind of clumping along. We are slowly getting the drift that maybe we should be taking a second look at us and the planet as coexisting organisms like cells are in a common tissue. Some of us look at it this way, while still the basic idea is that this is nature. You know this table here is holding all of this stuff up. It's nanotech. So how does it tie in? Here's how it ties in: We are about to be given, probably, the most powerful tools of creation. As far as we can see for the moment, this is the most powerful tool of creation.*

ROSALYN: *Nanotechnology?*

KENT: *Nanotechnology, right. So we are going beyond the limitations of traditional solution-based chemistry. We are about to reorganize energy and matter as we know it today.*

ROSALYN: *Yes.*

KENT: *Let me give you a little snapshot. This might be helpful. Right here at the National Lab we turned lead to gold. It's quite simple actually. Every element has it's potential to become an isotope. An isotope merely being where the nucleus has an uneven number of protons and neutrons and so then it radiates out material, both particulate matter and also different frequencies of energy, before it finally comes to a rest state. It is decaying through these different stages. So there is a table that shows all of the different elements and how they go through decay to become a different element. Think for a moment. Is it possible to take a heavier atom, smash into it some neutrons, to a slightly different form? So in this case we wanted to have a controlled reaction, where the material would reform into, guess what, gold. That was back in 1974, 1975.*

ROSALYN: *You got the basic element of gold?*

KENT: *Yeah.*

That's what we got, sure. Thank you my dear, because this was the ultimate laboratory trick of all times. But the point was, the point was it cost about $30,000 worth of energy to make.

ROSALYN: *We are still less efficient than the ultimate.*

KENT: *This is where I am trying to go.*

ROSALYN: *OK.*

KENT: *In a superconductive system where the cost of energy drops to essentially zero.*

ROSALYN: *Yes.*

KENT: *Then we can propel stuff and smash into other stuff and the cost would be almost nothing to do it.*

ROSALYN: *That's what's scary.*

KENT:	*Well, a different approach.*
ROSALYN:	*Alright, I am listening.*
KENT:	*A different approach.*
ROSALYN:	*I am listening.*
KENT:	*Right now it's not just that we have to set the world on the ground and drive our cars around and make plastics, we also have to go to various parts of the Earth, engage in all sorts of rather dubious political interactions and so forth.*
ROSALYN:	*Waging war?*
KENT:	*Yeah well, anyway, war is a business for us.*
ROSALYN:	*Hum.*
KENT:	*War has been for centuries.*
ROSALYN:	*Yes.*
KENT:	*But the point is to dig up and get out of the soil some minerals or some of that stuff that we are going to smelt off and turn to some kind of metal or whatever. What if we could just do an interval around the whole process?*
ROSALYN:	*Just do a what?*
KENT:	*An interval around the whole process. So, in other words, what if you could just make generic elements for whatever it is, instead of reorganizing the atoms, and crystal into stuff you want. That's what nanotech ultimately represents, up to and including the ability to actually change elements themselves. But we will need to sort of bring it back on Earth. Nanoscale material manipulation allows you to investigate new ways of printing things like superconduction materials. The ability to create things that don't require huge megalithic top down timing centralized manufacturing distribution systems, that's the real core of what's going on in our current situational paradigm. To support these large megalithic top down and often highly predatory manufacturing distribution systems you have to uproot massive amounts of material and move people around and cause all this mayhem. Most of which suddenly becomes unnecessary if you can sort of cross into this nanoworld which means that you are now manufacturing just as needed, just in time on a highly granular and low cost basis. It's a very different approach. Within the current associated limited systems that are in place it's extremely disruptive and there are many folks that would find it highly threatening and very unpleasant. So is it within the tangible scope to a purely technical precision that this could be deployed at some point, yeah probably. But from the geopolitical economic limitation perspective, is it possible?*
ROSALYN:	*Does it break down entire economic systems?*
KENT:	*It reorganizes them.*

ROSALYN: *Does it reorganize them justly, for equal access?*
KENT: *In theory it might.*

NOBLE REVISITED

Similar to the thinking of Noble, Stahl (1999) viewed technology as one of the "greatest ironies of our times," because the language we use to discuss technology is ideological, mystifying, magical, and implicitly religious. Stahl argued that we falsely claim to live in a rational, secularized society, and this claim strips technology of systems of meaning. He wrote that the "creed of technological mysticism" is fundamentally about "men, displaying mastery of their expertise, single-mindedly pursuing the challenge of "sweet" problems, seeking perfection through their machines, leading us into a progressive future in which life will be under our control" (p. 34). Stahl's view is that although implicit and hidden, technology is permeated with symbol and myth, giving it a power over its users by making people subject to manipulation and self-deception. The challenge and imperative, then, is to recover the meaning of technology through demythologization. To some extent, I agree. It appears to me that there are in fact mystical, mythological, religious elements embedded in conceptualizations of nanoscale science and engineering that are fundamentally about the perfection of machines toward the quest to bring life under control. I am not so clear, however, about the effects of manipulation and self-deception. What is apparent is the importance of recognizing those elements, as a means toward becoming conscientious about personal investments of meanings in the pursuit of nanotechnology.

Noble wanted technology resources to be used for the benefit of humanity. He was joined by others of like mind. Dyson, (1997) for example, commented that "pure scientists have become more detached from the mundane needs of humanity, and the applied scientists have become more attached to immediate profitability" (p. 199). He observed that market-driven applied science usually results in the invention of toys for the rich, and whereas fashionable research projects are supported, the unfashionable go without. Surely, nanotechnology falls into the category of fashionable research. Dyson further expressed particular concern over what he called the "new ages" flooding over humanity like tsunamis—the information age, the biotechnology age, and the neurotechnology age. To Dyson's thinking, the new technologies are profoundly disruptive:

> They offer liberation from ancient drudgery in factory, farm, and office. They offer healing of ancient diseases of body and mind. They offer wealth and power to the people who possess the skills to understand and control them. They destroy industries based on older technologies and make people trained in older skills useless. They are likely to by-pass the poor and reward the rich. They will tend, as Hardy said eighty years ago, to accentuate the inequalities in the existing distribution of wealth, even if they do not, like nuclear technology, more directly promote the destruction of human life. (p. 200)

Nanotechnology encompasses and further enables the development and convergence of each of these three technological ages. Dyson, Noble, and I seem to agree on the potential misdirection of newly emerging technologies. But whereas Noble wanted to disabuse us of religious belief and myth (where they are part of technological imaginings), Dyson wanted to use ethics to guide and direct technological undertakings. This is where he placed his hope. I, like Dyson, want to see ethics grounding and guiding the beliefs and visions, and especially in the searches for meaning reflected in both conceptualization and development of nanotechnology. But to my thinking, the only possible way to engage a conscientious evolution of nanotechnology is to acknowledge the moral significance and creative power of the mythical elements of it. Ethics is a matter of competing interests, resolved through agreement on rules and principles, often masking true feelings and desires, fear, and ambitions. As a system of governance, pseudoethics that mask authentic belief and feelings can be dangerous and deceptive. Authentic moral wisdom is born from honest, introspective engagement with the human heart, not just with control of the human mind. That means facing even the most frightening and baffling elements of the human psyche. It is an introspective process, which embraces mythical imagery as a means to understanding the depth of who we are. Only from that process can an ethics of nanotechnology development emerge.

The exponential increase in our ability to control matter that nanotechnology represents points toward the increasingly powerful ability of humans to experiment with, and even alter, the fundamental constitution of living organisms, matter, and human material experience. Where does the desire and ambition for such awesome power come from? What kinds of beliefs are held about who we are as humans living in an increasingly technological world? How is meaning being made about the changes that occur within the human psyche in response to that world? One critical source of answers to these questions is to be found in the symbolic processes of the human mind, such as in the myth, and particularly in the archetypes we humans perpetuate and share.

How might research scientists and engineers be instrumental in the momentum of the rapidly rising tsunami? What might they perceive to be their role in creating an increasingly technological world? What do they imagine it will take for humanity to withstand the profound changes to come?

Noble's critique may be useful for measuring and judging the social outcomes of technological pursuits. But he too quickly dismissed myth as an impediment to humanitarian ends. For example, Noble (1999) cited Fredkin as evidence of the entanglement of technology with religious myth: "I think our mission is to create artificial intelligence. It is the next step in evolution. One wonders why God didn't do it. Or, it's a very god-like thing to create a super intelligence, much smarter than we are. It's the abstraction of the physical universe, and this is the ultimate in that direction. If there are questions to be answered, this is how they'll be answered" (p. 163). But, there is another way to understand these mythical references. For one, fantasy about humans creating superintelligence is archetypal; as myth, its material facts are irrelevant. Its power and significance lie in the symbolism. Images of omnipotence leading to precise control and manipulation of material existence are very powerful in their ability to contain and control the terror associated with material death. Creation of superintelligence buffers human frailty from that which is perceived to be greater than us, yet out of our control—like God. As a human creation, superintelligence, such as AI, can be held within our control; we become god-like. What would it actually mean if we could master material control of the physical universe?

Nanotechnology visionary Drexler (1986) speaks to an ancient myth of a fountain of youth:

> Aging is natural, but so were smallpox and our efforts to prevent it. We have conquered smallpox, and it seems that we will conquer aging .... Aging is fundamentally no different from any other disorders; it is no magical effect of calendar dates on a mysterious life force. Brittle bones; wrinkled skin; low enzyme activities; slow wound healing; poor memory; and the rest all result from damaged molecular machinery; chemical imbalances, and misarranged structures. By restoring all the cells and tissues of the body to a youthful structure, repair machines will restore youthful health. (pp. 114, 115)

Jugs primary interest with myth was its ability to bring to consciousness some hidden elements of the psyche in resolving psychological and emotional conflicts for the individual, in freeing the archetype and permitting the formation of healthier association with the outside world. My interest with the presence of myth inside of nanotechnology development is to reveal and embrace it as a healthy element of cognitive processing, toward freeing ourselves from the mentality of moral neutrality and powerlessness that so often accompanies new technological development. I disagree with Noble; mythology does not necessarily inhibit humanitarian approaches to technology. Rather, if it is seen and understood, it can reveal core beliefs, and thus contribute to the moral guidance of technological development.

Noble suggested that myths are constructed to provide a sense of meaning and control to shared experience; they guide and inspire us, and enable us to live in an otherwise uncontrollable and mysterious universe. From that point, I suggest that in nanotechnology development, mythological elements of expression may serve a critical purpose of providing much-needed ethical guidance. What I hope is that once the archetypes at work are acknowledged and conscious engagement is made with the myths that are employed to construe and engage the imagination in the nanotechnology future, we can retrieve our conscious access to those things that are believed, dreamed of, hoped, and feared. Therein lays the path to a conscientious, ethical pursuit of nanotechnology.

Kents' discussion with me is unusual. Kent is atypical in my group of participating researchers in that he no longer has association with a university or national laboratory, nor is he part of a research and development effort inside of a corporation. Many more mainstream, "legitimate" researchers would deny him accountability or affiliation with the professions of science and research engineering. As such, a reader of this writing may be disinclined to count my conversation with him as valid or in any way an example of what nanotechnology researchers believe. I suggest the possibility that because of Kent's utter independence from the professions and any of its formal institutions, he may be freer with his words, and less guarded about sharing his thoughts.

What is to be understood by the promise of nanotechnology to improve on the "quality of life?" Although its meaning is vague, the notion of quality connotes a standard of something as measured against other things of a similar kind. It has to do with gradations and points to either increase or decrease in measurable terms. Quality invokes sentiments that are widely shared. The conceptualization of life, on the other hand, is much more difficult to define. For our purposes, consider "quality of life" to mean standards of living, as measured against either previous standards or that of others who are living in ways that appear to be more or less desirable or valuable than ones' own. In terms of the industrialized world, for most people, notions of quality in living are no longer about basic needs, such as access to food and shelter, health and prosperity. Rather, the idea points toward an unspecified but limitless improvement to our material lives, over and above the qualities we currently have.

Given the enormous prosperity enjoyed in these societies, what kind of appealing improvements might nanotechnology offer? These might include increasing longevity of human life, fewer diseases, less sickness, more food, more money, more property, more stuff, faster stuff, and smaller stuff. What is being sought in the reach for improved quality of life? Beliefs about quality of life are constructed from culturally shared, socially shaped perceptions. They are also formed from cognitive processes that reach well beyond the primary and instinctual thrust to survive, toward dreams and fantasies of freedom from want, need, and limitation. The pursuit of new knowledge has some of these myths at its core. As such, nanoscale science and technology researchers are engaged not only in the acquisition of new knowledge, but also in the negotiated and competitive sociocultural construction of meaning. As Fuller (1993) explained more generally,

> The so-called ultimate ends—such as peace, survival, happiness, and (yes) even truth—refer not to radical value choices for which no justification can be given, but rather to constraints on the manner in which other instrumentally justifiable ends are pursued. Thus, happiness in life is achieved not by reaching a certain endpoint, but by acquiring a certain attitude as one pursues other ends. (p. 16)

Similarly, improvements to the quality of living have no end point, only perpetual adjustments to attitudes, made possible by the allure of technological promise. Perhaps the nanotechnology initiative is one that offers not improvement on the quality of life, but maintenance of the quality we now have. Or, is it promising to improve the quality of life for those who have not yet achieved that high quality we have here in the United States? Or, perhaps as both social collectives and individuals, we are simply dissatisfied with current living and feel compelled toward the acquisition of new and novel things, and the changes to human lives that those things might bring. Planetary history certainly does point to the proclivity of all living entities to pursue change as a means toward self-improvement. That seems to be our human nature.

GEOFFREY

GEOFFREY: *The ideal project in a group is one which in a cycle of about 10 years goes from work with sufficiently fundamental work, really exploratory because that's what we do well, but no one can understand why we are doing it and hopefully gives us a very hard time for trying, to something which is really university advance engineering; the work that is done just before you transfer it into something that's commercial, the entire thing. Different students like to do different parts of that, but it's good to have all parts of it going on at one or another time somewhere in the laboratory.*

ROSALYN: *These are all at some level extensions of whom you are as a scientist. Every one of these projects, would you accept that?*

GEOFFREY: *Well, everything is an extension of what one is as a scientist. I mean, science covers a pretty broad spectrum. So yes, those are parts of what I am as a scientist, so are the companies, so is public policy, so is just general curiosity.*

ROSALYN: *OK, then who is it that is showing up? I mean, how would you define what it is, that is showing up?*

GEOFFREY: *Oh, it's me.*

ROSALYN: *Yes.*

GEOFFREY: *Right. I am a creature who is driven in large part by curiosity.*

ROSALYN: *Alright.*

GEOFFREY: *I find amusement in solving problems, which I think is a characteristic of many people who are scientists.*

ROSALYN: *I hear it in nearly every interview, actually—the problem solving and the curiosity in particular. I know, because I have read some of what you have written, that you do extend that curiosity to think about the actual implications of the work.*

GEOFFREY: *I think most people do in one or another sense. Now they may have more or less limited experience against which to think through a problem, because you can only think in the context of what you know.*

ROSALYN: *Of what you know.*

GEOFFREY: *So different people will, I think, people will always think as broadly as they can, depending on the circumstances.*

ROSALYN: *OK. Do you think, for whatever it is this word means to you, whatever nanoscience means to you, does it present us with new challenges that are somehow distinctive from other scientific challenges?*

GEOFFREY: *I don't think so.*

ROSALYN: *OK.*

GEOFFREY: *To me the greatest potential for something really new in terms of ideas is that one can imagine a future in which nanotechnology moves into something which is essentially applied quantum strangeness. You know these are the phenomena that everybody sort of knows that are there, but basically no one can understand. You know one really doesn't understand many of the phenomena of quantum mechanics.*

ROSALYN: *Yes.*

GEOFFREY: *You begin to see signs of people trying to figure out what that might be of quantum photography and quantum teleportation.*

ROSALYN: *Teleportation, interesting.*

GEOFFREY: *That's action to distance.*

ROSALYN: *I see.*

GEOFFREY: *And those are things which are actually new. They are only generally understood and they come out of quantum mechanics in a pretty straightforward way. But, most people don't understand quantum mechanics. We don't have any intuition and experience, really, at that level of the universe. Our understanding is largely rudimentary. It works, and we proceed onward on that basis. But anyway, that area has, I think, potential to be something really quite new. And then nanotechnology has some other things that go with it, which could make a very big difference, in terms of application. For example, if one has infinite memory for free, and very rapid acquiring, storing, and classifying data, then the whole concept of privacy may change in ways that may take us a while to grow accustomed to. But those are going to happen one or another way anyway.*

ROSALYN: *Why do you say that?*

GEOFFREY: *The evolutionary process toward loss of privacy seems to be part of packing more and more people on the planet. Privacy, as we know it, is probably substantially greater than privacy in lives of our great, great grandchildren.*

ROSALYN: *Is that simply as a function of the number of people who are here?*

GEOFFREY: *And the efficiency of communication between.*

ROSALYN: *OK, now that's what I was thinking you might be alluding to. Are you a proponent of this notion of synchronicity? Is that where you are going?*

GEOFFREY: *I am not sure that I know what synchronicity is.*

ROSALYN: *We can talk about it some other day, but for now, it's this concept that there is an ultimate reality in the universe which is simply information, and we are evolutionarily moving toward a recognition of that reality in a merging of all intelligence into one—that we humans are connected to one another and to that ultimate reality through intelligence.*

That's as best as I can explain it. Alright, so then, I am trying to find out whether your aspirations and curiosity have an end point, whether there is something that is there to be discovered or if it's an eternal process of simply satisfying the curiosity of the mind.

GEOFFREY: *Do you mean do I want to save the world?*

ROSALYN: *Save the world, I think we all want to save the world.*

GEOFFREY: *I mean the implication of wanting to save the world is that you are smart enough to know what would be involved in saving the world and I am not sure that I am anywhere near that smart. I mean, one of the few laws that I truly believe in is the law of unintended consequences.*

ROSALYN: *OK.*

GEOFFREY: *And, to save the world is terrific, but be a little cautious about the process, because it's not clear it's all going to work out the way you want it to work out. Now, I don't know why I work in these various things. There is always at the end a notion that there is a public utility.*

ROSALYN: *OK.*

GEOFFREY: *And so, when we were working in heterogeneous catalysis it was because, among other things, a large part of the energy in the world is produced through catalytic reaction. Better, cleaner ways of doing that seems like something that might come out of fundamental studies of how catalysis works. Material science, the virtue of material science to me, has always been that everything that you see around you is material. So, of course if you are curious about the world you want to know what everything is. And, if you know what everything is and how it works, then you can manipulate it.*

ROSALYN: *There is a key word.*

GEOFFREY: *Manipulate.*

ROSALYN: *Um hum.*

GEOFFREY: *Fine.*

ROSALYN: *Alright, let's back up for a minute. Is this stuff that's going across my hands material as well?*

GEOFFREY: *As far as I am concerned.*

ROSALYN: *OK.*

GEOFFREY: *I mean, I take it in the broadest possible sense. I think most people would say that to a first approximation materials are more likely to be things that are solids or lasting.*

ROSALYN: *Yes.*

GEOFFREY: *A condensed form of matter, say, an extended form of matter.*

ROSALYN: *So, what is the interest in manipulating the material world, then?*

GEOFFREY: *What does every one of us do?*

ROSALYN: *Well, let's think about that. Is that what we do?*

GEOFFREY: *Well, here you are manipulating the material world for information storage, and you are sitting on something, which is a manipulated form of material world, and you are dressed in it and you can see me because somebody has tinkered with glasses that have happened to work. I think we are all, perhaps unconsciously, profoundly interested in manipulating the material world.*

ROSALYN: *For our comfort or simply for our ends?*

GEOFFREY: *For our various ends. So, I am a material system.*

One of the interesting things that is happening now is the notion of biology as a materials problem beginning to emerge as something interesting to think about. So I am a machine in a certain sense, a mechanical machine, a fluid machine, a low grade computation machine.

ROSALYN: *Low grade computation?*

GEOFFREY: *There are things actually which I do quite well, but most things can be done better by higher grade computation machines. But as a computation machine, I am good at patterning, associative memory, sense of humor, things like that. I don't know how long that advantage is going to last. I think that's much more likely to be the kind of thing that turns the world on its ear than nanotechnology.*

ROSALYN: *Oh, I agree with you.*

GEOFFREY: *Because the people who tinker with these things I think make the case, which is not a bad case, that the collective intelligence of whatever is here and the sensors that go with it, whatever is in the mind, and the collective intelligence of what's in a modern sort of front end work station is probably a lower .... I think that in one of the last matches with Big Blue, whoever Big Blue was playing remarked for the first time he could not tell whether he was playing a computer or a human being .... This notion that a substantial fraction of the people in the world are not real, but they are just papier-màché put there to test the faithful to see if they can figure out what's real and what's not real. It's a neat notion. How do I tell that you are real?*

ROSALYN: *Do you have an answer to that question?*

GEOFFREY: *No. I have no idea of what test I would apply and I am willing to believe that the large numbers of people that I know are not real.*

ROSALYN: *Do you have a test for your own reality?*

GEOFFREY: *No, I don't. It is a matter of faith on my part. But certainly in many circumstances I behave in ways that I would not think of as human.*

ROSALYN: *OK.*

GEOFFREY: *So, from that point of view, I am not recognizable as human as humans should be.*

ROSALYN: *A scaffold.*

GEOFFREY: *It's a clever idea. In the absence of a test that at least gives one a starting point for conversation.*

ROSALYN: *That's a good point. I have been reading the Bible recently to try to locate the origins of some of our belief systems—there are all kinds of metaphors and myths that are embedded there.*

GEOFFREY: *Right.*

ROSALYN: *One of the things that is intriguing to me there is the concept of "I am that I am." It's a really very beautiful passage in response to "what are you?" and I really can't come up with a better answer.*

GEOFFREY: *It passes.*

ROSALYN: *It passes.*

GEOFFREY: *It would be interesting to know what it was in Aramaic.*

ROSALYN: *It would be interesting to know but I am afraid we have lost that translation.*

GEOFFREY: *I don't care, but I mean it would be interesting to know. You know the Bible was one of the courses in high school that I had, that I have always been grateful for. We studied it for 2 solid years: one on the Old Testament and one on the New Testament.*

ROSALYN: *Hum.*

GEOFFREY: *And, you are right about the number of references there are from it in the world, the Christian world that we live in—it is just staggering. And you know there is an enormous gap between people who would say, for example, "the burning bush" and they'd know immediately what you are talking about, and those that wouldn't. It's an interesting problem. It's going to make it harder to talk without those kinds of shorthand phrases, which summarize in fact thousands of years of practical human experience. Do you think that science fiction is a form of new theology?*

ROSALYN: *I think that science fiction is reflective of the human creative process and of our aspirations.*

GEOFFREY: *So, do you think about nuclear weapons?*

ROSALYN: *Yes, absolutely.*

GEOFFREY: *I am not sure I know how to put those on the table at the same time.*

ROSALYN: *Science fiction and nuclear weapons?*

GEOFFREY: *Um hum.*

ROSALYN: *Well, nuclear weapons came to be in science fiction before they came to be in material reality. I think science fiction gives us a way to imagine and to place our imagination into something that we can identify with. I don't think it's that far removed from who we are.*

GEOFFREY: *Um hum.*

ROSALYN: *It's just another expression of our fears, our desires to manipulate and control and our expressions of where we think we might be going, particularly expressions of the fear and struggle—very powerful material.*

GEOFFREY: *I am finding that. You don't hear Jung coming into conversations.*

ROSALYN: *Oh, actually I write about Jung—in particular about archetypes that I see embedded in some technology. They are beautiful. They are helpful.*

GEOFFREY: *I absolutely believe that. And I know my little children, I noted this at the time, were afraid of snakes long before they had any idea of what a snake was.*

ROSALYN: *It's such an important symbol. I happen to wear a snake bracelet on my wrist.*

GEOFFREY: *And spiders on bracelets?*

ROSALYN: *Not me, I am not ready for that one.*

GEOFFREY: *Why not have spiders on bracelets?*

ROSALYN: *It's a lot of fun; they did a really good job. So, I am afraid we could talk for a long time, but this is a really good beginning for me, because I am interested in who the person is behind this science/ policy/business interest in nanotechnology and what it means to you. I am getting little tiny pieces here and there. It's almost as if you have no choice. That it is just because this is what your mind is asking for.*

GEOFFREY: *I think to be perfectly candid; we do many things we would be perfectly happy doing.*

ROSALYN: *Why science then?*

GEOFFREY: *I like science. I am curious and these are problems …*

ROSALYN: *I want to know where we are going. I don't know where we are going.*

GEOFFREY: *Don't you think the world is a better place than it was in the dark ages?*

ROSALYN: *Do you think we are evolving to places of higher good in the human condition, is that what you are saying to me?*

GEOFFREY: *Yes, yes I am. I am saying that we are evolving to a world in which people have more time to think. I mean, there are lots of problems, which you know are more than we have time to discuss.*

ROSALYN: *So you mean more people have more time to think.*

GEOFFREY: *More people have more time to think. The life span is longer and we tend to think that once life is established, that it's a good idea to keep it going.*

ROSALYN: *Indefinitely.*

GEOFFREY: *I didn't say that.*

ROSALYN: *I am just asking.*

GEOFFREY: *Indefinitely, you know that's one of those hypothetical.*

ROSALYN: *OK. Well, 200 years?*

GEOFFREY: *If one could keep people happy and make an interesting life for 200 years and cut the birth rate by a factor of two.*

ROSALYN: *OK.*

GEOFFREY: *That might be an interesting thing to work through.*

ROSALYN: *Alright.*

GEOFFREY: *I don't know the answer to it.*

ROSALYN: *Just curious.*

GEOFFREY: *In some greater sense I don't know whether it would make an enormous amount of difference.*

ROSALYN: *OK.*

GEOFFREY: *Does the universe in the long term care whether we are here, the scum on this particular planet is here or not?*

ROSALYN: *I went to the planetarium in New York City for one of their shows and I walked in pretty self-assured and cocky feeling like I had a place of significance in the universe. The film started out with a shot of the planet Earth and I thought, "Oh yeah, that's where I am, on my home." The perspective then backed up, showing the solar system and then again further showing the galaxy. Then it backed up and showed the galaxy turning in on itself and I thought, "Wait a minute. Wait a minute! What just happened to the size of the universe to my relevancy? It's gone!" I walked out experiencing the classic crisis of meaning.*

GEOFFREY: *That's true; there is something I think that's even more alarming than that. Things that are made out of atoms comprise 0.5% of the matter of the universe. The other 99.5% is …*

ROSALYN: *Empty?*

GEOFFREY: *No, no.*

ROSALYN: *Just—other.*

GEOFFREY: *Other?*

ROSALYN: *Other.*

GEOFFREY: *And we know what about 50% of it is and the other 50% we don't even know what it is.*

ROSALYN: *Isn't that thrilling?*

GEOFFREY: *Oh yeah. How about that?*

ROSALYN: *But what if we did know everything about the physical material universe, then what would our focus become?*

GEOFFREY: *I don't think we have to worry about it.*

ROSALYN: *OK, theoretically speaking, it seems that's what keeps us alive, to some extent.*

GEOFFREY: *You know the nice thing is it's a game where you can never finish.*

ROSALYN: *You are right.*

GEOFFREY: *But you know what's required is a sense of how are you going to know everything about everything.*

ROSALYN: *We can't. We don't have the capacity.*

GEOFFREY: *You are on. Pretty slow processing capacity. It certainly isn't built into an individual, and we have some kind of collective process and some kind of collective memory, but its not enormously efficient. So I really am not …*

ROSALYN: *It's not a problem?*

GEOFFREY: *I am not concerned.*

ROSALYN: *OK. OK. So.*

GEOFFREY: *But the point that I was making on the original subject was I think it's important for science to be honest with society. I think it's important, very important, for everybody to be honest. If what one is doing is interesting science, it should be interesting science, not puffed up to be cosmic science. Because that particular business of wishing to appear larger than one is, is not, I think not very admirable.*

ROSALYN: *But it's not unique to science as an enterprise. I see it in entertainment, in sports.*

GEOFFREY: *But part of the value system in science is supposed to be honesty.*

ROSALYN: *Is it?*

GEOFFREY: *Um hum. It is.*

ROSALYN: *Perhaps we are sort of lost in a time when we have other values at work.*

GEOFFREY: *So you have now answered my question.*

ROSALYN: *You have answered the question?*

GEOFFREY: *No, you have answered.*

ROSALYN: *Oh, have I?*

GEOFFREY: *Right, so you are on.*

ROSALYN: *I am?*

GEOFFREY: *I mean, it just took a little while to get the answer to that question, And I think that could certainly be a part of the story, I don't disagree with that. But, you know one of the terms of science is that I think for those who genuinely love it and do it well, it is something like music. It's more than making a living. The fact of the matter is that there is something, because of the intrinsic interest in the subject and the subject is by itself, separate from the individual. I mean remember to look in the mirror. That makes the subject almost to me by definition, larger than people who study it and that's not something you can say about everything.*

ROSALYN: *And again, the fact that we are working at the scale of the nanoscale changes nothing or very little?*

GEOFFREY: *From a point of view of technology, nano makes things smaller, and that's good.*

ROSALYN: *Because we can control them better?*

GEOFFREY: *No, I mean actually for practical reasons. What's driven the recent microfabrication is not because it makes better early warning radar systems or Microsoft Word, just because it makes it possible to do some useful things more cheaply. If nanotechnology were all more expensive, it wouldn't have happened. And now that doesn't mean that it isn't great stuff to do, but as the ox made it easier and cheaper to pull the plow, and the tractor made it easier and cheaper than the ox, the microprocessor makes it easier and cheaper to do certain kinds of things. The question is, will nano do any of those things, you know, in a big way? There is presently no example that suggests it will. I am a believer in science, so I reserve judgment on that and if it turns out through nanotechnology one can go into this wonderful world of applied quantum strangeness, then that might be such a thing. Those are the things that are intrinsically not predictable. The changes that have the characteristic that you don't know what they are and don't know what their consequences are could make a big difference. So conversations about predicting the future are guaranteed to fail for one of two reasons. Either you predict the future and anything that you can predict is probably relatively speaking uninteresting, since you are predictive. Or, you can't predict the future in which the conversation hasn't gotten you very far.*

ROSALYN: *Well, the word* predict *is awfully decisive. I think we can muse over the future. We can consider possibilities within our frame of desire and ambition and we can, I hope, anticipate to some extent what challenges we might have if we do X, Y, it will lead to Z, possibly lead to Z.*

GEOFFREY: *I have a different algebra.*

ROSALYN: *Alright, please.*

GEOFFREY: *Which is, to go at it another way and not take what we are doing and try to* predict *where it will lead, but rather to look at sort of core assumptions in the world.*

ROSALYN: *Yes.*

GEOFFREY: *And then ask what is the probability that science will prove that assumption wrong.*

ROSALYN: *Yes.*

GEOFFREY: *Different way of doing things. I think it's a little bit more modest and in a very, very uncertain process you are a little bit more likely to understand where you are going. So, one we were talking about earlier, one of our assumptions is that we are mortal.*

ROSALYN: *That's the assumption we have all made.*

GEOFFREY: *Um hum, now what happens if it's wrong and you came to a subcategory of that? What happens if science allows us to live to be 200? I think the chance of science allowing us to live to be 200 is essentially good.*

ROSALYN: *Um hum.*

GEOFFREY: *Now, ask me whether it's going to be in 50 years or 200 years, I am a little bit uncertain about that. I would say more likely 50 than 200, but I don't know that to be sure.*

ROSALYN: *Um hum.*

GEOFFREY: *Live forever? I don't see how to do that part. Would a much-prolonged life make a difference to society? The answer is yes. You can put together a list of your own. E-mail connects us all the time. Our connections are the kind of collective consciousness that holds the hive together and who knows, we might have networks that begin to look like brain communications in 1,000 years. Maybe. I mean, you can trace a pathway that might lead off in that direction, but I can't see any of the technical details. I think one works ones' way through that. Human life is valuable, that's an assumption. That's one that is much more frail around the edges, because we already know human life is a commodity, that you spend in various ways for various purposes. But I think you can see the processes in science that are going to change that very substantially. As I say, it's interesting to make that kind of list. I find it more interesting and I think more productive to go that way than to ask, where is nanotechnology going? This is not a criticism of your project.*

ROSALYN: *No, actually, my project does more of what you are suggesting; it looks at the embedded assumptions and how those are challenged.*

GEOFFREY: *So this is what you are interested in, more in the sociology and psychology than you are in the ...*

ROSALYN: *In the ethics?*

GEOFFREY: *The output.*

ROSALYN: *In the technical output?*

GEOFFREY: *Ethics, that's just a complicated order.*

ROSALYN: *Oh yes. How true.*

GEOFFREY: *I know more about morality than I know about ethics.*

ROSALYN: *You know about morality. Ethics is a system of control and manipulation, so that we humans can live in relative peace on the planet and have some basic agreements about how we are going to do that.*

GEOFFREY: *Um.*

ROSALYN: *Like, you work as a professional on Wall Street, so ethically you can do this, but you don't do that. You are a physician at the bedside. You talk to your patient this way and not that way because these are the principles of ethics we apply, etc. I have more of a metaethical emphasis, asking questions about why we ask what's good, the significance of that word, and particularly to understand how the psyche plays into a conception of reality and the good, particularly in science.*

GEOFFREY: *You used a word which is a slightly slippery word.*

ROSALYN: *Which one, reality or good?*

GEOFFREY: *I'm not worried about reality. I can define it, reality I mean, but I don't know how to define good.*

ROSALYN: *Well, that's why I asked. What does it mean to define it, why is it important to us and where might answering it take us in the pursuit of nanoscale science? Those are the kinds of things I am curious about.*

GEOFFREY: *I had a scarifying experience some years ago, a long time ago. I was mugged. It was scarifying, but the scarifying part was not being mugged, but I was jumped in a parking lot and really beaten up thoroughly, so that I ended up with my face a mess and my cheekbone rearranged and my shirt looked like I had been rolling around in the gutter, and there was nothing left of my pants except my belt. I mean it was a real mess. I was taken down to the police department and there was a little young woman who was the second lieutenant, she was the booking officer. I walked in and she said it looks like you have been mugged. I said, literally, "Yes ma'am." She said, "You are a college professor aren't you?" Recognition, that's what we were actually pretty good at.*

ROSALYN: *It's a survival mechanism.*

GEOFFREY: *It was interesting. It resolves an issue. Michael Lucas, have you met Michael Lucas?*

ROSALYN: *No.*

GEOFFREY: *He is from Cal Tech. He is the only guy who makes true nanomachines. He is an interesting person. He is finding some new physics. Have you had enough?*

ROSALYN: *No, but I am concerned about your time and your energy. It's 11:10 p.m.*

GEOFFREY: *We can go a bit longer.*

ROSALYN: *OK.*

GEOFFREY: *I have got to figure what I am going to say tomorrow. Anyway, to go back to this issue of is there something different there? I mean quantum teleportation is different. We really don't understand action and distance and understanding how to use it, even if we don't understand it, that would be new. I have no idea what the consequences would be. There is that characteristic.*

ROSALYN: *But we pursue things without knowing the consequences, don't we?*

GEOFFREY: *Well, even in things in which you know the consequences, because the law of unintended consequences, you don't get it right. So, we don't know the consequences of anything.*

ROSALYN: *Not anything?*

GEOFFREY: *What are the consequences of this perfectly innocent conversation?*

ROSALYN: *I have no idea. I can make some assumptions, I can have some hopes, but I have no idea.*

GEOFFREY: *That's a question that is asked about lots of things.*

ROSALYN: *This is precisely why I get concerned when people aspire to make things "perfect" as they define it. For example, to get rid of disease as we know it. To map the genes so that we can master them and factor out those things that we don't want so that we can make the human machine optimal in its performance. All of this kind of lan-*

guage that I am hearing and I ask, can we wait a minute and think about the consequences? At what point in life do you let life be rather than to continue to desire to control it? That's why the language of nanotechnology perplexes me. To control matter with precision, to what end, to what end?

GEOFFREY: *You are interested in the world because of the interest in the world, not because of its effect on you, so it is curiosity. Understanding of oneself is one objective as understanding the world independent of oneself is another objective.*

ROSALYN: *Do you really believe we are able to do that through science?*

GEOFFREY: *I think there is a lot that one learns through science that has nothing directly to do with you or me.*

ROSALYN: *Do you think it's possible at all to use this human brain and not see oneself in what ones doing?*

GEOFFREY: *Yes.*

ROSALYN: *Or project oneself into ones work?*

GEOFFREY: *If you think about it for a bit, everything is you and your point of view, or whatever. I have no trouble with that. But, there is an article of faith that comes into this. One possibility is that there is no objective reality.*

ROSALYN: *Right.*

GEOFFREY: *And no objective truth. It's all just a question of our interpretations and whatever fog we happen to live in.*

ROSALYN: *Yes.*

GEOFFREY: *Or, there is an objective reality and then the fog is somewhere in between objective reality and us or us or something of that sort and I think it is probably difficult to hold both of those points of view simultaneously. I think that the personal interpretation of a black hole, sucking in neutron star, emitting light that I am seeing only 4 billion years after it happened, the sort of personal relevance of this is small. On the other hand, I am also willing to admit that virtually anything that happens in biology right now has social significance and actually a fairly straightforward social agenda in any way that you go after biological science and things that have to do with people. So there is probably a radiant from sociology or psychology, people interpreting the behavior of people through people with human minds, to black holes sucking in neutron stars, which I would argue is an activity that could be done probably as well without any people involved in the process as with people involved in the process. You could almost program the computer to work it out. There are the sensors—things looking at wave lengths that no human being will ever see—in a time so far past that there is no way that we can trace it, using phys-*

ics, there is not much humanity in it, to arrive at conclusions, inferences about the nature of the universe. It really has little place for humankind in it. So, you could go from that to this and there are no big steps in that process. This doesn't upset me at all, but I think there are differences in the process. What we are talking about in nanotechnology could have a big human component if it turns out to be like biology, biotechnology. I think that's the case. But I would argue right now that it's much more on the side of physics and abstraction.

ROSALYN: *Um.*

GEOFFREY: *In nano there are a lot of competing technologies going through that particular objective. The development of very, very large databases certainly will happen. It is happening.*

ROSALYN: *Um.*

GEOFFREY: *A friend of mine got a quote from IBM for 15 hetabits of memory. He is attempting the 16 bits. He had 20/10 for $200,000. That's so much information, I mean it's more information than one could imagine. I think that if one does look at any given time for ideas that have the potential to really be big, and one of the big ideas that is sort of floating around the edge in some unrespectable form is complexity and the genome—I think is going to turn out to be correlating genomics with behavior and personality is going to be a complex problem; complex in the real sense that knowing all of the pieces, knowing the data will not lead you to the answer, by any deterministic way. I don't particularly want people to be able to access all of the telephone calls that I have made and all of my bank transfers, not that there is anything interesting in that, except that I think that you could probably take that information and in a pretty straightforward way, predict my behavior. My group put on a birthday party for me a few years ago. Instead of doing the usual sort of trick of "this is what I have done in my research," and so on and so forth, one student asked me four or five questions from Myers Briggs Personality indicator. And then he simply went to the sheet, and read off the characteristics, and that was fine. I was really very amused by that. It was right on. But, afterwards he sort of took me aside and said look, you know, let me just give you what your page is.*

ROSALYN: *Um.*

GEOFFREY: *It was really very interesting, because it wasn't 100% right, it was 98.6% right. It's like a tooth. It loosens that sense of individuality, to recognize them on the basis of a limited number of questions you can figure out a lot of things that you thought were very complicated. So my statement about complexity may in fact be absolutely wrong, but it also may be that we don't need genomics to get there. But privacy is a big deal.*

ROSALYN: *Privacy is a big deal.*

GEOFFREY: *That's one of those big assumptions. We are individuals and as individuals we have a right to have ownership over knowledge concerning our lives.*

ROSALYN: *In our thoughts and our ...*

GEOFFREY: *In our thoughts.*

ROSALYN: *And fantasies.*

GEOFFREY: *We are forging into unfamiliar territory with other things and in the nano area, if you want to take part of this, I would focus on this area of strangeness. The trouble with that is it's actually very hard to understand. I think it would make your job difficult. Because a small something one can get a grip on and it's pretty easy to understand. The consequences are not complicated. But action and distance is hard to understand. It's straightforward to understand how wave functions collapse, but if you think about how you can have something happen instantaneously in a connected way, 10,000 light years apart, you know, you can think about that for a moment. It makes you queasy in the stomach. There is something going on there that doesn't make sense, and yet it happens.*

ROSALYN: *I spent a lot of years studying mystical writings in different traditions. If one wanted to explore this quantum strangeness from that realm, perhaps you could make sense of it. But most scientists that I am speaking with are not interested in mysticism, because it can't be repeated or tested under laboratory conditions. And it's not objective, right now.*

GEOFFREY: *What's the definition of mysticism?*

ROSALYN: *I think that it has to do with approaching questions from beyond the material realm. I don't know how else to explain it.*

GEOFFREY: *Well, you know we have a rule set in quantum mechanics, which gives good results, but we don't understand it. Now I am not sure that that doesn't fit perfectly well, with the mysticism .... There are no answers to these things and there can't be. It's not in the system.*

GEOFFREY: *The quantum strangeness is an interesting part of the nano story. Nanotechnology is a very legitimate place to start seeing new applications of quantum mechanics. There is one component of it which we haven't touched on which is actually in some ways the most interesting part of the nano story; machinery that makes the cell run.*

ROSALYN: *That makes the cell run?*

GEOFFREY: *Such as the mitochondria and the bacteria and all the rest of this stuff. It's just that machine intelligence is the emergence of a competitor species.*

ROSALYN: *Is that what you see, that we are creating machine intelligence as a competitive species?*

GEOFFREY: *A major part of our activity right now is to create our competitors. That's really interesting.*

ROSALYN: *Did you see the film,* The Matrix*?*

GEOFFREY: *I guess I did see that one. Do you ever read the collective works of Geoffrey Bullel? He made a collection of short stories called,* A Delicate Prey. *But one is relevant to this, "The Scorpion and the Frog." The scorpion and the frog are sitting by the banks of the creek and the scorpion comes to the frog and says, carry me across on your back, I can't swim. The frog says that's crazy, why should I do that, you can kill me at any point and the scorpion says, why would I possibly do that, because we get out in the stream, I sting you and you die and I die. And, so the frog thinks about that and says, OK, I am a decent person and you are right, I will do that. So off they go in the stream. The scorpion stings the frog and the frogs' dying remark is why on earth did you do that, to which the scorpion says ...*

ROSALYN: *I am a scorpion.*

GEOFFREY: *It's my nature.*

ROSALYN: *Yes.*

GEOFFREY: *That's our nature.*

GEOFFREY: *It's our nature to create and then there is the law of unintended consequences.*

ROSALYN: *But we project out onto that which we create what we are inevitably aiming to be.*

GEOFFREY: *We don't know that.*

ROSALYN: *We imagine it. We can.*

GEOFFREY: *You think we can control our fears? We have always felt we could beat it. But it's more complicated. Nuclear weapons are a great example.*

ROSALYN: *They are a perfect example.*

GEOFFREY: *There is a plot that you can make of the fraction of the worlds' population that has died, the civilian population that has died per year.*

ROSALYN: *OK.*

GEOFFREY: *Don't ask me where the numbers come from. From time immemorial this number has been 1%. OK?*

ROSALYN: *OK.*

GEOFFREY: *There was a spike in 1917/1918 or 1918/1919 which was due to influenza, the epidemic, and then a large bulge in 1935, 1944/1945, at which point it dropped to .1% where it has been, where we are now. So, the conclusion from that is the development of nuclear weapons have saved more lives than all the rest of a war, through all the rest of mankind. Because we haven't had a global war.*

ROSALYN: *Because we were afraid?*

GEOFFREY: *Yes.*

ROSALYN: *Um hum.*

GEOFFREY: *That's unintended consequences. Now, it's true that on three occasions we were close to sterilizing the Northern Hemisphere.*

ROSALYN: *Um hum, Hiroshima, Nagasaki and ?*

GEOFFREY: *The missile crisis and two other events.*

ROSALYN: *Oh, you mean near misses. OK, I see.*

GEOFFREY: *But we didn't and so you know life is like riding a bicycle. The existence of species is like riding a bicycle; we don't know how it is going to come out. I have no clue what will happen if we created a competitor or superior species. For one thing, I think it would take a while if I am trying to imagine this. If I were a silicon-based intelligence.*

ROSALYN: *Um hum.*

GEOFFREY: *How long would it take me to phantom that out there whatever there was, where my world consisted in streams and bits emerging from something, how would I infer from those streams and bits the existence of something which took other things that I couldn't imagine and put those things into this mouth whose function or existence I couldn't image, and chewed it, and mixed it with fluids, and dissolved it, and reincorporated it to make more of those things by a process or binary division. I mean, this would seem to be inferring existence. From that it would seem to be such a difficult job that it would keep it occupied for several hundred years, or forever. I mean for all we know, this is a white mouse. I don't know that there isn't some intelligence whose nature I absolutely can't infer pulling the strings right now. If we were to create a peer competitor of that sort, can it really compete? I don't know. But it's an interesting thing to speculate.*

ROSALYN: *One day I hope to understand why we are so ambitious about strong A.I. I just don't understand what it is that we are doing.*

GEOFFREY: *I think we are attracted by the idea of intelligence and we don't find it in humans, so that in a platonic sense, we look elsewhere. Therefore, we are not an intelligence species. We are an intuitive species. And so if we can't find it in us, we will construct it. But as far as complexity emerges, use a simple rule so that you can combine them when they cease to be simple and cease entirely to be predictable.*

ROSALYN: *And all we have to do is look inside of the nature that exists separate from the human to the extent that it does, at the immense complexity of interconnections of every ecosystem that there is on the planet. Amazing and inter- and co-dependents and sophistication and complexity and it's right in front of us and yet we try to …*

GEOFFREY: *The interconnection yes, but co-dependents we don't know about.*

ROSALYN: *Alright, I am making an assumption here.*

GEOFFREY: *For all I know, you can eliminate 90% of the species on the planet and things will get along just fine.*

ROSALYN: *There's a point well taken. We couldn't do that with the trees, but we could do it with ...*

GEOFFREY: *You could surely eliminate us.*

ROSALYN: *They were fine without us for how many eons?*

GEOFFREY: *Exactly. And so I think the lower down in the system you go, the more careful you want to be. E. O. Wilson makes a strong case for ants.*

ROSALYN: *Well, ants are critical for our survival.*

GEOFFREY: *But if you eliminate us, not a problem.*

ROSALYN: *No way.*

GEOFFREY: *It might be better, but ants would be a problem.*

ROSALYN: *They would be a problem, as with spiders by the way.*

GEOFFREY: *Spiders I am sure would be a problem.*

ROSALYN: *Major problem to us if they were gone, so I don't think we ...*

GEOFFREY: *Major problem to everything, if they were gone. Wilson makes the point that the biomass due to ants and the biomass due to humans is about equal and if you eliminated us, the world wouldn't even notice, and if you eliminated spiders and certain ants, everything would collapse.*

ROSALYN: *Gravity, we would have a problem with the rotational spin of the Earth because of the missing mass, or?*

GEOFFREY: *No I think that the issue is with the scavenging, turning over things, and the sort of cleaning up.*

ROSALYN: *Oh I see, I see.*

GEOFFREY: *Ants really are, according to him, absolutely crucial in the interconnecting sense of tying all of the birds of the ecosystem together.*

ROSALYN: *Interesting.*

GEOFFREY: *And we are not. We are just the top food chain and so we are basically parasitic on the system.*

CHAPTER EIGHT

Conscientious Moral Commitments

I have tried to show how technology is developing completely independently of any human control. Carried away in some Promethean dream modern man has always thought he could harness nature whereas what is happening is that he is building an artificial universe for himself where he is increasingly being constrained. He thought he would achieve his goal by using technology but he has ended up its slave. The means have become the goals and necessity a virtue. We have become conditioned in such a way that we take on every new technology without once wondering about its possible harmfulness. There is nothing worrying about technology as such but our attitude toward it is very worrying.

—Ellul (1998, p. 184)

All scientific research, including nanoscale science, is governed by recognized and well-established research ethics. It is built on a foundation of trust, to assure that results are valid, safety is a factor, and the observable world is being described without bias. For example, great care is given to the use of human subjects, to inform those subjects, and to minimize any known harms that could come to them as a result of the research. Likewise, authenticity of authorship, honest documentation, respect for intellectual property, accurate reporting of findings, proper detailing of protocol, and allocation of credit are standards of ethics in the professions of scientific and engineering research and development. This is no different in nanoscale science. Engineers who work as researchers, taking the results of science into the development of new devices, machines, and techniques for human use, are also guided by detailed, well-developed, and evolving professional codes. These codes govern issues such as integrity and safety of design, the intention of anticipating and then minimizing potential harm to humans and their environment. As a group, engineers are entrusted by the public to design, build, and assemble products and materials, which will serve and benefit the best interests of the common good, while rigorously testing against known harms, and communicating all known risks. This holds true in nanoscale engineering.

We, who are the beneficiaries of those professions, rarely are cognizant and appreciative of the relative ease with which we are able to live, due in large measure, to science and engineering. The world we know and live in is robustly reflective of technological ingenuity. Technological humans no longer live in natural surroundings—virtually all that we do is technologically dependent and technologically formed. Perhaps it is because of this ubiquitous nature of technology that we are generally unaware of the profound ways in which technology regularly alters and changes very fundamental qualities of material life, as well as its meanings. Perhaps it is because Ellul is correct that we technologically dependent people have given science (and technology) over to a soteriological role in our lives.

If, however, there is a tragedy, or a major disruption connected to technological access, then for a moment we become more conscious of our complete and utter dependence on science and technology. This is when discourses critical of technology proliferate, seeking a source for blame of responsibility for those technological failures. Hurricane season is an example, when power outages and rapidly rising waters bring grave inconveniences to human living, and a rise in human agitation and frustration with the powers that be. Often, aspersions are cast at scientists and / or engineers who are responsible for discovery, design, building, or maintenance of the process, structure, or device in question. This is not to deny that scientists / engineers are morally accountable for their creations. However, others join them in culpability, such as those who manage, market, distribute, and ultimately we who consume technologies. Scientists / engineers are no more capable of orchestrating the outcomes of technological developments than they are capable of controlling the material world and filling the existential emptiness we all share (despite intentions or beliefs holding the contrary). As one researcher queried during our conversations, "Did the fact that Einstein discovered that $E = mc^2$ make him morally responsible for the atomic bomb?" The seeds of inquiry may be fertilized in the minds of individual scientists, but without the germination process of public or other research funding, the socio / cultural nutrients of want and desire and consumption that encourage growth, or the yielding and distribution of its fruits, technology would not flourish in our world and science would be in want of a true purpose. The process of nanotechnology development is growing up from a complex social construction evolving from narratives about how we wish to live in relation to the tools and devices we use, other people, and ourselves.

What is it that we, in various societies around the globe, might rightfully ask and expect of our researcher scientists and engineers regarding nanotechnology? Is it to avoid any harm that may come to humanity as a result of the knowledge we gain and the nanotechnology technologies we develop, while bringing forward envisioned improvements in our material existence? Is it to take heed of the values that are implicitly a part of their technological designs, and to be more aware of the moral responsibility that comes with them? Or, is it to avert any unintended consequences of nanotechnology that may adversely affect the public? Some research scientists and engineers respond to those expectations by placing moral responsibility for the outcomes of technological development squarely back on the public and on its policymakers. It may be that because of the unknowable elements of nanotechnology development, the researcher has very particular obligations with regard to precautions and safety. Yet, for the purposes of the ethical development of nanotechnology, where else might members of society look for leadership and responsibility? Who is in the position to provide ethical direction for its outcomes, to establish its directions, to articulate its purposes?

As experts, and by virtue of their training and capacity to reshape our world, research scientists and engineers do have a particularly high level of moral responsibility for the development of nanoscale science. But what power do they really have as individuals, what responsibility can they be assigned for the use, application, and direction of nanoscale technology development? Perhaps they hold some, but none alone. Those responsibilities have to be shared widely, not just placed primarily in the laps of those who by training and cultural induction are in a position to foster and pursue the new discovery and understanding, which makes possible our continued evolution as an increasingly technological species. Nor should responsibility be given over to policymakers who may or may not be scientifically or technically trained, but who nevertheless write the laws that may or may not avoid or ameliorate the harmful consequences of nanotechnology development. And despite what some of the researchers themselves are saying, there is no such willful entity or force as a "society" in which to place responsibility for the ethical and societal implications of nanotechnology. Technological society is a collection of individuals, not an entity unto itself. Science, which leads to technology, is a reflection of who we as individuals perceive and wish ourselves to be. The two are linked to our identity as individuals, families, and communities, and intrinsic to our politics, beliefs, and values formed through and in response to the narratives we weave.

Ellul (1990) explained that "in spite of all the progress, we still have an uneasy conscience," claiming that science leaves behind those who have the "scruples of conscience." He continued, "It [*science*] goes its inexorable way until it produces the final catastrophe" (pp. 186, 187). He blamed science. As much as I would like a force under which moral leadership of nanotechnology can be held, I am afraid that there is no such moral entity as the "science" Ellul spoke of so despairingly. Questions of ethics and social responsibility for nanoscale science and technology belong with individual scientists in dialectic with individual consumers, politicians, capitalists, agency sponsors, graduate students, and anyone else with the intellectual and emotional capacity to hold up a conscientious moral commitment, which insists that we proceed with perpension toward the humanitarian development of nanotechnology. Responsibility for nanoscale science and engineering development belongs to us all.

PROPOSAL FOR THE HUMANITARIAN, CONSCIENTIOUS PURSUIT OF NANOTECHNOLOGY

I have two primary interests in nanotechnology research and development: what it may mean for the well-being of the humanity, and what effect it may have on the stewardship of our earthly home. These concerns are over whether or not it is possible to have conscientious control and ethical guidance of nanotechnology development—a commitment that necessitates assessing the possible cultural, social, environmental, and moral outcomes of this particular perturbation in human technological pursuit. One challenge in doing so is the resistance to clearly see and ascertain our living relationships with technology, and with our own beliefs and desires about it. Ellul spoke of the world in which we live as "increasingly a dream world as the society of the spectacle changes bit by bit into the society of the dream." This dream, according to Ellul (1990), is a "dream of a science which is plunging us into an unknown and incomprehensible world":

> This will no longer be the world of machines. In that world we had a place. We were at home. We were material subjects in a world of material objects. The new world is no longer the familiar world of prodigious electronic

> equipment. In that world we were in a setting that was astonishing from many standpoints but that was still accessible and could be assimilated. What is changing in an incomprehensible way is the very structure of the society in which we find ourselves. This is a direct effect of science. But the average person has no awareness of it, does not know what it is about, cannot understand the change that is taking place, but is aware only of being on the threshold of a great mystery. (p. 184)

That dreamlike world was captured in the science fiction film, *The Matrix*, in which highly intelligent, humanlike machines take over and control the human race. In the film, humans believe themselves to perceive actual conscious reality, while in fact they live in isolated, confined pods, unaware of their true condition as energy producing slaves to the very superintelligent technologies humans themselves created. Neurologically connected to a virtual reality, the human mind has no consciousness of what is real, of self-knowledge, or knowledge of the truth.

I fully agree with Ellul that our world is becoming more dreamlike to human perception and experience. But science is only one element of influence in a larger cultural evolution in which the modern, Western technological culture has multiple levels of relationship with scientific pursuit and technological use. At one level, understandings that come from science are cherished for their ability to create technological solutions for addressing basic human needs. Pumps for retrieving clean drinking water from deep under the ground, instrumentation for measuring body temperature and substances for detecting the presence of bacteria, electronics for transmitting information over hundreds of thousands of miles are at one level simply about human basic survival. At another level, science and the technologies derived from it are coveted for their ability to address personal needs that fall a bit higher on Maslow's hierarchy of needs, such as entertainment, community, nurture, and ever-increasing economic gain. At yet another level, science and technology both express and reflect implicit human longings such as a sense of meaning and purpose, and feelings of immortality, worthiness, and power. For example, the former Twin Towers of New York City, and their representation of ultimate power, challenged human notions of scale, symmetry, gravity, and limitation.

When the word *conscientious* is used here, it is about wishing for us to do what is right. It is about wanting for there to be a conscious awareness of which levels of relationship are being pursued in the development of nanotechnology and at what planetary, social, economic, and moral costs.

In agreeing with Elull on his assessment of our increasingly lacking awareness of the changing structure of the world, I also use the word *conscientious* to call for an awakening. In the dream world of contemporary technological living, we slumber under our own beliefs. What is believed about nanotechnology, desired from it, and by whom? What meaning may those beliefs and desires have to human society? Conscientiousness—being alert and committed to knowing what is right, embracing what is right and acting on it—comes in part from being awake to the motives, desires, and beliefs that trigger one's actions. Conscientiousness in the development of nanotechnology, then, recognizes both the tacit and explicit elements of belief, which are at work in its pursuit, toward seeing more clearly how those elements of the mind are motivating and directing research and development at the nanoscale into appropriated technologies.

When the word *humanitarian* is used here it is in calling for an approach to nanotechnology research and development, which advances the belief that there is a connection humans have to one another, and to the Earth that is our home. Consider, for a moment, the human body. Mentally healthy individuals do what is feasible to care for it. They nourish it, bathe it, adorn it for protection and beauty, seek to cure it when ill, and generally respect it as one's very being. Whether or not one believes in a soul, self and body are integral. Suppose one were to view that self and body as inclusive of all others, wherein self and humanity are integral. In other words, by an affluent resident of Tokyo or London, an 8-year-old Sudanese girl who daily searches for water with her sisters are somehow also perceived to be part of one's own self. Then, all feasible efforts might be made both individually and collectively to care for, nourish, cleanse, and protect that self, which is also other: the human family. Now, what happens if the self that is extended to include humanity, also includes the Earth as home? So when a grove of redwoods is cut down, one feels inside that something in one's self is also being destroyed. What might nanotechnology development look like with that frame of reference? How might it be used and for what purposes? If self is body / humanity and home is Earth, then I imagine we might seek to develop nanotechnology to assure that all human beings are fed, clothed, and have access to potable water. It might mean that economic systems worked in such a way that market opportunities were global, and basic health care was universally accessible. Perhaps it would mean as well that waste products were part of a stream of reuse and toxins to the planet and human health were never produced.

But it is not my personal, utopian vision that is of importance here. That's my own and it may or may not even be materially possible, given humanity's relative lack of knowledge about the material universe, and lack of compassion about other. Instead, what is being put forth here is the ideal of humanitarianism, acknowledgment of other as worthy of concern, and a recognition of the body (one's own and those of other human beings) and that of the planet that sustains human life, as worthy and deserving of utmost care. Without that acknowledgment, nanotechnology will most likely follow the normal paths of technology development, which tend to flow along the courses of national competitions and market demands. Then, trickle down effects would be the best we could hope for in using nanotechnology for tending to the needs of humanity. If, instead, humanitarian concern were to be placed at the root of nanotechnology development, it is more likely that novel, innovative devices will emerge from the ingenuity of the human mind, toward a world with less suffering from material need and more universal prosperity.

Various narratives suggest that the congruence of nanoscale science and nanotechnology will mean for radical changes to humanity and other forms of life. It has been my intention to make plain that conscientiously directing those changes requires our conscious recognition of the values embedded in the narratives we construe about it, and in the meanings ascribed to the nanotechnology quest. Nanoscale science and engineering, like all scientific revolutions in human history, is a process. That process includes, among other things, the internal, often tacit, distinctively human search for meaning. If it is to be a humanitarian pursuit, then it must also involve deliberation over certain questions of ethics, deliberations over which could have a profound impact on nanotechnology public policy and application. Such questions are implicit in some researcher's narratives about nanotechnology, but rarely, if ever, explicitly addressed in the public domain. They include the following:

1. How is it we wish to live, and at what cost?
2. What is it we hold to be most true and most important in living together, not just in this particular society, but also as a human community living on this planet?
3. What might it mean to society, the globe, and the individual for the modern, technological world to acquire the knowledge required to further refine its abilities to manipulate the material world?

4. What does it mean to better and improve on that material existence? Is there an end to that process?
5. To what ends are we pursuing nanoscale science and engineering?

HUMANITARIAN HOPES AND PRACTICAL LIMITATIONS

Akhil, a graduate student from South Asia commented:

> Many of us see nanotechnology as a boon to the Western world. We are working on problems related primarily to Western conditions, such as diabetes and heart disease. There is a lot of research money available for problems to be researched at the nanoscale. As scientists, we are mostly driven by curiosity, so we can get interested in any scientific problem. But some problems just don't get funded, so they don't get our attention. Like malaria. All over the world millions of people are affected by malaria. We are not given the resources to study that. So, we just hope that there will be some kind of trickle down effect.

As with others like himself, who are from developing countries and in the U.S. on academic scholarships in the sciences and engineering, there is a sense of helpless hopefulness that suffering people in his own country will somehow feel a difference as a result of his work. Akhil wishes to actively seek to make a difference, but geopolitical and economic factors seem insurmountable. Researchers who are now U.S. citizens and already established as principal investigators running their own labs express similar sentiments, (see chap. 2, Luis), and yet despite their relatively high status as internationally recognized scientists, generally they too are not sure what to do other than to hope for a secondhand donation of goods and services to those who are most in need. Luis feels compelled to go a bit further than that, and is speaking out as an individual to his university deans, to associates in his professional society, and to me.

It remains to be seen what will come of those societies, which have no nanotechnology initiatives, for whom nanotechnology is unlikely to have any direct benefit at all. It may not seem to be a matter of grave social injustice to exclude the better part of the technologically developing world from the revolution that is embarking. But it is. At the time of this writing, except for South Africa, the African Continent had no national nanotechnology initiatives. In South America, Brazil is involved, but at this time, Central America and the Caribbean have no formal programs. Some would argue that free market economics, not socioeconomic justice, have always driven technological innovation, and nanotechnology can be no exception. There are those researchers, like Luis, who hope that all of the world will in some way be able to participate after the first world reaps its initial benefits. There is a more likely scenario, however. One is that for the most part Africa, Central America, the Caribbean, and much of South America will simply be left out of the military developments, new world markets, and improvements to human health that come to the "developed" world as the result of nanotechnology. But if nanotechnology turns out to be an unprecedented health and environmental horror, those regions of the world that do not participate may offer safe havens of protection. On the other hand, if things get seriously out of control in the nanotechnology-developed world with such products as synthetic self-replicators[1] there may be no safe haven possible in the world.

Human suffering has many sources, one of which is material neediness. It is often mistakenly believed that eventually, and with enough financial and intellectual resources, technology can cure all the ills that afflict humanity. This is not true. Technology changes the human condition but does not necessarily always improve on it. For example, existential emptiness and spiritual longing are sources of suffering that come from conditions of human consciousness that can only be fulfilled by sources other than material ingenuity. True human intimacy, silent contemplation, compassion, peace, and belonging in community are the more likely remedies for these states of mind. Seeking to enrich experiences of empty sexuality with gadgets, mechanisms, and video entertainment may be pleasurable and exciting for some, but leaves empty sexuality as empty sexuality. Despite the feelings of connection that appear to be there, enrolling in online chat rooms with strangers does not abate loneliness. Sony's Aibo gives the illusion of owning a pet, but the feelings of nurture evoked by it can only be painfully artificial in that people truly need to feel needed. And they truly need to feel mutual affection in connection to other living beings. Although a very sophisticated, seemingly intelligent robot dog can serve as a source for projected feelings from the imagination, it is not a living being,

---

[1]Devices that can make copies of themselves, which could have many valuable applications. Substances made of microscopic self-replicators could heal themselves by producing replacements for damaged parts. For example, molecular scale self-replicators could combine to form self-maintaining paint or spacecraft skins that can repair damage caused by space debris. In addition to replacing damaged parts, self-replicators can scale up production exponentially, as each new product is at the same time a new factory for more products. This could be a solution for accurately and inexpensively producing useful quantities of novel nanoscale materials. See Keck Futures Initiative for complete description and definition at: http://www7.nationalacademies.org/keck/Keck_Futures_Nano_Conferences_Synthetic_Self_Replicator_Summary.html

and therefore unable to have empathy, desire, an appreciation for suffering and an ability to play, or any of the other attributes living animals express with humans.

We need to realize that human relationships with living beings are not the same as human interactions with technology. Attempting to replace one for the other is pointless. Looking to nanotechnology to solve world hunger or to cure AIDS or to correct for the damage of CO_2 emissions may be as well-intentioned as the examples of false technological fixes referred to earlier, but equally ineffectual. Whereas technology may change the situations, and have utility, when applied to the ills of human life, it is ineffectual unless accompanied by a humanitarian perspective of self and other love and respect. Only then are the changes sought through technology authentically healing and improving to human conditions. Nanotechnology can and will touch human life, and the life of the planet in supportive and enriching ways, if those who have any influence or contribution to make in its development are conscientious and humanitarian in their aims. Speaking generally for the human race in using the word "we," I want to assert that as "we" proceed to refine our capacity to manipulate matter, and the capacity to restructure the material world we perceive to be our domain as we wish, we also inherit a corresponding increase in responsibility for the Earth and its inhabitants. Does humanity possess the collective wisdom, which is requisite for taking on that responsibility? Are we mature enough to handle the potency of the devices and processes we now aim to develop, toward precise refinement and control of matter?

It is a worthy and plausible aim to guide the science and development of nanotechnology ethically, and in existentially fulfilling ways, by directing it toward authentically humanitarian and "planet-arian" aims (which includes fiduciary responsibility for the Earth and its inhabitants). But to do so, to make that kind of conscientious determination about the goals and intentions of nanotechnology development, requires a continuing, honest appraisal of the competing values and interests that are present in its inner workings—the internal, personal struggles for meaning, which lay at the core of the nanotechnology quest. It is through such assessments that we might be able to recognize how our human sense of finiteness and frailty lends itself so adeptly to aggressive technological development. Furthermore, we of highly technologically developed societies increasingly disregard our utter dependence on the Earth that nurtures and supports our lives, and proceed to develop new and increasingly powerful technologies with only superficial consideration of their effects on the Earth. In our pursuit of notions of "progress" through technological development, our impact on human communities, animal life, and ecosystems grows as well. Therefore, we are increasingly, morally responsible for attending to genuine forethought, reflection, and care over the purposes of that technological progress. If the meanings we evolve for the future are explicitly expressed and authentically valued, then those meanings can be used to ethically guide and direct human and financial resources in the enterprises of nanotechnology.

It has been argued that although science enables us to predict natural events, it is remarkably unpredictable as far as its own future is concerned (Richter, 1972). Although he recognized that the technological implications of science have stimulated drastic and disruptive transformations of society, he saw that because of its very nature, the course of science cannot be determined by society. (Because it is shaped by social forces, science is, by definition, not science.) I am suggesting that because in the nanotechnology initiative science and technology are absolutely linked, its course can and should be decided and guided. Too much is at stake here to accept our usual laissez-faire attitudes about the uncontrollable evolution of scientific knowledge and technology. The common assumption and claim that the future of nanotechnology, or any technology for that matter, is both unpredictable and uncontrollable is a myth. I concede that the belief systems that frame the nanotechnology initiative and the institutional structures and market forces, which support its pursuit, are unwieldy, tacit, and supremely powerful. And further, to stake a moral claim on controlling and guiding the future of nanotechnology is much like standing before a tsunami and yelling, "No, go the other way!"

It may seem to be an impossible task to bring conscientious commitment and humanitarian leadership to the direction and outcomes of nanotechnology. Obviously, there are limitations to what elements of the future can be determined in the present. Who could know that nuclear power in the U.S. would fail so miserably, while becoming a major source of energy in Europe? No one could have known. And, its fate could not have been controlled; there were too many unpredictable factors at work. What I am asking for, the responsibility I am speaking about, is the moral responsibility to be clear about our intentions, honest about our motivations, rooted in our virtues, and committed to those commonly shared values we can agree to uphold. That is the only real means we have to control and direct the nanotechnology future.

Here is an analogy. If one seeks to lose excess body fat, there are technological fixes to do so such as stapling the stomach and liposuction. But real, authentic change toward the well-being of the person must include introspection. It has to address questions about one's relationship to food, what one eats, what particular kinds of foods are selected, what the body represents, lifestyle choices, and what the self is perceived to be relative to the body. Only when these kinds of questions are considered can fat reduction happen in a lasting way that is respectful and nurturing of the body. Even if the technological procedure appears to fix the problem, the efforts will be superficial if they fail to also engage the inner self. The same goes for nanotechnology. The work to be done on the interior—with values, beliefs, meanings, aspirations, fears, myth, and so on—is essential.

In the public domain, great claims have been made by its most vocal proponents that nanotechnology has the tremendous capacity to revolutionize our world. Interestingly, when individual researchers have spoken with me about the work of their groups and labs, a very different kind of claim is made. Rather than making great proclamations, they tell stories about inspired curiosity, humility in the face of elusive scientific knowledge about the material universe, desires for personal happiness, and the pursuit of their own individual dreams. Yes, they want to make a difference in our world, and they dream of breakthroughs. But they speak mostly about small steps, and shun grandiose claims of major changes to come.

Florman (1994) lamented the end of what he saw to be the Golden Age of Engineering. He explained how during that time, engineers could feel their work to be existentially pleasurable. In those days, according to Florman, engineers felt fulfilled as men, loved their work, and felt it was inherently good. But no more, because today's technologists work in a time of criticism, even hostility toward their works. He elaborated:

> It is being said that engineering, no matter how clever, is destructive. It is being said that engineering, no matter how well-intentioned, is pernicious. Engineers are being called charlatans, fools, and devils. And such things are not being said by a single eccentric philosopher sitting by Walden Pond, but by myriads of people in every walk of life. Even engineers, to judge by their journals, have become uncertain, self-critical, and defensive. It should be apparent that engineering Golden Age ended abruptly about 1950, and that

> the profession, for all its continuing technical achievements, finds itself at the present time in a Dark Age of the spirit." (p. 11)

Yet he also acknowledged that engineers, who are long said to be obsessed with materials and machines, are increasingly thinking introspectively. "It could hardly be otherwise," he wrote, "considering the social turmoil of the 1960s, the environmental evolution of the 1970s, the political upheavals of the 1980s, and the evolution of a totally transformed global economy" (Florman, 1997, p. xi). If Florman were writing today he would have to include the mapping of the human genome, cloning, and the evolution of nanotechnology as reasons for engineers to become more inward looking, to examine their personal thoughts and feelings, and to see "under the surface of things." Florman placed great emphasis on the value of introspection for the engineer. But, he indicated, the implicit conviction of every engineer is that thought will lead to action.

My conversations with researchers suggest a particular pattern of reasoning that is at work inside of nanotechnology development. That is, that although introspective as individuals, the work of research scientists and engineers is focused externally, on understanding processes in order to acquire more refined technological capabilities toward solving technological problems. That is what they are trained to do. What is interesting is how they speak about that ambition in terms of what they explicitly value. Writing down a recipe for making materials speaks of research in terms of cooking, or magic, where a recipe or concoction has to be created before it is possible to take the individual ingredients of material substance and change them into what one wants. Striving for enjoyment in work and having fun point to personal satisfaction in the creative and challenging process of answering elusive questions about how things work, and solving scientific problems through that knowledge. The ability to manage otherwise fatal diseases reinforces the commonly held ideal that the body processes can kill us if they are not somehow managed. Understanding what is going on at the nanoscale is a highly prized outcome of the daily grind of searching for new knowledge, while understanding how a functional device can be made is the payoff. Those are obvious sources of meaning for the individual. Given the opportunity to speak introspectively, as these conversations provide, to express thoughts and feelings about their work and its implications to the larger society, researchers of the nanoscale have a great deal to share.

During the industrial revolution, not everyone simply accepted that material "things" should completely rule people's lives, and several groups experimented with alternative ways of using technology to shape society. One group that most effectively linked technology and society were commonly known as the Shakers. Carlson (2005) explained:

> The Shakers believed that everything around them should be a true reflection of the inner spirit .... All of the virtues that were good for their souls—honesty, utility, simplicity, purity, order, precision, economy—they felt should be part of the things they made and the way they lived each day. As one Shaker explained, "Heaven and Earth are threads of one loom." As a result, the Shakers laid out simple but wonderfully ordered villages, they built plain but elegant buildings, and they made remarkably useful and beautiful objects to use in daily life. Shaker villages and products stood in marked contrast to many of the goods of American industrialization whose design was driven by the limits of mechanization and the competitive forces of the marketplace .... For the Shakers, technology was not to be rejected but rather carefully shaped to advance their spiritual beliefs. (p. 86)

It's true; the Shakers didn't even survive their ideology. But, they are nevertheless a remarkable example of how values can be conscientiously placed at the center of technological intent.

It seems to me that nanotechnology can be correctly viewed as another marvel of human ingenuity and curiosity, but it is also a potentially very powerful new capacity in human hands. Therefore, its development must be guided with great care, through loving concern about humanity, and the Earth on which we have our existence. But that kind of development will be possible only if the leadership of this "next technological revolution" can somehow envision and embrace the intention of directing its development conscientiously, toward humanitarian aims, through open, public discourse, with honesty about our struggles over the meanings nanotechnology holds for us, our beliefs, and ethics. And as this happens, let the voices of the research scientists and engineers, whose individual and collective minds and efforts continue to move humanity into ever more refined ways of knowing and living, be heard and included in that discourse. Then, although it may not be possible to know what's going to be next, we might actually be able to direct, apply, and use nanotechnology in ways that allow multitudes of people all over the globe to move closer to fulfilling the human potential of universal well-being and prosperity. Keen's is one of the many researchers' voices that surely ought to be heard:

ROSALYN: *Do you think that the physical universe is something which simply needs to be discovered, or is the purpose of scientific inquiry to discover its properties so we can control, manipulate, change, and recreate our experience in the physical universe?*

KEEN: *I think that I really don't understand that question.*

ROSALYN: *I guess it's a value statement on my part. I just am curious where we are going, and why. In other words, is there any end to this inquiry for new knowledge? Is there an end point?*

KEEN: *It's hard for me to imagine.*

ROSALYN: *Why, because of the physical nature of the universe or because of the human being?*

KEEN: *Because of the human being. The human being will consistently and always ask deeper questions. Maybe the way biology has evolved is the best example. Look how specialized it is. It's specialized because we have come to know a heck of a lot more about the hand and about the foot. So there are hand surgeons and there are foot surgeons, rather than general surgeons as they were 50 years ago. The depth of knowledge of each aspect is becoming so great that only one person can handle a very small segment.*

ROSALYN: *Soon we will have pinky surgeons.*

KEEN: *Yes. I think that's right, specialized. And who would you go to as a surgeon? The person who has operated on many pinkies, and has as detailed a knowledge of pinkies as there possibly could be.*

ROSALYN: *Then, when we get to highly interdisciplinary work, such as you are doing, we want to know how does the pinky relate to the thumb, and can I find a surgeon who can do surgery on the pinky and also understand how it relates to the thumb?*

KEEN: *Yes.*

ROSALYN: *I guess that takes us to a new level of inquiry when we go back up from the detail.*

KEEN: *I have a quibble with that, because it doesn't mean that a person who is into science is able to grasp any larger range of knowledge. But they become very specialized. For example, we have a small program in which we make organic materials interact with metals. It involves some chemistry. It involves some physics and surface analysis. The graduates do become very knowledgeable in that very thin slice of a problem. In this particular case, that affects organic displays. It is interdisciplinary for sure. He needs to take some chemistry knowledge and he needs to take some physics knowledge. But I think on the whole, that we have become more and more specialized, not less specialized. "Interdisciplinary" just means that we are grabbing a little of this and grabbing a little of that and making it work.*

ROSALYN: *Is that your specialty?*

KEEN: *That's right; I am defining a new specialty.*

ROSALYN: *So, does the acquisition of new knowledge necessarily mean going down into sub, sub, sub, subspecialties?*

KEEN: *No. You may acquire new knowledge by going into sub, subspecialties, but you also could look more deeply into an existing specialty. You don't have to form a specialty. But the acquisition of new knowledge means finding out the obvious, finding out something new, some scientific system. It seems to me that it would be a hard thing to prove, but it seems to me that the cleverest are those who have figured out the most interesting questions. They haven't answered the questions necessarily, surely they have in some cases, but even fewer have figured out good questions. Clever people will consistently ask deeper and deeper questions. But everything still belongs to basic knowledge.*

ROSALYN: *I am finding fewer people are interested in the basic knowledge, at least among the people I am talking to, who have put a nanoscience tag on their work. Most seem very interested in moving quickly to application.*

KEEN: *That's good.*

ROSALYN: *Is it?*

KEEN: *Well, we need some people to think along those lines. Certainly you want to be able to see some impacts in the near future because otherwise how are you going to convince the funding agencies that this is valuable work? Not only do they have funds, but they are not that patient. You want to know this is going to be important, so then more fundamental science work will be supported.*

ROSALYN: *Oh, that's interesting.*

KEEN: *Nowadays it is very hard to argue that we should continue to use taxpayers' money to do fundamental research. Because, normally people don't understand fundamental research, or what's involved, and it's hard to explain and you always have to explain it using everyday examples. So if you have something that is making an impact, this is good. Otherwise, we would not know how important the laser is. That comes from fundamental research. When the first person recognized lasers, they were not looking for something that would become a CD writer. They were looking to understand transistors at molecular levels. It's very, very fundamental science and it turns out that the laser could be used for many things, such as laser surgery. I think a country like the United States has to invest in the science, fundamental science. It's for the future because*

> *you don't know what you are going to get. It cannot be planned, normally. You have to let people create, according to their interests. Some people will recognize very early on their research will have a great impact, but it's not the original plan that you are going to research a goal, that you do this, this, this. In scientific research, you don't know what's going to be next.*

FINAL THOUGHTS

My conversations with Keen and other researchers have instilled in me an appreciation for the beauty of basic science research toward the unknowns of discovery and the intrigue of unanticipated application. They have also confirmed what I suspected: that whereas the unknowns of application are valued, the basic nanoscience researchers are pressured to focus on their applications, largely as a matter of practicality. Like the engineers who are by nature interested in solving practical problems, the scientists are caught in the emerging culture of nanotechnology that says, make your work do something useful, soon, and more money will come to support your continued research. Nanoscience and engineering present a number of nodes of ethical concern. Among them is the external pressure on research scientists and engineers to move as quickly as possible through basic research to marketable application. More highly valued by sponsors and investors than discovery for learning sake is the promise of potentially high returns on financial investments. The system of rewards makes that obvious. It seems to be especially true that in nanotechnology, "the pure scientists have become more detached from the mundane needs of humanity, and the applied scientists have become more attached to immediate profitability" (Dyson, 1977, p. 199). It would appear that under current conditions of the nanotechnology initiative, basic research for its own sake gets devalued and applied science is artificially accelerated. This condition alone could stymie the critical reflection and deliberation over ethics, meanings, and beliefs about nanotechnology, which is essential to its humanitarian development. Other than the scholars of such things, who has the time, really?

Surely there is another, more conscientious, way for nanotechnology research and development to proceed.

APPENDIX A

Methodology and Preliminary Research Findings

I began meeting with researchers for this study in summer 2002. Thirty-five research scientists and engineers whose work takes place at the nanoscale were among those who agreed to participate. All have been interviewed once. Twenty-three have been interviewed twice. By the time this book has been published, about 18–20 should have been interviewed three times. The interviews last an average of 1 to 1.5 hours. On occasion, the researcher is happy to continue, but most have very demanding schedules and are pressed to give even 1hour of their time. On large part, this work is intuitive, on my part. My hope has been to evolve a theory of how particular values and themes comprise the structural framework for meaning making and beliefs about nanotechnology. Having only recently been introduced to it, I have come to see that my intuitive approach to the conversations, and my interpretations of them, closely resemble the evolving Grounded Theory approaches of Glaser and Strauss. Using the methodology of Grounded Theory (which refers to theory developed inductively from a corpus of data; in this case, the conversations themselves), I can take a discourse-oriented perspective that assumes variables interact in complex ways. Grounded theorists are concerned with or largely influenced by emic understandings of the world, using categories drawn from respondents themselves, toward making implicit belief systems explicit.[1]

Following that approach, I have begun to identify some of the emergent categories inside of these interviews, and their properties. Although they may change, currently there are 17 such categories.

[1]For further, detailed explanation, see *Introduction to Grounded Theory* by Steve Borgatti, available at http://www.analytictech.com/mb870/introtoGT.htm

EMERGENT CATEGORIES AND PROPERTIES OF ANALYSIS

I. Categories

1. Matters relating to reporting of research results
2. Matters relating to grant writing and other elements of gaining support
3. Personal responsibility
4. Political perspectives
5. Personal aspirations
6. Beliefs about science
7. Perceptions about nanotechnology generally
8. Conceptual blocks to ethics considerations
9. Ethics in nanotechnology generally
10. Personal values pertaining to nanotechnology research
11. Collaboration issues
12. Problems, concerns, and fears
13. Notions of failure and success
14. Issues pertaining to financial profit and personal fame in the profession
15. Pure science versus engineering or application
16. Future directions and applications of nanoscale science and research
17. The role of the government

II. Properties Identified

A few preliminary properties have been identified in most of the 17 categories. Those are as follows:

1. Matters relating to reporting of results

All of the researchers I have formally interviewed are principal investigators. Some have spoken of there being a great deal of time pressure on them to report findings. They seem to feel that the time is too short to do so adequately, too many interim reports are required, and there are pressures to get results out prematurely. A few have found this process to be compromising to their work, but acknowledge its critical role in assuring their continued financial support. Other reporting pressure comes from the competition to get journal articles out quickly, and before others do. Although laboratory work can take a great deal of time, and be difficult to control, they commonly feel that there are external, professional pressures to gain and hold high status. The way to do this, primarily, is to get results into the top journals first. High status, they say, brings the rewards of more grant money, which means larger labs with better and more equipment and more graduate students to do the work, which in turn means more work can be done to produce faster results. Unfortunately, this is a real challenge for the junior professors, except for the "hot shots," and to women, most of whom are junior professors.

2. Matters relating to grant writing and other elements of gaining support

A few have spoken cynically of nanotechnology as a way to get more money for what they were already doing before nanotechnology became "hot," but that was not recognized before the initiative came along. They feel that there is a language game that must be played to assure fundability of their projects, and they find themselves adapting their primary research questions to fit the goals of the nanotechnology initiative. Some, especially those who are senior level scholars at the top of their fields, and internationally recognized, express no such criticism. They speak optimistically and with enthusiasm about their prospects for new findings and, particularly, for the creation of new processes, devices, and applications. The smaller group PI's (Principle Investigators) have on occasion referred to the "big guys" with established nanocenters as being the ones who "always" get the federal grants, as opposed to themselves, who they perceive are out of the "in" group of highly recognized and therefore politically attractive for funding.

3. Personal responsibility

Ever since a scientist got caught and widely popularized last year with falsified, published data, the subject of reporting integrity has come up regularly. Most expressed some empathy, and were anxious and nervous about what happened. They seemed to see themselves as vulnerable to the same mistake, given the enormity of the pressures they feel to compete for the place of "first" in reporting new findings in the literature. The issue of far-reaching effects of what scientists may learn from their work and its possible unintended consequences has also been a theme in these discussions. Almost without exception, researchers have emphasized that they cannot be held responsible for what someone else does with the new knowledge they themselves gain and report. For example, I was asked, "Just because Einstein discovered $E = mc^2$, was he responsible for its application in the atom bomb?" Whenever we talk about what happens to the new knowledge they acquire, or the new devices and applications they contribute to, the researchers say that their only responsibility really is in accurate reporting. Otherwise, most feel they would be immobilized. Another concern is their responsibility for graduate students. Nearly all researchers spoke of their graduate students as their primary focus. The sense they gave is that there is a family structure where the principal investigator functions in the parental role (i.e., teaching, guiding, and caring for their student), whereas the students, much like children in a family, become a source of personal pride by carrying the investigator's "name" out into the world. There is also some deep commitment to responsibility to the profession or to science generally, in terms of the quality of research and adherence to the principles of the scientific method. Finally, because most of these researchers are receiving public money, the theme of responsibility to the public emerges periodically as well.

4. Political perspectives

A few senior researchers have talked about nanotechnology as a politically driven initiative. In those discussions, there have been concerns raised about the use of nanotechnology to increase power and wealth in the developed West, and particularly for a few already wealthy people. Health care issues are given as examples of politically motivated funding for nanotechnology; whereas one of the biggest sources of human suffering in the world is still malaria, cancer research gets priority because this hits most closely to home for the politicians who are making the funding decision. For those researchers who have concerns about this, there seems to be some frustration about putting their efforts and resources forward in these areas, as opposed to the areas they believe warrant greater social effort. They hope there will be some trickle down effect from their work to the developing world, but they do not see the mechanisms for making it truly effective. International graduate students in the labs of PI's have also raised this issue. They feel badly that at home people have no access to potable water, yet they are doing research here on technical questions pertaining to increased wealth for the developed world, such as how to beat Moore's law. Otherwise, most do not discuss politics at all.

5. Personal aspirations

When speaking about their personal aspirations, most researchers seek to make a change in our world, to have an impact on the people. Most of these changes deal with new ways of manipulating matter, which would provide us with new vaccines, new scientific tools, new consumer products, or new ways of living. One of the scientists quotes his friend, the inventor of the supermarket bar code reader, saying: "Every time I go to the grocery store, I feel like I did something important." This type of feeling and personal recognition appears to be what the researchers are truly aspiring toward. Peer recognition also seems to be of great importance. Most scientists and engineers in my group talk about the importance of being a leader in the field, of making a breakthrough, and of being recognized for their discoveries. A lot of the researchers project these goals onto their graduate students. In fact, a considerable part of the researchers direct their team, but some admit that their students do the core of the work, and those are the ones who will make the breakthroughs. In the interviews, the researcher–student relationship is often portrayed as a parent–child relationship, in which parents live their dreams and aspirations through the success of their children.

6. Beliefs about science

My entire group agrees that science is a social good and that scientific research is a morally neutral enterprise. Most believe that the material world is out there, waiting to be discovered. In this context, science takes the form of a search tool, devoid of human values, used to dig a way into the mysteries of the objective world. As a consequence, researchers believe that their discoveries are in themselves neutral, but they agree that the applications of these discoveries carry moral values. A lot of science and engineering examples are usually discussed, going from the invention of a hammer or a knife, to the invention of the atomic bomb. The consensus seems to be that you can use a hammer to hit a nail on the head, or you can use it to hit a person on the head. Hence, science is neutral, but its applications are double-sided. In our conversations, the notions of progress, scientific advancement, and human progress are often spoken of as synonymous. Indeed, most scientists will use these notions interchangeably. Science, being progress, is thus portrayed as an entity with its own specific direction and momentum, whose course we cannot and should not stop.

7. Perceptions about nanotechnology generally

There seem to be very distinct visions of nanoscale science and engineering research within the group I interviewed. First, a few researchers disagree with the fact that it is a field of its own. They argue that "it's just chemistry," or that all they do is material science, physics, or biology. Hence, there seems to be a belief that there is nothing truly different about nanotechnology, because nano is just the continuation and evolution of already existing fields. Who actually does nano and who doesn't is also a source of debate. Although some scientists adhere to and enjoy the multidisciplinary aspect of nanotechnology research, others tend to separate their work from that of physicists, or material scientists, or biologists, or computer scientists. Eric Drexler, among others, is often placed into a completely different field of knowledge. When referring to a particular aspect of the field some researchers would reply "that's Drexler," or "that's physics," reflecting their belief that this isn't truly nanoscience. Hence, there are different degrees of belief in the existence of the nano field itself, and in the multidisciplinary aspect of this field also. On the other hand, when the researchers are asked to talk about their work, what they do, and what they invent, there is a unique element that comes up. Indeed, the discourse suggests another revolution in terms of economics, social implications, laws of physics and chemistry, and devices soon to be created. There are those researchers who do claim that the field instills never-before-seen collaboration between disciplines. Some researchers do claim, or at least do not deny, that there is something new about nanoscience. This leads to the conclusion that nanoscale science and technology is not really different in terms of the disciplines themselves, or the knowledge associated with it. Rather, nanoscience appears to be a revolution in terms of the collaborative research structure built around it, its incredible potential to alter our material experience, and the degree of control over natural elements (including ourselves) that it provides us with.

8. Conceptual blocks to ethics considerations

Almost all of the researchers are highly concerned with issues of ethics. This concern, however, is accompanied by a feeling of powerlessness. As stated previously, they see the danger of nanoscale science and technology in its applications, not in its discovery or in the conception of nanodevices. Therefore, a few reject any obligation to make ethical decisions, often placing this responsibility on the shoulders of policymakers. This separation between nanoscience and ethics becomes apparent in the interviews. For instance, when I ask a question about the technical side of the researchers' work, followed by a question regarding concerns about their work, a few researchers pause for an instant and ask, "Have you asked a philosophical question?" or "Are we talking about ethics?" This clearly defines a line between the scientific nature of some of the researchers' work and matters of ethics. Hence, the belief that their work is neutral, they have no power over its applications, and their work and ethics are two distinct fields, is the primary roadblocks to a careful consideration of ethics.

9. Ethics in nanotechnology generally

It is interesting to note that researchers do view nanoscale areas other than their own bearing a big ethical weight. For instance, when I ask biologists, chemical engineers, or material scientists what they think about the social and ethical implications of nanoscale computer engineers or physicists, they usually see a load of moral dilemmas that should be addressed. When asked about their work, however, the computer engineer or the physicist would reply that, for instance, they are only working on the theoretical level, and the real concerns are found in biochemical engineering. From this example, it seems clear that although they do spend a significant amount of time thinking about ethics and morality, the researchers have a really hard time viewing their own work as a source of ethical concern. Nevertheless, a very few researchers demark themselves from the rest in that they are deeply concerned about their personal work and its consequences. One researcher expressed how deeply affected she feels when hearing scientists publicly claim that they use stem cells for their research solely for the sake of science and personal curiosity. She acknowledges her responsibility to the public and affirms that she makes every effort to pursue worthy goals while employing ethical research methods.

10. Personal values pertaining to nanoscale research

A number of the researchers have spoken about their childhood-born interest in science and engineering. Most talk in terms of wanting to make a difference in the world, toward improving the quality of living, alleviating suffering, curing diseases, and the like. Others are frank about simply being curious. Every one means to be conscientious and to do what's right, but the most consistently expressed and deepest values pertain to the acquisition of new knowledge; the contributions each might be able to make to "the literature." This, beyond all else, seems to be the most significant value for the researchers. In a few cases, personal experience with tragedy, such as losing a loved one to cancer or suffering from it oneself, point to the value of human life. It is also expressed as a source of motivation, one that has largely determined the direction and purpose of the research.

11. Collaboration issues

Most are excited about the new opportunities offered through collaborations. No one has expressed any hesitation to collaborate. In fact, they often say that their work at the nanoscale could not be done without the help of people from other fields of expertise. Overall, they are both stimulated and challenged by having to learn and understand the technical language of fields of expertise outside of their own. Although there is a real sense that collaborations are expected by such agencies as the NSF and DARPA, they are motivated by the apparent financial opportunities that collaborative efforts represent, and by the fact that their research is more likely to be successful when collaborated.

12. Problems, concerns, and fears

Most problems stem from the financial costs of running a laboratory and the exceptionally high prices for equipment and supplies needed to do research at the nanoscale. Keeping graduate students funded is a related and very serious concern for those who are running relatively small labs on short-term, soft money. Problems related to international graduate student visas are mounting. No one answered my question about what fears they might have. In fact, it was generally seen to be a strange question.

13. Failure, success, and the competitive race

In our conversations, scientists and engineers clearly expressed what it means for them to succeed or fail in their research efforts. For most of the researchers, success is associated with a breakthrough. One of the scientists mentioned that just one true breakthrough in a lifetime would be enough to meet his notion of success. Recognition in the field is also an indication of achievement. This notion of success primarily based on making an impact in the world and obtaining recognition in the field parallels the researcher's aspirations mentioned previously. At the same time, scientists acknowledge that not everyone can be successful. One of the scientists believes that there are only about 3% of leaders in the field. The rest of the scientific and engineering community belongs in the remaining 97%. For that particular scientist, the notion of success means to be in this leading 3%. This idea is paraphrased in other interviews.

Numerous researchers speak of nanotechnology research as being a race in which one has to make a finding first. It is a very difficult race in that the person who gets second place obtains very little credit. As a matter of fact, for some researchers, making it to the "finish line" late, or not making it to the finish line at all, is what constitutes failure. Through these interviews, one can sense a strong, perhaps tacit, desire to compete against the rest of the research community, and to come out first. Whereas most research teams are openly cooperative and multitalented, from a global perspective, these teams are competing in an arena with teams from all over the world. In other words, this pressure to "be the first to" suggests a mind war involving teams from all over the globe. Failure in the athletic world is represented by giving up, being weak, or arriving last, however, failure in the nanoresearch world is to give up a research project, not come up with significant results, or being outdone by another research group, or to lose ones' grant to non-renewal. At least one of the researchers does believe however, that this competitive structure is beneficial. He, in fact, claims that this pressure to succeed brings more results in the scientific community. Moreover, he believes competition / imitation to be the finest form of admiration, and suggests that having several teams work on a specific research topic will help investigate all aspects ("holes") of the subject, and in doing so, will build a stronger foundation for science to move on.

14. Issues pertaining to financial profit and personal fame in the profession

Nearly everyone of the participating researchers has expressed a longing for professional recognition for their work. Most seem hopeful that their research will culminate in some product that will be taken up by a business venture of some sort. A few have started their own small development companies, or joined efforts with existing for-profit companies.

15. Pure science versus engineering or applied science

There is a distinctive difference in the way scientists who are working at the nanoscale and the engineers speak about the nature of their own research. At the same time, the theoretical, philosophical, and practical divisions of science and engineering are blurred at the nanoscale of inquiry. Whereas the scientists (i.e., physicists, chemists, biochemists, etc.) tend to speak about basic research and answering questions simply for reason of their own curiosity and contributing to the body of existing knowledge, the engineers (i.e., bio-medical, material scientists, mechanical, etc.) are very clear about wanting to get something to work in order to solve a specified problem. With an increasing focus on collaborations between formerly distinctive disciplines, and with the focus of nanoscience research on very specific nationally stated goals and objectives, the notion of pure science for science's sake is somewhat obscured. With only a few exceptions, nearly all speak more in terms of tasks and problems than in terms of knowledge.

16. The direction of nanoscale science and technology

There is great hesitation on the part of most to answer the question of where nanotechnology is leading. Some feel that this cannot be known. Others feel ill-equipped to think in those terms. A few, who are key public proponents of nanotechnology, are very clear about the possible applications of their own work and of nanoscience generally. All have been willing to project 10 years out about their own research developments, but with caveats about the unpredictability of science research. Interestingly, when the subject of science fiction comes up, and the respondents are given the freedom to think fantastically and creatively without their ideas being judged, then they offer many possibilities about the futures of nanotechnology. But always, they qualify their statements.

17. The role of the government

There exists a conceptual and perceptual tension over whether or not the scientists serve the interest of the government, and private business, or some otherwise neutral, universal quest for knowledge. Although there are those who adamantly defend their role as independent, others acknowledge the source of their financial support as inextricably linked to the determination of their academic freedoms.

APPENDIX B

History of the Nanotechnology Initiative in the United States[1]

As nanoscience has advanced and discoveries in the field applied, the potential contributions of nanotechnology to future economic growth has brought increasing government attention. Today, nanotechnology is a top research priority of the Bush administration.

Attempts to coordinate federal work on the nanoscale began in November 1996, when staff members from several agencies decided to meet regularly to discuss their plans and programs in nanoscale science and technology. This group continued informally until September 1998, when it was designated as the Interagency Working Group on Nanotechnology (IWGN) under the National Science and Technology Council (NSTC).

The IWGN sponsored numerous workshops and studies to define the state of the art in nanoscale science and technology and to forecast possible future developments. Two relevant background publications were produced by the group between July and September 1999: *Nanostructure Science and Technology: A Worldwide Study,* a report based on the findings of an expert panel that visited nanoscale science and technology laboratories around the world; and *Nanotechnology Research Directions*, a workshop report with input from academic, private sector, and government participants. These documents laid the groundwork and provided the justification for seeking to raise nanoscale science and technology to the level of a national initiative.

In August 1999, IGWN completed its first draft of a plan for an initiative in nanoscale science and technology. The plan went through an approval

[1]See http://www.nano.gov/html/about/history.html for more detailed information on the history and evolution of the National Nanotechnology Initiative.

process involving the President's Council of Advisors on Science and Technology (PCAST) and the Office of Science and Technology Policy. Subsequently, in its 2001 budget submission to Congress, the Clinton administration raised nanoscale science and technology to the level of a federal initiative, officially referring to it as the National Nanotechnology Initiative (NNI).

Once the NNI had been set up, the IWGN was disbanded and the Nanoscale Science, Engineering, and Technology (NSET) Subcommittee was established as a component of the National Science and Technology Council's (NSTC) Committee on Technology (CT). The CT is composed of senior-level representatives from the federal government's research and development departments and agencies, provides policy leadership and budget guidance for this and other multiagency technology programs.

The NSET is responsible for coordinating the federal government's nanoscale research and development programs. The NSET membership includes representatives of departments and agencies currently involved in the NNI and OSTP officials.

The National Nanotechnology Coordination Office (NNCO) was established to serve as the secretariat for the NSET, providing day-to-day technical and administrative support. The NNCO supports the NSET in multiagency planning and the preparation of budgets and program assessment documents. It also assists the NSET with the collection and dissemination of information on industry, state, and international nanoscale science and technology research, development, and commercialization activities.

The NNCO serves as the point of contact on federal nanotechnology activities for government organizations, academia, industry, professional societies, foreign organizations, and others. The NNCO facilitates outreach through the planning, organizing, and conduct of workshops, as well as through reports and development and maintenance of the NNI Web site (www.nano.gov).

This information is taken in part from *Small Wonders, Endless Frontiers: A Review of the National Nanotechnology Initiative*, by the National Research Council (2002).

See also Nanoscale Science and Engineering R&D Extend Frontiers of Scientific Knowledge, Lead to Significant Technological Advances, Supplement to Presidents' FY 2004 Budget 2003, Oct.

APPENDIX C

Interview Protocol

Ethics and Belief inside the Development of Nanotechnology

1. Please tell me something about your background.
2. When did you first get interested in science? (Engineering?)
3. How did you first get involved in nanotechnology?
4. What is it that interests you about the field?
5. Please describe and explain, in lay terms, the nature of your current research.
6. What are you ultimately hoping to accomplish?
7. What would it mean if you were successful?
8. To scientific understanding?
9. To yourself as a scientist/engineer?
10. To people in general?
11. What makes this project most interesting to you?
12. What elements of your research are most frustrating to you? Worrisome?
13. Do you have other concerns about your work in nanotechnology?
14. How about the exciting elements? What are those?
15. Where does your funding come from?
16. What expectations do you think that (those) funding agencies have for you?
17. How about moral support for your work? Do you need that? Where does it come from?
18. Who are you working with? Collaborations?
19. What do you imagine could go wrong with your project?
20. How, if at all, do you envision your research resulting in changes to the human body?
21. What areas of human life could be affected by your work?

22. What are you learning in your work, or might you learn, that is a surprise to you?
23. What do you still hope to learn?
24. What do you wish you knew, that would help you to be successful?
25. How important is it to you that your project is successful? Why?
26. What do you think about Eric Drexler's ideas and predictions about nanotechnology?
27. What moral/ethical factors are part of your work?
28. Do you have a religious orientation or practice?
29. What do you believe about the existence of a human soul?
30. If the soul is real for you, what do you believe happens to it after bodily death?
31. What do you think about the idea of extended, even indefinite human life?
32. Do you think nanotechnology offers any promises for that possibility?
33. What would it mean if we could use technology to eradicate all forms of bodily suffering?
34. What is the best outcome you can imagine from your research? The worst?
35. Have you ever/always had a religious faith? Explain.
36. Do you feel any connection between your professional work and your personal beliefs about life's meaning?
37. What do you care about most in your research?
38. What would it mean to fail?
39. To whom might the results of your work really matter?

Note: Interview questions are asked as open-ended questions, with follow-up responses to solicit deeper reflection:

Is that important?
What do you mean by that?
How do you explain that concept?
Why did you use that word?
Whose idea is this?
How would others feel about that?
Why do you care?
What do you believe about that?
Is there another way to understand that?

How might you explain that further?
What does that tell you?
Is there a personal reason for your enthusiasm? Concern?
What if that's not true?
Is it possible that's not true?

APPENDIX D

21st Century Nanotechnology Research and Development Act[1]

P.L. 108-153 (S. 189/S. Rept. 108-147 and H.R. 766/H. Rept. 108-89)

IMPACT OF PUBLIC LAW

P.L. 108-153, the 21st-Century Nanotechnology Research and Development Act, authorizes programs for nanoscience, nanoengineering, and nanotechnology research. Final provisions of this Act deleted authorization language and funding levels for the National Institutes of Health (NIH), but it is expected that NIH will continue to "be an active participant in the National Nanotechnology Program."

The Act establishes in statute a National Nanotechnology Research Program with a National Nanotechnology Coordination Office, and through the authorized partners (National Science Foundation, U.S. Department of Energy, National Aeronautics and Space Administration, National Institute of Standards and Technology, and Environmental Protection Agency), requires the awarding of grants and the creation of nanotechnology research centers on a competitive basis. It also provides for a research program to identify the ethical, legal, environmental, and other societal concerns related to nanotechnology. Also included in the final version by the sponsor, Senator Ron Wyden (D-OR), was a reference to biotechnology. The legislation is not intended to limit research and development to the physical sciences but rather is intended to include a wide variety of research, including the biotechnology–nanotechnology

[1]{2004 #278}

interface, with applications ranging from industrial manufacturing to advances in medicine to breakthroughs in defense against bioterrorism.

LEGISLATIVE HISTORY

Two bills on nanotechnology research were introduced in the first session of the 108th Congress. Only the Senate bill included reference to NIH.

S. 189 was introduced by Senator Wyden on January 16, 2003, and it is identical to the bill that he introduced in the 107th Congress. The bill was referred to the Senate Committee on Commerce, Science and Transportation, which held a hearing on May 1. NIH submitted a statement for the record. The bill was reported out of the Senate Committee on September 15.

H.R. 766 was introduced by Representative Sherwood L. Boehlert (R-NY) on February 13, 2003, and was referred to the House Committee on Science. On March 19, the House Committee on Science held a hearing to examine Federal nanotechnology research and development activities and to consider H.R. 766, which would statutorily authorize these programs. Questions raised at the hearing were based on a general concern as to whether the National Nanotechnology Initiative (NNI) would facilitate technology transfer and provide the United States with a competitive advantage in the global market. Although NNI is focused on basic research, the witnesses emphasized that NNI already has structures in place that would facilitate interaction between the research community and the private sector. The bill was reported out of the House Committee on May 1 and passed the House on May 7.

On November 18, the Senate passed an amended, preconferenced version of S. 189, proposed by Senators Wyden and George Allen (R-VA), a version different from that which had been reported out of Committee on September 15. The House passed S. 189, as amended, on November 20. Provisions of this Senate- and House-passed bill deleted authorization levels for NIH at the request of the House, as indicated by Senator Ted Stevens (R-AK) in his floor statement during debate.

On December 3, the President signed into law S. 189, the 21st Century Nanotechnology Research and Development Act, as P.L. 108-153.

References

President Clinton's Remarks at Science and Technology Event, Retrieved from http://www.mrs.org/pa/nanotech/clinton.html. 2000.

(2003). http://www.house.gov/science/hearings/full03/apr09/charter.htm. House Committee on Science. Washington, D.C.

Abram, D. (1996). *The spell of the sensuous.* New York: Vintage.

Baum, R. (2003). Nanotechnology: Drexler and Smalley make the case for and against "molecular assemblers." *Chemical and Engineering News, 81*(48). Retrieved from http://pubs.acs.org/ cen/coverstory/8148/8148counterpoint.html

Becker, E. (1973). *The denial of death.* New York: The Free Press.

Carlson, B. (2005). *Industrialization of America* (p. 86). New York: Oxford University Press.

Crichton, M. (1990). *Jurassic park.* New York: Knopf.

Crichton, M. (2002). *Prey.* New York: Harper Collins.

Dowling, A. P. (2004). *Nanoscience and nanotechnologies: Opportunities and uncertainties, Working Group on Nanotechnology and Nanoscience.* The Royal Society and the Royal Academy of Engineering.

Drexler, K. E. (1992). Drexler writes Smalley open letter on assemblers. Retrieved from http://www.foresight.org/Updates/Update52/Update52.2.html

Drexler, K. E. (1986). *Engines of creation.* Garden City, NY: Anchor/Doubleday.

Dyson, F. J. (1997). *Imagined worlds.* Cambridge, MA: Harvard University Press.

Dyson, F. J. (1999). *The sun, the genome, and the Internet.* Oxford, England: Oxford University Press.

Ellul, J. (1990). *The technological bluff.* Grands Rapids, MI: Eerdmans.

Ellul, J. (1998). *Jacques Ellul on religion, technology and politics: Conversations with Patrick Troude-Chastenet.* Atlanta: Scholar's Press.

ETC, G. (2003). *The big down: Atomtech—technologies converging at the nanoscale.* Retrieved from http://www.etcgroup.org/documents/TheBigDown.pdfETC

Feller, I. (2001). *Societal implications of nanoscience and nanotechnology: NSET workshop report.* Dordrecht: Kluwer.

Feynman, R. P. (1959). *There's plenty of room at the bottom.* Paper presented at the American Physical Society, annual meeting, Pasedena, CA. Retrieved from http://resolver.caltech.edu/CaltechES:23.5.0

Florman, S. C. (1994). *The existential pleasures of engineering.* New York: St. Martin's.

Florman, S. C. (1997). *The introspective engineer.* New York: St. Martins.

Frankel, V. E. (1992). *Man's search for meaning: An introduction to logotherapy.* Boston: Beacon.

Fuller, S. (1993). *Philosophy, rhetoric and the end of knowledge.* Madison: University of Wisconsin Press.

Gerson, E. M. (1976). On "Quality of Life." *American Sociological Review, 41*(5), 793–806.

Hauerwas, S., & Burrell, D. (1989). From system to story: An alternative pattern for rationality in ethics. In S. Hauerwas & G. Jones (Eds.), *Why narrative* (pp. 158–190). Grand Rapids, MI: Eerdmans.

Hauerwas, S. J., & Jones, G. L. (Ed.). (1989). *Why narrative?* Grand Rapids, MI: Eerdmans.

Habermas, J. (1990). *Moral consciousness and communicative action.* Cambridge, MA: MIT Press.

James, E. (1994). *Science fiction in the 20th century.* Oxford, England: Oxford University Press.

Jaspers, K. (1978). *Man in the modern age.* New York: AMS Press.

Johnson, M. (1994). *Moral imagination: Implications of cognitive science for ethics.* Chicago: University of Chicago Press.

Joy, B. (2000). Why the future doesn't need us. *Wired, 8*(4). Retrieved from www.wired.com/wired/archive/8.04/joy_pr.html

Jung, C. (1933). *Modern man in search of a soul.* New York: Harcourt, Inc.

Jung, C. (1967). *The symbolism of evil.* Boston: Beacon Press

Jung, C. (1981). *The archetype and the collective unconscious.* Princeton, NJ: Princeton University Press.

Kaku, M. (1997). *Visions: How science will revolutionize the 21st century.* New York: Anchor.

Lakoff, G., & Johnson, M. (1980). *Metaphors we live by.* Chicago: University of Chicago Press.

Latour, B. (1987). *Science in action: How to follow scientists and engineers through society.* Cambridge, MA: Harvard University Press.

Marburger, J. (2003). *Workshop on Societal Implications of Nanoscience and Nanotechnology.* Arlington, VA: National Science Foundation.

Marburger, J. (2004). *Nanotechnology policy for the 21st century.* National Nanotechnology Initiative: From Vision to Commercialization, Washington, DC.

Marcuse, H. (1964). *One dimensional man: Studies in ideology of advanced industrial society.* Boston: Beacon.

Merchant, C. (1980). *The death of nature.* New York: Harper Collins.

Mikulski, B. (2001). *Societal implications of nanoscience and nanotechnology: NSET workshop report.* Dordrecht: Kluwer Academic.

Mitcham, C. (1994). *Thinking through technology: The path between engineering and philosophy.* Chicago: University of Chicago Press.

Mnyusiwalla, A., Daar, A., & Singer, P. (2003). Mind the gap: Science and ethics in nanotechnology. *Nanotechnology, 14,* R9-R13.

Mukerji, C. (1989). *A fragile power.* Princeton, NJ: Princeton University Press.

National Science and Technology Council. (2002). *National nanotechnology initiative: The initiative and its implementation plan.* Washington, DC.

Noble, D. (1999). *The religion of technology.* New York: Penguin.

Norris, P. (2004). *Aerogel Research Lab,* Home page. Retrieved from http://fourier.mech.virginia.edu/~microhx/home1.html

Parks-Daloz, L., Keen, J., Keen, C., & Parks-Daloz, S. (1996). *Common fire.* Boston: Beacon.

Popper, K. R. (1992). *In search of a better world: Lectures and essays from thirty years.* London: Routledge.

Rhodie, S. (2003). *Charles fears science could kill life on earth.* Retrieved from http://www.scotlandonsunday.com/uk/cfm?id=481682003 Scotland on Sunday

Richter, J. M. N. (1972). *Science as a cultural process.* Cambridge, MA: Schenkman.
Ricoeur, P. (1975). *The rule of metaphor.* Toronto: University of Toronto Press.
Ricoeur, P. (1976). *Interpretation theory: Discourse and the surplus of meaning.* Fort Worth: Texas Christian University Press.
Ricoeur, P. (1995). *Figuring the sacred: Religion, narrative and imagination.* Minneapolis: Augsburg Fortress.
Roco, M. C. (2003). *Government nanotechnology funding: An international outlook.* National Science Foundation. Retrieved from http://www.nano.gov/html/res/IntlFundingRoco.htm
Roco, M., (Ed.). (2004). *The coevolution of human potential and converging technologies.* Annals of the New York Academy of Sciences. New York: The New York Academy of Sciences.
Rouke, M. (2002). *Understanding nanotechnology.* New York: Warner.
Sarewitz, D., & Woodhouse, E. (2003). Small is powerful. In A. Lightman, D. Sarewitz, & C. Desser (Eds.), *Living with the genie* (pp. 63–84). Washington, DC: Island Press.
Shelley, M. W. (1992). *Frankenstein, or, The modern Prometheus.* New York: Knopf.
Shelley, M. W. (1831). *Frankenstein.* Electronic Text Center, University of Virginia Library.
Smith, N. D. (Ed.). (1982). *Philosophers look at science fiction.* Chicago: Nelson-Hall.
Society, A. P. (1997). 1997 Nicholson Medal for Humanitarian Service to Henry W. Kendall, MIT. Retrieved from http://www.aps.org/praw/nicholso/97winner.cfm
Stahl, W. A. (1999). *God and the chip: Religion and the culture of technology.* Ontario, Canada: Wilfrid Laurier University Press.
Stephenson, N. (1995). *The diamond age, or, young lady's illustrated primer.* New York: Bantam.
Stevens, A. (1990). *On Jung.* London: Routledge.
Toulmin, S. G. J. (1962). *The architecture of matter.* New York: Harper & Row.
Toumey, C. P. (1996). *Conjuring science.* New Brunswick: Rutgers University Press.
Urh, J. Personal conversation, December, 2004
Warner, M. (2002). *Fantastic metamorphoses, other worlds: Ways of telling the self.* Oxford, England: Oxford University Press.
Weil, V. (2001). *Ethical issues in nanotechnology.* Societal implications of nanoscience and nanotechnology: NSET workshop report, M. Roco (pp. 193–198). Dordrecht: Kluwer Academic.
Whitesides, G. (2001, September). The once and future nanomachine. *Scientific American, 81.*
Winner, L. (1986). *The whale and the reactor: A search for limits in an age of high technology.* Chicago: University of Chicago Press.
Winner, L. (1997). *Technomania is overtaking the millennium.* Retrieved from http://www.rpi.edu/~winner/How%How20Technomania.html

FURTHER READING

Altmann, J. and Gubrud, M. (2004). Anticipating military nanotechnology, *IEEE Technology and Society Magazine,* winter 2004, pp. 33–40.
Baldi, P. (2001). *The shattered self: The end of natural revolution.* Cambridge, MA: MIT Press.
Bierlein, J. F. (1994). Parallel myths. New York: Ballantine.
Borgmann, A. (1999). *Holding on to reality: The nature of information at the turn of the millennium.* Chicago: University of Chicago Press.
Crandall, B. C. (1996). *Nanotechnology: Molecular speculations on global abundance.* Cambridge, MA: MIT Press.
Crandall, B. C., & Lewis, J. (1992). Nanotechnology: Research and perspectives: Papers from the first Foresight Conference on Nanotechnology. Cambridge, MA: MIT Press.

Davis, E. (1998). *Technosis: Myth, magic, mysticism in the age of information*. New York: Harmony.

Denton, M. (1998). *Nature's destiny: How the laws of biology reveal purpose in the universe*. New York: The Free Press.

Drexler, K. E. (1992). *Nanosystems: Molecular machinery, manufacturing, and computation*. New York: Wiley.

Flynn, M. (1991). *The nanotech chronicles*. New York: Baen Books.

Frankel, V. E. (1992). *Man's search for meaning; An introduction to logotherapy*. Boston: Beacon.

Götz, I. L. (2001). *Technology and the spirit*. Westport, CT: Praeger.

Gerhart, M., & Russell, A. M. (1984). *Metaphoric process: The creation of scientific and religious understanding*. Fort Worth, TX: Texas Christian University Press.

Gendron, B. (1977). *Technology and the human condition*. New York: St. Martin's.

Herken, G. (2002). *Brotherhood of the bomb: The tangled lives and loyalties of Robert Oppenheimer, Ernest Lawrence, and Edward Teller*. New York: Henry Holt.

Danne, J., & Dozois, G., (Eds.). (1998). *Nanotech*. New York: Ace Books.

Kaku, M. (1997). *Visions: How science will revolutionize the 21st century*. New York: Anchor.

Kurzweil, R. (1999). *The age of spiritual machines: When computers exceed human intelligence*. New York: Viking.

Lakoff, G. (1987). *Women, fire and dangerous things*. Chicago: University of Chicago Press.

Lakoff, G., & Johnson, M. (1980). *Metaphors we live by*. Chicago: Chicago University Press.

Crow, M. M., & Sarewitz, D. (2001). *Nanotechnology and societal transformation*. In AAAS Science and Technology Policy Yearbook 2001. Washington, DC, American Association for the Advancement of Science.

MacKenzie, D., & Wajcman, J. (1999). *The social shaping of technology*. Philadelphia: Open University Press.

Miller, F. (1982). *Philosophers look at science fiction*. Chicago: Burnham.

Naylor, T. H., Willimon, W. H., & Naylor, M. (Eds.). (1994). *The search for meaning*. Nashville, TN: Abingdon.

Pacey, A. (1983). *The culture of technology*. Oxford, England: Blackwell.

Pool, R. (1997). *Beyond engineering : How society shapes technology*. New York: Oxford University Press.

Postman, N. (1992). *Technopoly: The surrender of culture to technology*. New York: Knopf.

Prahalad, C. K. (2004). *The fortune at the bottom of the pyramid: Eradicating poverty through profits*. Philadelphia: Wharton School Publishing.

Ratner, M. A., & Ratner, D. (2003). *Nanotechnology: A gentle introduction to the next big idea*. Upper Saddle River, NJ: Prentice-Hall.

Regis, E. (1995). *Nano: The emerging science of nanotechnology: Remaking the world—molecule by molecule*. Boston: Little, Brown.

Roco, M. C., & Montemango, C. D. (Eds.). (2004). *The coevolution of human potential and converging technologies*. Annals of the New York Academy of Sciences. New York: New York Academy of Sciences.

Sarewitz, D. R. (1996). *Frontiers of illusion: Science, technology, and the politics of progress*. Philadelphia: Temple University Press.

Sarewitz, D. R., Pielke, R. A., & Byerly, R. (Eds.). (2000). *Prediction: Science, decision making, and the future of nature*. Washington, DC: Island Press.

Shale, M. H., & Shields, G. W. (Eds.). (1994). *Science, technology, and religious ideas*. Lanham, MD: University Press of America

Shrader-Frechette, K. S., & Westra, L. (1997). *Technology and values.* Lanham, MD: Rowman & Littlefield.
Suzuki, D. (1989). *Inventing the future.* Toronto: Stoddart.
Webster, A. (1991). *Science, technology, and society: New directions.* Basingstoke: Macmillan.
Wolfson, J. R. (2003) Social and ethical issues in nanotechnology: Lessons from biotechnology and other high technologies. *Biotechnology Law Report, 22*(4): 376–396.

Author Index

Note. Page number followed by *n* denotes footnote.

Subject Index

Note. Page number followed by *n* denotes footnote.

D

E

F

G

H

I

J

O

P

Q

R

S